A PRIMER OF PROBABILITY LOGIC

Publisher's Note

The author, Ernest W. Adams, passed away in 2009 before he could make corrections for a second printing. Thanks are due to the late Theodore Hailperin, Professor Emeritus at Lehigh University, and to the late Lewis Henry LaRue, Professor of Law at Washington and Lee University for stepping in to make the corrections that the author would have made. Thanks are also due to Greg Stokley for his careful checking and uncovering other problems and to Alistair Isaacs for his help with the bibliography. The second printing reflects these corrections; the publisher assumes full responsibility for remaining problems.

2016

CSLI Lecture Notes
Number 68

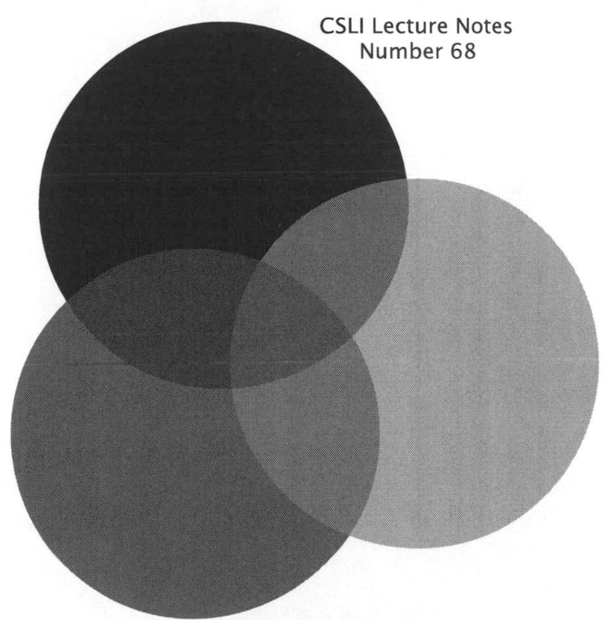

A PRIMER OF PROBABILITY LOGIC

ERNEST W. ADAMS

Foreword by Brian Skyrms

CSLI PUBLICATIONS
Center for the Study of
Language and Information
Stanford, California

Center for the Study of Language and Information
Leland Stanford Junior University
Printed in the United States
20 19 18 17 16 5 4 3 2

Library of Congress Cataloging-in-Publication Data

Adams, Ernest W. (Ernest Wilcox), 1926–
A primer of probability logic / Ernest W. Adams ; foreword by Brian Skyrms.

p. cm.

(CSLI lecture notes ; no. 68)
Includes bibliographical references and index.

ISBN 1-57586-067-8 (hardcover : alk. paper).
ISBN 1-57586-066-X (pbk. : alk. paper)

1. Algebraic logic. 2. Probabilities. I. Title. II. Series.

QA10.A34 1997
511.3—dc21 97-20385

CIP

ISBN 978-157586-067-1 (hardcover : alk. paper).
ISBN 978-157586-066-4 (pbk. : alk. paper)

∞ The acid-free paper used in this book meets the minimum requirements of the American National Standard for Information Sciences—Permanence of Paper for Printed Library Materials, ANSI Z39.48-1984.

CSLI was founded in 1983 by researchers from Stanford University, SRI International, and Xerox PARC to further the research and development of integrated theories of language, information, and computation. CSLI headquarters and CSLI Publications are located on the campus of Stanford University.

CSLI Publications reports new developments in the study of language, information, and computation. Please visit our web site at
http://cslipublications.stanford.edu/
for comments on this and other titles, as well as for changes and corrections by the author and publisher.

Contents

Foreword

An adequate logic of science cannot rest solely on the concept of deductive validity. It must also deal with error and uncertainty—matters best addressed in terms of the theory of probability. This was already taken as a first principle in the 19th century by LaPlace, Venn, de Morgan, Peirce and Poincare. And the logic of probability has been seen as central to practical reasoning since its inception. (And perhaps before its inception on a generous reading of Aristotle.) But contemporary logic texts tend to treat probability as an addendum or afterthought rather than as an integral part of the subject.

This book is different; probability is at its center. The very structure of the language investigated is pragmatically based on the probability concepts involved. The conditional is construed as the bearer of conditional probability, along the lines of Adams' eye-opening research on the logic of conditionals. Arguments containing such conditionals are evaluated from a pragmatic probabilistic perspective. The connections to epistemology and practical reason are evident throughout.

As a Primer, this text is written so that it is accessible to a student new to the subject matter. In another sense it can serve as a primer for professional philosophers who are not already familiar with Adams' original contributions to probability logic. Here these are given an elementary systematic exposition. Readers will be introduced to a new and fascinating perspective on the function of language.

Brian Skyrms

Preface

This book is an experiment. With an important exception to be noted below, the subject of probability logic as it exists today exists only in scattered articles in inaccessible journals, for the most part written in arcane terms dealing with mathematical topics of no concern to nonspecialists.[1] Nevertheless the author's opinion, based on classroom experience, is that not only does this subject have important applications to the kinds of problems that elementary logic deals with, but beginning students with no more than a background in high school algebra can easily master its rudiments and learn to make these applications. This book aims to present the writer's approach to teaching this subject at the level of elementary logic.

There are two main motivations for introducing the subject into the elementary curriculum at the present juncture. First and foremost, the writer believes that the insights into human reasoning that probability logic provides constitute a very important advance beyond those provided by modern nonprobabilistic logic. In large part this derives from the fact nonprobabilistic logic does not take into account the *fallibility* of human reasoning, if not in the steps that lead from 'givens' to conclusions, almost always in the givens themselves, and it is as important to deal with 'error' and uncertainty in this domain as it is, say, to deal with it in the domain of physical measurement. This will be the subject of the first five chapters of this text. But chapter 6 will also suggest that there is at least one vitally important kind of human reasoning that involves uncertainties that cannot be conceptualized as *errors*, because there are no 'ideal truths' that correspond to them. This involves *conditional*, if–then propositions, which are ubiquitous in everyday and in scientific reasoning, but which are treated badly by traditional logic. And, when we consider why the traditional 'truth-conditional' approach fails in application to conditionals, we will find the answer in traditional logic's radical separation of pure and practical logic, and its failure to take into account the way in which con-

[1]The references at the end of this book give some indication of this.

clusions arrived at in 'pure reason' influence practical decision and action. This brings us to the second main motivation for introducing probability logic into the elementary curriculum.

Though probability logic is a subject that is barely thirty years old, with hardly any tradition in either philosophy or science, it is a rapidly developing research specialty with practitioners in widely diverse fields, and with a potential for much wider ramifications. Listing some of these, we may begin with philosophical logic, a rapidly developing specialty in logical theory, which, among other things, has sought to redress the imbalance that existed in the past between abstract mathematical logic and its application to the real world. Another is decision theory, especially probabilistic decision theory such as is now very actively studied in philosophy of science, philosophy of action, economics, and various branches of psychology, and which probability logic impinges on because of its unification of pure and practical reason. Another is linguistics and the philosophy of language, in which theories of action have already played important parts and in which conditionals are a recognized specialty, but in which until recently probabilistic ideas have played a very small role. Still another is computer science and Artificial Intelligence, which attempts to model human reasoning processes in computational terms, and in which the analysis of the *nonmonotonic reasoning* that is closely related to conditional logic has become very important in the last decade.[2] And, last but not least, although statistics and probability theory have very long and distinguished traditions, they have not as yet considered the new problems that arise within the framework of probability logic, particularly those that arise because this subject does not accept the idealized randomness and independence assumptions that are central to almost all of traditional statistical theory. This suggests that an introduction to probability logic may bring the student in on the ground floor of a discipline with a potential for rapid development in many directions, and which some at least may wish to pursue professionally.

Let us now comment briefly on the prerequisites required for reading this text, its intended audience, and its scope and limitations.

This book is written to be read independently by persons with interests in all or any of the subjects mentioned above. It presupposes only such knowledge of sentential logic as is ordinarily acquired in a one semester course in formal logic. In particular, except for a few starred exercises and three appendices, quantificational logic not presupposed. No prior knowl-

[2]Nonmonotonic logic recognizes that beliefs held at one time may be fallible and therefore may have to be given up when new information comes to hand, and it rejects traditional logic's idealized picture of reasoning as a process in which the acquisition of information only *increases* our store of knowledge. Pearl and Goldzmidt (1991) gives an excellent survey and exposition of this and other recent advances in this area, from the point of view of computer science and Artificial Intelligence.

edge of mathematical probability is required, and the little that is used in this book is introduced in the text itself. This is standard elementary probability, and therefore while previous study of probability and statistics can be helpful it is not essential.

For assistance, especially with independent study, most chapters of the text include glossaries of the key terms introduced in them, and these serve to some extent as capsule summaries of the theories that involve them. Most sections include numerous exercises, answers to approximately half which are worked out, in some cases at length, at the end of the text. As said, unlike most writings on probability logic, the present text will concentrate on applications. Technical mathematics is therefore kept to a minimum consistent with reasonably concise exposition of the core materials. Problems of primarily mathematical interest will be pointed out as we go, and commented on in starred remarks and exercises. On the other hand, because applications are highlighted, currently open problems that relate to them and to foundations are pointed out, which technically unsophisticated students may consider with profit.

Another respect in which this text is untypical is that it does not state basic principles as unquestioned 'axioms', and this is especially true of the treatment of conditionals in chapters 6–8. As it happens, the principles of conditional logic are presently very much in dispute, and we will not seek to disguise this—in fact, chapter 8 sketches elements of an approach that differs from ours. It is therefore only because our approach to conditionals is probabilistic that its inclusion in this book is justified. But, that an important application of probability logic is based on controversial principles is an indication of the fact that the field has not yet developed a stable, accepted body of concepts and doctrine paralleling that which has evolved in 'classical', nonprobabilistic logic over the past century. The field is not a 'mature science' in Kuhn's sense, and no textbook can pretend to present its 'foundations'. As a consequence, the reader should be aware that the approach and outlook of this text are in some respects those of the author, and they are not necessarily shared by others who may with equal right be styled 'probability logicians'. As said, this is particularly true of chapters 6–8 of this work, and its relation to 'practical reason', which is the topic of chapter 9, and which links 'pure' probability logic to probabilistic theories of decision.

A word should be said about the relation between this primer and current research and writing on probability logic. Though a relatively new subject, the literature on it is extensive and rapidly expanding, but our concentration on elementary aspects precludes making more than passing references to its more advanced topics. Nevertheless an attempt is made in footnotes, starred sections and exercises, and brief appendices, to point to these topics. Representative items in the literature on the subject are cited in the references, but, especially in the case of more recent additions, these do not pretend to be encyclopedic. Because of our 'philosophical' bias we have tried to give some indications of the 'philosophical roots' of the probabilistic approach to logic. In this regard, however, we strongly recommend Professor Hailperin's new book *Sentential Probability Logic* (Hailperin, 1997), which gives excellent accounts of the contributions of such prominent historical figures as Leibniz, de Morgan, Boole, Peirce, Keynes, Carnap, and others to the development of this subject, as well as of more recent developments.

The author would like to acknowledge the help and encouragement of Brian Skyrms, Vann McGee, Dorothy Edgington, students in his elementary logic class at the University of California, Berkeley, and especially an anonymous referee who read a draft of this book with painstaking care and recommended improvements on almost every page. The good qualities of the result owe a great deal to her or him, though its faults are those of the author alone. Finally, the author is greatly indebted to Tony Gee, Maureen Burke, and Dikran Karagueuzian of CSLI Publications at Stanford University for their unstinting help and patience in seeing this work through the press, and to Gerome Vizmanos for creating the numerous diagrams.

1

Deduction and Probability: What Probability Logic Is About

1.1 Deduction and Certainty

It is often said that deductive logic is concerned with what can be deduced with logical, mathematical certainty, while if probability has a part in logic at all it falls into the province of Inductive Logic. This has some validity, but it also overlooks something important. That is that deductive logic is usually supposed to apply to reasoning or arguments whose premises aren't perfectly certain, and whose conclusions can't be perfectly certain because of this. For example, a currently widely used textbook asks the student to determine the validity of the following inference:

> If Ed wins first prize then either Fred wins second prize or George is disappointed, and Fred does not win second prize. Therefore if George is not disappointed then Ed does not win the first prize. (Copi, 1965: 25)

This is formally valid, but even so the premise "If Ed wins first prize then either Fred wins second prize or George is disappointed" isn't the sort of thing that a person can be perfectly sure of, and if that is so then "if George is not disappointed then Ed does not win the first prize" shouldn't be 'deductively certain' either.[1]

An important thing to notice about the 'problem' of uncertainty in deductive reasoning is its similarity to problems that arise in applying 'exact' mathematics in calculating numerical measurements. For instance, a person who has to calculate the area of a rectangle whose sides are $2\frac{1}{2}$ and $3\frac{1}{2}$ feet will most likely multiply $2\frac{1}{2}$ by $3\frac{1}{2}$ to get an area of $8\frac{3}{4}$ square feet, even though she knows that the 'premises' of $2\frac{1}{2}$ and $3\frac{1}{2}$ feet aren't absolutely exact, and therefore the area of $8\frac{3}{4}$ square feet that she calculates can't be

[1]Most contemporary philosophers agree with the Scottish-English philosopher David Hume, 1711–76, who maintained that no one can be perfectly, rationally certain of 'matter of fact statements' about the future. This is related to the so called *Problem of Induction*, which will be returned to in chapter 4.

exact either. But without more exact information she has to *idealize*, and calculate as though the $2\frac{1}{2}$ and $3\frac{1}{2}$ foot lengths were exact, assuming that if they are close to the 'true' lengths, then the $8\frac{3}{4}$ square feet will be close to the true area. Similarly, a person who deduced the conclusion of the 'Ed, Fred, and George' inference would idealize by ignoring the fact that its premises aren't perfectly certain, but, supposing that they are 'certain enough', the conclusion ought to be 'reasonably certain'.

But, to get back to measurement, it is also important that you can't always safely ignore errors and inexactness in numerical calculations. For instance, if you add too many inexact lengths you can arrive at a *very* inexact sum, and if you divide by small differences between inexact lengths, you can arrive at a very erroneous ratio. By the same token you can't always ignore uncertainties in deductive 'calculations'. The famous lottery paradox is an example (Kyburg, 1965).

A lottery with 1,000 tickets labelled "Ticket #1", "Ticket #2", etc., up to "Ticket 1,000", will have only one winner, to be chosen in a random drawing. Any one ticket's chance of winning is 1/1,000, and its chance of losing is a near certainty of 999/1,000 = .999, which would be certain enough that someone speaking loosely would be justified in saying any of "Ticket #1 will lose", "Ticket #2 will lose", and so on, up to "Ticket #1,000 will lose". But it would be absurd to deduce the *conjunction* of these 1,000 'premises', "Ticket #1 will lose and Ticket #2 will lose and ... and Ticket #1,000 will lose", even though this is a valid inference in idealized deductive logic.

Now, a scientific theory of errors of measurement has been developed over the past 300 years that accounts for such things as that even though individual measurements have very little error, they can 'accumulate' when a large number of measurements are added together, and current scientific practice requires numerical data to be reported with margins of error. But it has only been in the past 30 or so years that a theory of *uncertainty in reasoning*, which is the analogue of error in measurement, has begun to be developed, which accounts for the lottery paradox and other aspects of deductive reasoning where it is important to take uncertainty into account. The following chapter will begin the discussion of this, and introduce certain laws of *uncertainty accumulation* that explain how small 'increments of uncertainty' in the premises of inferences can 'accumulate' as premises are added, until they become 'absurdly large' in certain conclusions that are deduced from them, such as in the lottery paradox. The same laws also explain why there must be *many* premises for this kind of accumulation to be serious.

But a complication that has no direct parallel in the case of measurement enters the picture in the case of deduction.

1.2 Inference and Probability Change: Nonmonotonicity

The error involved in saying that the distance from Los Angeles to New York is 3,000 miles when it is really 2,794 miles is not something that changes over time, but uncertainties can change very quickly (*Rand McNally Road Atlas*, 1986: 105). Last night you might have been pretty certain that it wouldn't rain today, but looking at the sky this morning can change your mind and make you pretty certain that it *will* rain. This kind of change is important in logic, because deductive reasoning in real life involves it. In the 'Ed, Fred, and George' example, for instance, if you finally deduce "If George is disappointed then Ed does not win first prize" you become more certain of something you weren't certain of before, and probability logic, which is concerned with certainty, is concerned with this.

Perhaps the thing that most clearly shows the importance of taking changes in probability into account is the light this throws on the rule anything follows from a contradiction, which may have seemed paradoxical when you first learned of it. In fact, in real life, often what look like contradictions when you formalize them are actually things that are either asserted by different persons, or by the same person at different times.[2] That's what might have happened if last night you had said that it *wouldn't* rain today, but looking out and seeing the sky this morning leads you to change your mind and say that it *would* rain today. Obviously, whether or not this is a contradiction, it would be irrational to deduce anything whatever from it. Probability logic accounts for this, and it explains why formal logic can't account for it; namely because it assumes that premises are perfect certainties which will never have to be retracted in the light of new information.

Nonretraction in reasoning, which follows from formal logic's idealizing and treating all premises as though they were perfectly certain, is called *monotonicity*, which means that the more 'data' or premises you acquire, the more things you can deduce from them without giving anything up. But one of the most striking things about probability logic, which allows for the fact that it doesn't treat all premises as certainties, is its *nonmonotonicity*.[3] And, once you allow that things can be subtracted as well as added to your store of beliefs, you see that this happens in many contexts besides contradictions. One of these involves *conditionals*.

[2]These aren't the only contradictions that you encounter outside of logic books, and another important type of contradiction occurs in the context of *indirect reasoning*, where you assume that something you want to prove is false, in order to deduce a contradiction—an absurdity—from it.

[3]In fact, it is sometimes called *nonmonotonic logic* in the branch of Artificial Intelligence (AI) that is concerned with uncertainty in reasoning, cf. Pearl (1988).

1.3 Conditionals

Two famous fallacies of material implication,

> It won't rain today. Therefore, if it rains today it will pour.

and

> It won't rain today. Therefore, if there is a terrific thunderstorm it won't rain today.

are so called because the "if... then..." propositions that they involve are symbolized as *material conditionals*, which assumes that they are true if either their antecedents are false or their consequents are true. Thus, "If it rains today it will pour" can't be false when "It won't rain today" is true, because its antecedent is false. Similarly, when "It won't rain today" is true, "If there is a terrific thunderstorm it won't rain today" also has to be true, because its consequent is true.

Now, the fact that formally valid inferences like the ones just noted seem intuitively invalid or irrational is connected with a controversy of long standing that had its origins in ancient Greece. For instance, the Greek poet Callimachus (b. about 330 B.C.) is quoted as saying "Even the crows on the rooftops are cawing over the question as to which conditionals are true".[4] In fact, many theories of which conditionals are true have been proposed, one famous one being by Diodorus Cronus, the teacher and rival of Philo, the originator of the material conditional theory, who held that a true conditional is one that "neither could nor can begin with a truth and end with a falsehood". The difference between this and the material conditional theory is the inclusion of the *modal* clause "neither could nor can be", which anticipates quite modern theories, including one that will be commented on in section 8.3. But this is only one among many rivals to the material conditional theory that are being debated as of the present writing.

As you might guess, however, probability logic approaches the controversy over conditionals not by considering the question of which of them are *true*, but by considering which of them are *probable*. The key to explaining why you wouldn't want to deduce "If it rains today it will pour" when you are 'given' that it won't rain today, is that you can't be perfectly certain of what you are given.[5] If you think that it won't rain, a conditional "If it *does* rain today then it will pour" would tell you what to expect in the unexpected situation in which it rains. But just because rain is unexpected, you wouldn't necessarily expect it to pour if it did rain. Because formal logic idealizes and treats premises as certainties, it doesn't seem that the

[4]Cf. Kneale and Kneale (1962), p. 128, and section 12.1 of Mates (1965).

[5]This is another example of 'Hume's thesis' that you can't be perfectly, rationally certain about matters of fact relating to the future.

question of what will happen if they are false has to be taken seriously, and the rule that, no matter what the 'then' part is, an 'if–then' statement is true if the 'if' part is false reflects that. But probability logic doesn't idealize in this way, and it takes seriously the question of what the 'then' part should be when the 'if' part is false.

Obviously the fact that it isn't rational in probability logic to accept any conditional statement just because you think its 'if' part is false is related to the fact that it isn't rational to deduce anything from a contradiction. These are both aspects of nonmonotonicity, that at later times you may have to give up beliefs that you had held earlier, and once you recognize this you begin to question other rules of formal logic. This will be discussed in detail in chapters 5 and 6, but the example of Ed, Fred, and George illustrates one important thing that we will find very commonly. This reasoning would be valid in *most* situations, but it is possible to imagine *exceptions*: situations in which it would be reasonable to accept the premises but not the conclusion. In fact, we will see that most rules of reasoning involving conditionals have exceptions like the ones that arise in the Ed, Fred and George case. A practical aim of probability logic is to *identify* the exceptions, to determine when it is 'safe' to apply the rules of formal logic, and when following them can run you into trouble. That is analogous to a practical aim of theories of error in measurement of telling you when you can safely ignore errors in calculating with exact mathematics, and when using these methods can lead you into trouble.

The next section brings in an even more practical aspect of probability logic.

1.4 Decision and Action: The Advantage of Being Right

Here is an example of what is sometimes called a *practical inference*:

> You *think* that your friend Jane is going to take logic, and you want to be in the same class as she is. Therefore you act, by signing up for logic.

This *practical syllogism* involves two 'premises': (1) your *thought* that Jane will take logic, and (2) your *wanting* to take the same class as Jane takes. Formal logic only deals with the first kind of premise—with a thought, or belief. The second premise, wanting to be in Jane's class, isn't a thought, and neither is the 'conclusion'—signing up for logic. That is something you *do*. It doesn't make sense to say that wanting to be in Jane's class or signing up for logic are true or false, so it doesn't make sense to ask whether the practical reasoning is valid, in the formal sense that if its premises are true, its conclusion must also be true. Nevertheless, signing up for logic is the conclusion of a mental process that wanting to be in Jane's class is a part of, and it is just as important to evaluate this reasoning as it is to evaluate inferences whose conclusions are things that can be true or false.

Probabilities come into practical reasoning because persons often have to balance them when they choose what action to take, i.e., when they make decisions. For instance, if you aren't certain what class Jane will take, you are likely to weigh the chances that she will take logic before you decide whether to take it yourself. The general principles of 'weighing chances' are formulated in the theory of *decision making under risk*, which is part of the broad picture of practical reason that will be outlined in chapter 9 of this book. Although these theories go beyond 'pure' probability logic by taking value or utility into account, there is a special reason for considering them in the present context. That is because they bring in the *value of being right*.

Suppose that your 'thought' that Jane will take logic is itself deduced from 'premises', e.g., that Jane will either take logic or ethics, and she won't take ethics. Probabilities aside, this would be valid reasoning, and therefore the conclusion must be true if the premises are true. But while formal logic distinguishes between the 'values' of truth and falsity, it doesn't say why truth should have a *higher* value than falsity. Practical logic does explain this. The practical reason why it is important for your conclusion, that Jane will take logic, to be right is that if you act on it by taking logic yourself, and you are right in thinking she *will* take logic, you will get what you want, to be in the same class with Jane. But if you are wrong, you will end up not getting what you want. More generally, very often the practical motive for aiming to be right in reasoning is that if you are right and you act on your conclusions then you will get what you want, but if you are wrong you won't.[6]

But, ending this section, it is important to note that you can be right or wrong not only about truth, but also about probability. That is because how you weigh probabilities determines how you *act*, and whether you act 'rightly or wrongly' determines whether you get what you want. This is something that isn't usually considered in the theories of decision making under risk, but there is a special reason for considering it here because it has an important bearing on the controversy concerning conditionals mentioned in the previous section. Celebrated *triviality results* due to the contemporary philosopher David Lewis and others,[7] seem to imply that an assumption that will be fundamental to the final chapter of this work, namely that the probabilities of conditionals are so called *conditional probabilities*, can't be the probabilities of these propositions being *true*. More

[6]This is related to the so called *Pragmatic Conception of Truth*: roughly that beliefs are true if they are practically 'useful', different versions of which were advanced by C. S. Peirce, William James, and John Dewey. These views will be returned to briefly in sections 8.6⋆ and 9.9⋆⋆ of this book; see also A. J. Ayer (1968).

[7]Lewis's (1976) and related triviality results are discussed in section 8.5 and in Appendix 3 of this work.

than any other, this discovery has split the logical community into opposing camps in the last decade, one holding that truth is the key 'logical value' of a conditional, and the other that probability is. The question of which is 'right' will be discussed in chapter 8⋆ and section 9.9⋆⋆,[8] which consider what, pragmatically, the best judgments about conditionals like "If it rains it will pour" are. However, this brings us to the fact that many topics that will be covered in this work are controversial at present, and one of our objectives will be to acquaint the reader with the controversies.

1.5 Summary and Limitations of This Work

As said, our primary concern in this work will be with applications of probability to deductive logic, and especially with the fact that the premises of reasoning in everyday life are seldom certainties and what this implies about the certainty of the conclusions that are deduced from them. But this leads to technical questions like asking, for instance, how probable the premises of an inference have to be if you want to assure that its conclusion is at least 99% probable. This is analogous to asking how accurate the measurements of the height and width of a rectangle have to be if you want to estimate its area with 99% accuracy. The minimal knowledge of formal probability needed to answer this question in the probability case will be given in chapters 2 and 4,[9] which will differ from the material in ordinary courses in mathematical probability mainly in not dealing with the kinds of 'combinatorial' problems that typically arise in games of chance and in the theory of sampling.[10]

The probabilistic background given in chapters 2 and 4 is applied in chapters 3 and 5 to answering the question of how probable the conclusion of a deductively valid inference must be, given the probabilities of its premises.[11] The chapters differ in that while chapter 3 looks at things 'statically', and leaves out of account the fact that probabilities can change as a result of gaining information or 'adding new premises', chapter 5 introduces the 'dynamical dimension' that is associated with the nonmonotonicity that was commented on in earlier sections. This brings in the first

[8]Sections and chapters that are marked with a star, '⋆', discuss subjects that are less elementary and less central than other topics discussed in the text. These sections can be omitted in studying the more central topics.

[9]Students who have already had courses in probability theory need only skim the basic probability ideas introduced in these chapters, though they will need to pay attention to their applications to deductive logic.

[10]More generally, courses in mathematical probability tend to focus on applications to certain kinds of reasoning that are *inductive* in a broad sense, and to make assumptions about probabilities that are themselves not valid *a priori*, the not making of which actually simplifies our theory.

[11]In spite of the obviousness of this question, it has not to the writer's knowledge been raised before in texts either on probability or deductive logic.

of the controversial principles that will be involved in our discussion, because the dynamical theory is based on *Bayes' Principle*, which is a general law describing how probabilities change when information is acquired. This principle is 'critiqued' in section 4.7⋆⋆, and it is the first of three fundamental but controversial principles that will guide the theoretical developments with which this text is concerned.

Chapters 6 and 7⋆ focus on the logic of conditional propositions, and this discussion is based on the second major controversial assumption of this text: namely that the *probabilities of conditional propositions are conditional probabilities.* Not only do these probabilities not conform to the laws presupposed in chapters 3 and 5, but they lead to a 'logic' of conditionals that differs radically from the logic that is based on the assumption that the truth value of an 'if–then' proposition is the same as that of a material conditional. This leads to an argument that is set forth in chapter 8⋆, which is a variant due to Alan Hájek of David Lewis' fundamental 'Triviality Argument', that shows that, given certain plausible 'technical assumptions', if the 'right' measure of a conditional's probability is a conditional probability then its probability cannot equal the probability of its being true, no matter how its truth might be defined. Chapter 9 gives a brief introduction to the theory of *practical reason*, showing how conclusions about propositions and their probabilities influence actions, and how the success of the actions in attaining the goals of the persons who perform them depends on the 'rightness' of the probabilities acted upon. This chapter ends with the most 'philosophical' section of this work, which concerns the 'pragmatics of probability'. This has to do with how probabilities ought to be measured, assuming that persons are guided in their decisions by them and 'right probabilities' provide good guidance. This is a question of 'the meaning of probability', and, as would be expected, it is the most controversial question of all. Our equally controversial assumption is that good or 'real' probabilities are frequencies with which human beings can correctly predict things, with the corollary that they are best guided in the long run if their estimates of probabilities correspond to these frequencies.

Finally, something should be said about the limitations of this text. It is meant to be a *primer*, that is, as an introduction to a subject that even now, in what appears to be its infancy, has many ramifications of both a mathematical and a philosophical character. But the fact that the subject has both a philosophical and a mathematical side means that, unlike other introductory works on probability and its applications, our primer will introduce formal mathematical problem-solving techniques only to the extent that they contribute to the understanding of matters that are not fundamentally formal.[12] In fact, significant mathematical developments in

[12]In fact, in the author's opinion, emphasis on formal techniques tends more to obscure

probability logic are passed over because their bearing on applications is less direct than the materials that will be covered here. Some of these, e.g., the extension of the theory to infinitesimal probabilities, are sketched briefly in appendices, some are brought up in exercises, and some are alluded to only in footnotes. It is hoped that by this means, students will be made aware of some of the important ramifications of the subject, current research in it, and they will have some indications to recent literature on it.

1.6 Glossary

Conditional A sentence of the form "If P then Q," where P and Q are themselves sentences. *Indicative conditionals* are of the form "If P is the case then Q is the case," and *subjunctive conditionals* (often called *counterfactual* conditionals) are of the form "If P were the case then Q would be the case."

Bayes' Principle The principle that the acquisition of new information, ι, results in changing probabilities according to the rule that the new probability of a proposition P equals the old probability of P, conditional on ι.

Hume's Thesis That there can be no rational certainty concerning future matters of fact.

Lottery Paradox The paradox that according to standard logical theory "No lottery ticket will win" is a logical consequence of the premises "The first ticket won't win," "the second ticket won't win," "The third ticket won't win," and so on for all of the tickets in a lottery with 1,000 tickets, although each premise is sufficiently probable to be accepted on its own, while the conclusion is certainly false.

Monotonicity; Nonmonotonicity That standard laws of logical theory imply that adding new premises or items to a store of information can only increase the class of conclusions that can be deduced from them. Nonmonotonic theories reject this by allowing the possibility that adding new information can actually result in giving up conclusions previously arrived at.

Paradoxes of Material Implication Paradoxes that follow from the assumption that the indicative conditional "If P then Q" is true in all cases except when P is true and Q is false—i.e., its truth value is given by the material conditional. The two best known of these paradoxes are that any inference of the form "Not P; therefore, if P then Q," or the form "Q; therefore, if P then Q" is necessarily valid.

than to clarify the very phenomena that theory is meant to help us understand, which in this case have to do with human reason.

Practical Reason The aspect of logical theory that concerns the connection between *beliefs*, which may be conclusions arrived at by 'pure reason', and *practical actions* that may be taken as a result of arriving at these conclusions.

Practical Syllogism; Practical Inference A pattern of practical reasoning of the form **want+belief→action**, exemplified in the case of a person who wants candy and who believes that he can buy some in a particular store, who acts accordingly by going to the store.

Pragmatic Conception of Truth Roughly, the view that the truth of beliefs consists in their being useful. Various versions of this doctrine have been advanced by the American philosophers Charles Sanders Peirce, William James, and John Dewey.

Triviality Results Mathematical results proved by David Lewis and generalized by others, that show that a particular measure of the probability of a conditional proposition—that it should be a conditional probability—is inconsistent with the assumption that it satisfies fundamental laws that the probabilities of nonconditional propositions are assumed to satisfy. An implication is that if conditionals' probabilities are measured in this way then they cannot be said to be true or false in senses normally taken for granted in formal logic.

2

Probability and Logic

2.1 Logical Symbolism and Basic Concepts

This chapter will review the concepts of formal symbolic logic that are presupposed in the rest of this text, and introduce basic concepts of probability. Both the concepts and the symbolism of formal logic that we will be using are illustrated in application to an inference that was commented on in chapter 1, which we will eventually consider from the point of view of probability:

> Jane will either take ethics or she will take logic. She will not take ethics. Therefore, she will take logic.

Symbolizing the inference in the way that will be used throughout this book, we will write its premises above and its conclusion below a horizontal line, as follows:

$$\frac{\mathrm{E} \vee \mathrm{L}, {\sim}\mathrm{E}}{\mathrm{L}},$$

or, equivalently, we may write "$\{\mathrm{E} \vee \mathrm{L}, {\sim}\mathrm{E}\} \therefore \mathrm{L}$".[1] The atomic sentences "Jane will take ethics" and "Jane will take logic" are symbolized by the sentential letters E and L, and "either... or..." and "not" (disjunction and negation) are symbolized by "$\vee$" and "$\sim$", respectively. "And", "if... then", and "if... and only if..." (conjunction, conditional, and biconditional) will be symbolized likewise, as "&", "$\rightarrow$", and "$\leftrightarrow$", respectively, though chapter 6 will introduce a new symbolism and 'probabilistic interpretation' of the conditional. Later we will use Greek letters for *sentence variables*, for instance when we say that for any sentences ϕ and ψ, ϕ and $\phi \rightarrow \psi$ logically

[1] Except in Appendices 7, 8, and 9 we will use a *sentential symbolism* such as "E" for statements like "Jane will take Ethics", and not a *predicate calculus* symbolism like "Ej", where "j" stands for "Jane" and E stands for "will take Ethics". We can do this because, except in the appendices, we will not be dealing with quantified sentences that have to be symbolized with variables.

Henceforth we will usually omit quotation marks around formulas like "E" and "E∨L," and write E and E ∨ L instead.

entail ψ. This implies that for all particular sentences such as E and L, E and E → L entail L, and the same thing holds for any other formulas that might be substituted for ϕ and ψ.

Formulas that can be formed using the above symbolism are assumed to satisfy the usual logical laws, and the student should be able to construct truth-tables like the one below, to determine the validity of an inference, the logical consistency of its premises, and so on:

	truth values		formulas: state descriptions				other formulas		
cases	E	L	E&L	E&~L	~E&L	~E&~L	E∨L	~E	L
1	T	T	T	F	F	F	T	F	T
2	T	F	F	T	F	F	T	F	F
3	F	T	F	F	T	F	T	T	T
4	F	F	F	F	F	T	F	T	F

Table 2.1

This table shows that the premises of {E ∨ L, ~E} ∴ L are consistent and the inference is valid, since L is true every case in which E ∨ L and ~E are both true. In the next section we will see that this implies that if E ∨ L and ~E are perfectly certain then L must also be certain, but our problem will be to find out what can be expected if the premises are only probable, and not completely certain. We can't say that the conclusion must be certain,[2] but we can ask: must it at least be probable, and if so, how probable? These questions will be answered in the next chapter, but that will require more background in probability.

We will end this section by introducing a special bit of technical terminology. The formulas E&L, ~E&L, E&~L, and ~E&~L are sometimes called *state-descriptions* (SDs),[3] and these have two properties that will turn out to be very important when it comes to considering probabilities. (1) Each SD is true in just one 'case', or line in the truth-table. (2) If you know which SD is the true one, you can deduce the truth of any other formula formed just from E and L. For instance, if you know that E&~L is the true SD, you can deduce that E ∨ L must be true because E&~L is only true in line 2 of the table, and E ∨ L is true in that line.

[2]At least, assuming that it *depends* on the premises. If it is independent of them it must be a logical truth and therefore perfectly certain. But that is not usual in applied logic.

[3]The term 'state-description' is due to R. Carnap (1947), and it is appropriate because a state-description describes the 'state of the entire world', at least so far as concerns propositions formed from E and L.

SDs are also pictured in a special way in *Venn diagrams*, like diagram 2.1 below, which the following section will connect with probabilities:

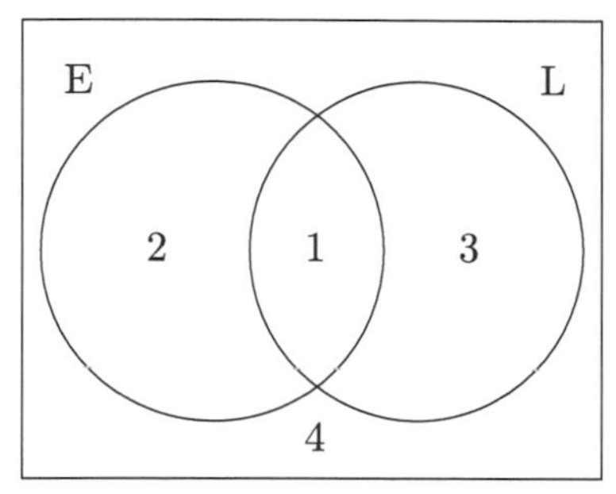

Diagram 2.1

Given a diagram like the above, the student should be able to identify the regions in it that correspond to formulas like E, L, $\sim$E, $E \vee L$, and so on. For instance, here the *atomic formulas* E and L correspond to circles, and compound formulas like $\sim$E and $E \vee L$ correspond to more irregular regions. $\sim$E corresponds to everything *outside* of circle E, while $E \vee L$ corresponds to the 'union' of circles E and L. More complicated formulas like $E \rightarrow L$, $(E \vee L) \leftrightarrow \sim E$ correspond to more complicated regions, which the student should also be able to identify. This can always be done in terms of unions of the 'atomic *regions*' 1–4, which themselves correspond to SDs.[4] For instance, $E \rightarrow L$ corresponds to the union of regions 1, 3, and 4 because it is logically equivalent to the disjunction of the SDs E&L, $\sim$E&L, and $\sim$E&$\sim$L, which correspond to regions 1, 3, and 4, respectively. The following review exercises will give you practice in doing this.

Exercises

1. Give the numbers of the atomic regions in diagram 2.1 that correspond to the following formulas (if a formula doesn't correspond to any regions write "none"):
   a. $E \leftrightarrow L$
   b. $\sim$(E&$\sim$L)
   c. $\sim(L \rightarrow E)$
   d. $E \vee$ (L&$\sim$E)
   e. (E&L) $\rightarrow \sim$E
   f. $(E \vee L) \leftrightarrow \sim E$

[4]Note the difference between atomic *formulas* and the regions they correspond to, and atomic *regions*, which are *parts* of the regions that atomic formulas correspond to. Thus, the atomic formula E corresponds to the left hand circle in diagram 2.1, whose atomic *parts* are regions 1 and 2. It will turn out later that atomic regions stand to probability logic somewhat as atomic formulas stand to standard sentential logic.

g. $(E \leftrightarrow \sim E)$
h. $(E \vee L) \& (E \rightarrow L)$
i. $(E \rightarrow L) \rightarrow L$
j. $(E \rightarrow L) \rightarrow (L \rightarrow E)$

2. Determine the atomic regions in the following Venn diagram that correspond to the formulas below the diagram:

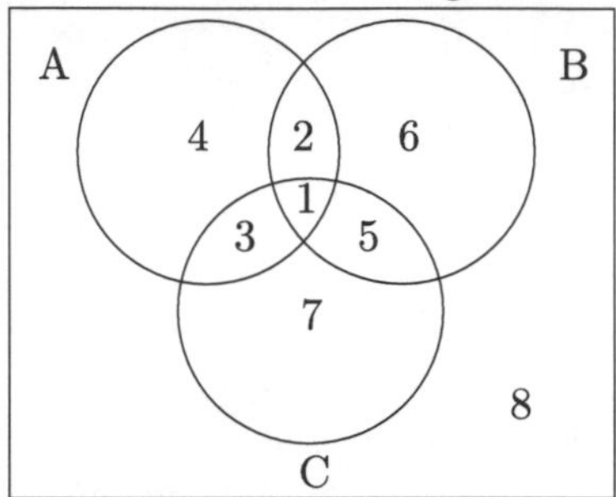

a. A&B
b. C&B&$\sim$A
c. $A \leftrightarrow B$
d. $A \rightarrow (B\&C)$
e. $(A\&B) \rightarrow C$
f. $A \rightarrow \sim(A\&B)$
g. $(A \vee B) \rightarrow (\sim B \vee C)$
h. $(A \rightarrow B) \vee (B \rightarrow C)$
i. $(A \leftrightarrow B) \vee (A \leftrightarrow C)$
j. $A \leftrightarrow (B \leftrightarrow C)$

2.2 Fundamental Connections between Logic and Probability

In order to apply probability to the analysis of inferences like $\{E \vee L, \sim E\} \therefore L$ we will add a geometrical interpretation to diagrams like diagram 2.1. This is illustrated in the following three diagrams, the first of which is simply a copy of diagram 2.1:

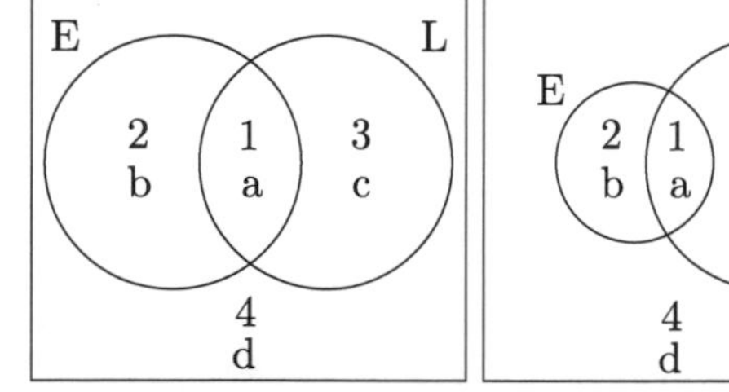

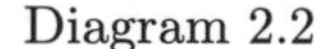
Diagram 2.2

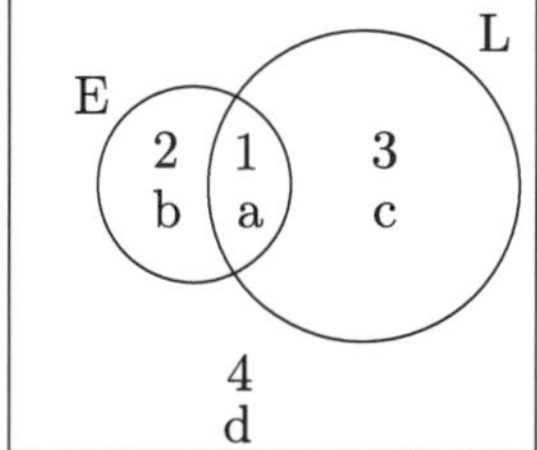

Diagram 2.3

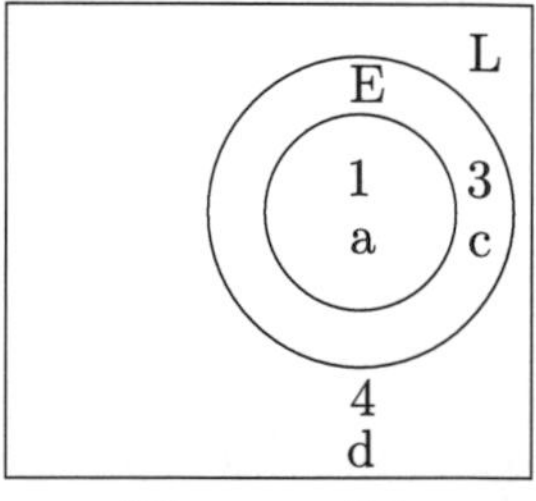

Diagram 2.4

Now we are going to assume that the *areas* of the regions in the diagrams represent the probabilities of the formulas they correspond to (assuming that the area of the 'universe' equals 1). The reason for having more than one diagram is that although all of the diagrams represent the same formulas, the probabilities of these formulas differ in the different diagrams. For instance, while E and L correspond to circles in all three diagrams, and L's probability is actually the same in all three, E's probability is larger in diagram 2.2 than it is in diagrams 2.3 and 2.4. These *possible probabilities* are in some ways similar to and in some ways dissimilar to the possible truth-values that E and L might have.

An extremely important similarity is that the diagrams represent not only the probabilities of E and L but the probabilities of the compound formulas that can be formed from them. For instance, the probability of ~E is represented by the area of the region outside of circle E and the probability of E ∨ L is the area of the union of regions E and L. But there is a fundamental dissimilarity between probability and truth.

Note that while E and L have the same probabilities in diagrams 2.3 and 2.4, the probability of E&L is smaller in 2.3 than it is in 2.4. What this shows is that E&L's probability is not determined by the probabilities of E and L, which is not like truth, since its truth-value is determined by the truth-values of E and L. This is also the case with E ∨ L, since while E and L have the same probabilities in 2.3 and 2.4, E ∨ L is more probable in 2.3 than it is in 2.4. In the terminology of logical theory, while conjunctions and disjunctions like and E&L and E ∨ L are *truth-functional* in the sense that their truth values are *functions* of the truth-values of E and L, they are not *probability functional* because their probabilities are not functions of the probabilities of E and L.[5] This means that we can't do for probabilities what truth-tables do for truth-values: that is, to calculate the probabilities of compound sentences like E&L and E ∨ L from the probabilities of their parts.

But it is equally important that while the probabilities of formulas like E&L and E∨L are not functions of the probabilities of the atomic *sentences* they are formed from, they are functions of the probabilities of the atomic regions they correspond to and the state-descriptions that correspond to them. For instance, E ∨ L corresponds to the union of atomic regions 1, 2, and 3, whose probabilities are a, b, and c, respectively, hence E ∨ L's probability is a+b+c. It is 'determined' by the probabilities of the state-

[5]This means that probabilities are not 'degrees of truth' as in *many-valued logic* (cf. Rosser and Turquette, 1952) or in the now widely discussed subject of *fuzzy logic* (cf. Zadeh, 1965). There are connections between probability and fuzzy logic, but we shall not enter into them here. Note, incidentally, that the negation of E, ~E, *is* probability-functional, which follows from the fact that its probability *plus* the probability of E must always add up to 1.

descriptions, which themselves form a *probability distribution*, as illustrated in table 2.2, which illustrates several other points as well.

There are ten important things to note in table 2.2, as follows:

(1) As previously pointed out, each of the atomic regions corresponds to an SD, and to a case or line in the truth-table, which occupies the black-bordered part of the entire table 2.2.[6] The student should be able to determine which SD and line any given atomic region corresponds to.

(2) As just noted, the set of probabilities or areas of all of the atomic regions constitute a probability distribution, in which probabilities adding up to 1 are 'distributed' among the atomic regions. Another way to picture a distribution is to write the probabilities directly into the corresponding regions in a Venn diagram. For instance, distribution #2 in table 2.2 can be pictured by a diagram like the following, in which areas are not drawn to scale because probabilities are already written into them:

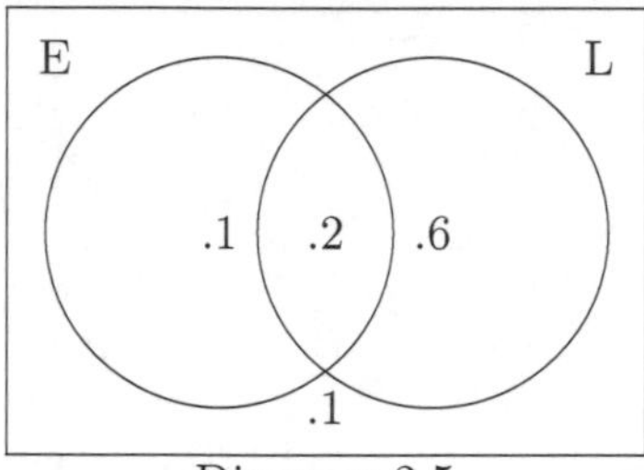

Diagram 2.5

(3) The left-hand columns of table 2.2 list four distributions, namely distributions #1, #2, and #3 plus a 'variable distribution', distribution #4, and the variety of these distributions brings out the fact that the probabilities that form them are entirely arbitrary except for the fact that they must be nonnegative and add up to 1.[7]

(4) Probability distributions are fundamental to probability logic, and possible distributions are to it as possible combinations of truth-values are to standard logic. In particular, they can be used to show that certain combinations of probabilities are *consistent* in the same way that truth-tables can be used to show that certain combinations of statements are consistent. For instance, distribution #2 'generates' probabilities of .9, .7, and .8, for $E \vee L$, $\sim$E, and L, respectively, and this proves that it is possible for these propositions to have these probabilities.

[6]Usually the converse is true, i.e., every line in the truth-table corresponds to an atomic region, although diagram 2.4 shows that this isn't always so. When a line doesn't correspond to a region we can imagine it corresponding to an 'empty region', whose area is 0.

[7]Assuming that a distribution can be *any* set of nonnegative numbers that add to 1 is what distinguishes probability *logic* from ordinary probability *theory*, which usually restricts itself to special classes of probability distributions.

probability distributions					truth values		formulas							
particular			#4				state descriptions				other formulas			
#1	#2	#3	variable	case	E	L	E&L	E&~L	~E&L	~E&~L	E∨L	~E	L	
.25	.2	0	a	1	T	T	T	F	F	F	T	F	T	
.25	.1	0	b	2	T	F	F	T	F	F	T	F	F	
.25	.6	1	c	3	F	T	F	F	T	F	T	T	T	
.25	.1	0	d	4	F	F	F	F	F	T	F	T	F	
							1	2	3	4	1,2,3	3,4	1,3	truth regions
							a	b	c	d	a+b+c	c+d	a+c	dist. #4 prob
							0	0	1	0	1	1	1	dist. #3 prob
							.2	.1	.6	.1	.9	.7	.8	dist. #2 prob
							.25	.25	.25	.25	.75	.5	.5	dist. #1 prob

Table 2.2

(5) However, there are obviously infinitely many possible probability distributions, and distributions #1–#3 are just three of them, while the variable distribution, distribution #4, includes #1–#3 and all other possible distributions as special cases. That there are infinitely many distributions means that you can't list all possible ones in a 'probability distribution table', and use that to determine whether any combination of probabilities is consistent, in the way that you can make a truth-table that lists all combinations of truth values that E and L can have, and use that to determine whether any combination of sentences is consistent.

(6) Distributions #1–#3 have special properties, as follows: distribution #1, which corresponds roughly to diagram 2.2, is a *uniform distribution* since it attaches the same probability to all SDs. It represents 'total uncertainty' as to which of E and L and their combinations is the case, and, as we will see, E and L must both be 50% probable in this distribution.

Distribution #2 is nonuniform, though it still embodies a high degree of uncertainty, and it corresponds roughly to diagram 2.3. It represents a state of affairs in which E is only 30% probable, while L is 80% probable.

Distribution #3, which doesn't correspond to a diagram, is not of a kind that is ordinarily considered in probability theory since it involves only certainties and not probabilities. You could say that ordinary formal logic pretends that these *certainty distributions* are the only possible probability distributions, and the only question is which of them is 'right'.

(7) There is a sense in which distributions #1 and #3 are at opposite ends of a continuum, going from total *un*certainty to complete certainty, with distribution #2 in between.[8]

(8) There is also a sense in which distributions #1–#3 could represent *evolving* probabilities, i.e., distribution #1 could give the 'right' probabilities at one time, distribution #2 could give the right ones at a subsequent time, and distribution #3 could represent the final stage of an inquiry in which all uncertainty is removed. As was noted in chapter 1, the fact that the probability of a proposition can change is another fundamental difference between it and truth. For instance, at one time it can be only 50%

[8]It is even possible to measure this 'distributional uncertainty' in terms of Information Theory's concept of *entropy* (cf. section 6 of Chapter 1 of Shannon and Weaver 1949). The entropic uncertainty of the variable distribution is defined as

$$-(\mathrm{a}\log\mathrm{a}+\mathrm{b}\log\mathrm{b}+\mathrm{c}\log\mathrm{c}+\mathrm{d}\log\mathrm{d}),$$

where logarithms are usually taken to the base 2 (the base determines the *unit* of uncertainty measurement). Given this, distribution #1's uncertainty is 2, which is the highest possible uncertainty of a distribution of this kind, distribution #2's uncertainty is 1.626, and Distribution #3's uncertainty is 0, which is the lowest possible uncertainty. Except for footnote comments, *distributional uncertainties* of this kind, which should not be confused with the *propositional uncertainties* that will be discussed in section 2.3, will not be discussed in this book.

probable that Jane will take logic, later it can become 80% probable, and finally it can become certain that she will take it, but the *truth* of "Jane will take logic" remains the same during all that time.[9] How probabilities should change will be returned to, but here the change looks as though it could come about as a result of acquiring *increasing information*, ultimately leading to complete certainty.

(9) The parts of table 1 outside the black-bordered part have to do with how probability distributions determine the probabilities of *formulas*, such as the ones at the heads of the columns on the right. The basic rule is very simple: the probability of any formula formed from E and L is the sum of the probabilities of the cases in which it is true. For instance, $E \vee L$ is true in cases 1, 2, and 3, so its probability is the sum of the probabilities of those cases, which are a, b, and c in the variable case. Similarly, because ~E is true in cases 3 and 4 its probability is the sum of the probabilities of those cases, hence it is c+d.

Of course, each SD, E&L, ~E&L, E&~L, and ~E&~L, is true in exactly one case, and therefore its probability is equal to the probability of the case in which it is true.

(10) Looking ahead to chapter 3, something may be said about the light that table 2.2 throws on the inference $\{E \vee L, {\sim}E\} \therefore L$. Distribution #3 makes its premises certain, and, as ordinary formal logic would lead us to expect, its conclusion has become certain in consequence. But distribution #2 may be more realistic, since it makes the premises probable but not certain. It also makes the conclusion probable. We would like to know, however, whether it is necessarily the case that conclusions are probable whenever premises are probable but not certain. This question will be answered in chapter 3.

Exercises

1. Calculate the probabilities of the formulas in exercise 1 of section 2.1, in distributions #1–#4. To simplify, you may want to write out the answers in a table, as follows:

		probabilities in distributions				
	formula	#1	#2	#3	#4	
a.	E↔L	.50	.3	0	a+d	probabilities
b.	~(E&~L)					of formulas
						↓

[9]More exactly, ordinary modern logic considers truth to be 'timeless'. However, in colloquial speech we sometimes speak of something as being true at one time and false at another, and it can be argued that such a view underlies Aristotle's famous 'sea fight tomorrow' argument (cf. McKeon (1941: 48)).

2. a. Can you construct a distribution in which $E \vee L$ and $\sim E$ have probabilities of .9 and .7, respectively, but the probability of L is less than .7?

   ⋆b. What is the smallest probability that L could have, if $E \vee L$ and $\sim E$ have probabilities of .9 and .7, respectively?

3. Fill in the blanks in table 2.3, which gives possible probability distributions for formulas formed from A, B, and C, corresponding to the regions in exercise 2 of the previous section.

prob. distributions particular					truth values			formulas				
#1	#2	#3	#4 variable	region	A	B	C	A&B	$A \leftrightarrow C$	$A\&(B \vee C)$	$(A \rightarrow B) \vee C$	
.125	.01	0	a	1	T	T	T					
.125	.02	0	b	2	T	T	F					
.125	.03	0	c	3	T	F	T					
.125	.04	0	d	4	T	F	F					
.125	.1	0	e	5	F	T	T					
.125	.2	0	f	6	F	T	F					
.125	.3	1	g	7	F	F	T					
.125	.3	0	h	8	F	F	F					
				→								truth regions
			→									Dist. #4 prob.
		→										Dist. #3 prob.
	→											Dist. #2 prob.
→												Dist. #1 prob.

Table 2.3

4. a. Construct a probability distribution for A, B, and C in which all of these formulas have a probability of .9, but A&B&C has a probability less than .8.

   ⋆b. What is the smallest probability that $A \vee B \vee C$ can have if each of A, B, and C have probability .9?

   ⋆c. What is the smallest probability that $(A\&B) \vee (B\&C) \vee (C\&A)$ can have if each of A, B, and C have probability .9?

2.3 Probability Functions and Algebra

Now we will start to use a more compact way of talking about probabilities, and instead of writing "the probability of $E \vee L$", "the probability of $\sim E$", "the probability of L", and so on, we will use *probability function expressions*, "p(E$\vee$L)", "p($\sim$E)", "p(L)", etc.[10] This allows us to express relations involving probabilities much more compactly than before, so that, for instance, instead of saying that the probability of L is less than the probability of E $\vee$ L we can write $p(L) < p(E \vee L)$. This also allows us to state principles of the *algebra of probability* in a simple way.

As we have seen, the probabilities of compound formulas are not generally *functions* of the probabilities of their parts, but they are in certain special cases. For instance, the probability of $\sim$E is a function of the probability of E, and this can be expressed algebraically by the equation:

$$p(\sim E) = 1 - p(E).$$

More generally, for any formula, ϕ, we can write:

$$p(\sim\phi) = 1 - p(\phi).$$

And, while we can't say that p(L) is always less than p(E$\vee$L) (see diagram 2.4), it is always true that $p(L) \leq p(E \vee L)$, and, more generally, for any formulas ϕ and ψ,

$$p(\phi) \leq p(\phi \vee \psi).$$

The equation and inequality above are particular cases of *general laws of probability* which, taken together, constitute a general *theory of probability*, and, though it is somewhat beside the point for present purposes, it is of some interest to describe some of its features.

2.4⋆ Deductive Theory of Probability

All of the general laws of the theory of probability can be deduced from a very simple set of 'axioms', known as the Kolmogorov Axioms,[11] which state fundamental relations between probability and logic. These are: for all formulas ϕ and ψ

K1. $0 \leq p(\phi) \leq 1$.
K2. If ϕ is logically true then $p(\phi) = 1$.
K3. If ϕ logically implies ψ then $p(\phi) \leq p(\psi)$.
K4. If ϕ and ψ are logically inconsistent then $p(\phi \vee \psi) = p(\phi) + p(\psi)$.

Certain theorems of probability that can be deduced from these axioms will be proved below, but first there are two important things to note about the axioms. One is the close connection between the concepts of pure logic

[10] When you read the expression "p(L)" out loud you will still *say* "the probability of L", but, technically, the "p( )" that is common to "p(E $\vee$ D)", "p($\sim$E)", and "p(L)" is a *probability function*, and E $\vee$ D, $\sim$E, and L are some of its possible arguments.

[11] Cf. Kolmogorov (1950: 2). Kolmogorov's axioms are stated in terms of *sets* rather than formulas and there are other differences of formulation, but for our purposes they are not essential. Sometimes the axioms are stated to include an 'infinite generalization' of K4, which is essential to most advanced mathematical probability theory but not to its present application.

and the laws of probability that is brought out by axioms K2, K3, and K4, which connect probabilities with logical truth, logical consequence (entailment), and logical consistency. Pure logic is a prerequisite to probability logic because the probability axioms, and the theorems that follow from them, depend on the concepts of logical truth, consequence, and consistency.[12] The other thing to note is the connection between logical truth, or *logical certainty* as it is sometimes called, and 'practical' or 'probabilistic' certainty, which follows from it according to axiom K2. Some philosophers have held that the only things we can be *really* certain of are 'truths of pure logic', which would amount to saying that the converse of axiom K2 ought to hold as well, i.e., that $p(\phi)$ can *only* equal 1 if ϕ is a logical truth.[13] But the fact that we consider that certainty distributions like distribution #3 in table 2.2 are allowed shows that we consider it to be possible for a proposition to be practically certain without being logically true. And, in this work when we say that a proposition is certain we will mean that it is practically, or probabilistically certain, in the sense of having probability 1, and not necessarily that it is logically certain.[14]

Now, though we will not develop the theory systematically, we will derive some *theorems of probability* from axioms K1–K4. Both of the laws stated above, that $p(\sim\phi) = 1 - p(\phi)$ and $p(\phi) \leq p(\phi \vee \psi)$, can easily be deduced from these axioms, as follows:

Theorem 1. $p(\phi) \leq p(\phi \vee \psi)$.

Proof[15]

1. ϕ logically implies $\phi \vee \psi$. By pure logic.
2. If ϕ logically implies $\phi \vee \psi$ then $p(\phi) \leq p(\phi \vee \psi)$. From K3.
3. $p(\phi) \leq p(\phi \vee \psi)$, from 1 and 2. By *modus ponens.* QED

[12]Certain writers, notably K. R. Popper, have developed probability theory independently of concepts of pure logic (cf. New appendices iv and v of Popper (1959)), but this approach is more complicated than the present one. Aspects of this approach are described in appendix 2.

[13]Probability functions that satisfy the converse of K2, i.e., ones for which $p(\phi) = 1$ holds only if ϕ is logically true, are sometimes called *regular* probability functions.

[14]Another thing to note about axiom K2 is that although it implies that logical truths have probability 1, this does not mean that they are certain *a priori*, before they are proved. It was pointed out in section 2.1 that probabilities change with circumstances, and one thing that can bring about this change is proving a proposition. But the Kolmogorov axioms only concern 'static probabilities', that apply to propositions at one time, and beyond that they idealize by presupposing that all of the purely logical properties and relations among the propositions they apply to are known. Thus, if ϕ is a logical truth that is assumed to be known, and in those circumstances $p(\phi)$ should equal 1. Similarly, if ϕ logically implies ψ that is assumed to be known, and therefore $p(\phi) \leq p(\psi)$.

[15]The proofs given here will be more informal than derivations that the student may have learned to give in courses in logic, but they are not entirely informal. Specifically, the steps in the proofs are numbered, and each step involves a *statement*, followed by

Theorem 2. $p(\sim\phi) = 1 - p(\phi)$.

Proof

1. $\sim\phi$ and ϕ are logically inconsistent. By pure logic.
2. If $\sim\phi$ and ϕ are logically inconsistent then $p(\sim\phi \vee \phi) = p(\sim\phi) + p(\phi)$. From K4.
3. $p(\sim\phi \vee \phi) = p(\sim\phi) + p(\phi)$. From 1 and 2 by *modus ponens.*
4. $\sim\phi \vee \phi$ is logically true. By pure logic.
5. If $\sim\phi \vee \phi$ is logically true then $p(\sim\phi \vee \phi) = 1$, from K2.
6. $p(\sim\phi \vee \phi) = 1$. From 4 and 5, by *modus ponens.*
7. $1 = p(\sim\phi) + p(\phi)$. From 3 and 6, by pure algebra.
8. $p(\sim\phi) = 1 - p(\phi)$. From 7 by pure algebra. QED

Theorem 3. If ϕ is logically false then $p(\phi) = 0$.

Proof

1. Suppose that ϕ is logically false.
2. $\sim\phi$ is logically true. From 1, by pure logic.
3. If $\sim\phi$ is logically true then $p(\sim\phi) = 1$. From K2.
4. $p(\sim\phi) = 1$. From 2 and 3 by *modus ponens.*
5. $p(\sim\phi) = 1 - p(\phi)$. From theorem 2.
6. $1 - p(\phi) = 1$. From 4 and 5 by pure algebra.
7. $p(\phi) = 0$. From 7 by pure algebra. QED

Combined with axiom K3, theorem 3 has an interesting consequence in application to the law that anything follows from a contradiction. The law is that if ϕ is logically false then it logically implies any proposition ψ, and therefore according to axiom K3, it follows that $p(\phi) \leq p(\psi)$. On the other hand, theorem 3 says that if ϕ is a contradiction then $p(\phi) = 0$, which, combined with the fact that $p(\phi) \leq p(\psi)$, simply says that $0 \leq p(\psi)$ for any proposition ψ. But this really tells us nothing at all, since axiom K1 already tells us that $0 \leq p(\psi)$. Therefore while it may seem paradoxical to say that anything follows from a contradiction, this has no serious consequences in probability logic because it tells us nothing about the probabilities of the propositions that follow from them. This will be returned to in chapter 4, in which possible changes in probabilities are considered, and which will bring out another peculiarity of the principle.

Three more theorems are very important for our purposes, though only the first will be proved:

---

the *reason* or justification for making it. For instance, the statement in step 2 in the proof of theorem 1 is

If ϕ logically implies $\phi \vee \psi$ then $p(\phi) \leq p(\phi \vee \psi)$. From K3.

and the justification for making it is that it follows from axiom K3—in this case by replacing the Greek letter "ψ" in the axiom by "$\phi \vee \psi$". Most mathematical proofs are much less formal than this, but the student is urged to use this 'statement-reason form' as much as possible.

Theorem 4. If ϕ and ψ are logically equivalent then $p(\phi) = p(\psi)$.

Proof

1. Suppose that ϕ and ψ are logically equivalent.
2. Then ϕ logically implies ψ and ψ logically implies ϕ. By pure logic.
3. If ϕ logically implies ψ then $p(\phi) \leq p(\psi)$. By K3.
4. $p(\phi) \leq p(\psi)$. From 2 and 3 by *modus ponens.*
5. If ψ logically implies ϕ then $p(\psi) \leq p(\phi)$. By K3.
6. $p(\psi) \leq p(\phi)$. From 2 and 5 by *modus ponens.*
7. $p(\phi) = p(\psi)$. From 4 and 6 by pure algebra. QED

Theorem 5. $p(\phi) + p(\psi) = p(\phi \& \psi) + p(\phi \vee \psi)$.

Proof left as exercise.

Theorem 6. Probability Sum Theorem. If $\phi_1,\ldots,\phi_n$ are mutually exclusive (any two of them are logically inconsistent) then:

$$p(\phi_1 \vee \ldots \vee \phi_n) = p(\phi_1) + \ldots + p(\phi_n).$$

The proof of this theorem requires mathematical induction because it involves arbitrarily many formulas, $\phi_1,\ldots,\phi_n$, and we will give it in detail only the special case in which n = 3. The student should be able to figure out how to prove cases n = 4, n = 5, and so on.

Theorem 6.3. If ϕ_1, ϕ_2, and ϕ_3 are mutually exclusive (any two of them are logically inconsistent) then:

$$p(\phi_1 \vee \phi_2 \vee \phi_3) = p(\phi_1) + p(\phi_2) + p(\phi_3).$$

Proof

1. Suppose that ϕ_1, ϕ_2, and ϕ_3 are mutually exclusive.
2. If ϕ_1 and ϕ_2 are mutually exclusive then $p(\phi_1 \vee \phi_2) = p(\phi_1) + p(\phi_2)$. By K4.
3. $p(\phi_1 \vee \phi_2) = p(\phi_1) + p(\phi_2)$. From 1 and 2 by *modus ponens.*
4. $\phi_1 \vee \phi_2$ and ϕ_3 are mutually exclusive. From 1, by pure logic.
5. If $\phi_1 \vee \phi_2$ and ϕ_3 are mutually exclusive then $p((\phi_1 \vee \phi_2) \vee \phi_3) = p(\phi_1 \vee \phi_2) + p(\phi_3)$. By K4.
6. $p((\phi_1 \vee \phi_2) \vee \phi_3) = p(\phi_1 \vee \phi_2) + p(\phi_3)$. From 4 and 5, by *modus ponens.*
7. $p(\phi_1 \vee \phi_2 \vee \phi_3) = p(\phi_1) + p(\phi_2) + p(\phi_3)$. From 3 and 6, by pure algebra. QED

Theorems 4 and 6 are important because they justify the method of calculating the probability of a compound formula by breaking it up into a disjunction of SDs, and then adding the probabilities of the SDs. In fact, if any formula, ϕ, is logically equivalent to $\phi_1 \vee \ldots \vee \phi_n$, where $\phi_1, \ldots, \phi_n$ are mutually exclusive, then theorem 4 tells you that $p(\phi) = p(\phi_1 \vee \ldots \vee \phi_n)$, and theorem 6 tells you that $p(\phi_1 \vee \ldots \vee \phi_n) = p(\phi_1) + \ldots + p(\phi_n)$, hence $p(\phi) = p(\phi_1) + \ldots + p(\phi_n)$. This is particularly useful when the component probabilities, $p(\phi_1), \ldots, p(\phi_n)$, are themselves easy to calculate, as they are in the case of *probabilistic independence*, to be discussed in section 2.5⋆. However, because the probabilities of the SDs can be arbitrary numbers that sum to 1 they will not usually be independent in the probabilistic sense. This is one of the consequences of the final series of theorems of this section, some of which are negative in that they show that things like independence are not always the case. The proofs of these theorems are shortened in some cases by combining two or more steps into one.

Theorem 7. $p(\phi\&\psi) \geq p(\phi) + p(\psi) - 1$.

Proof

1. $(\phi\&{\sim}\psi) \vee (\phi\&\psi)$ is logically equivalent to ϕ. By pure logic.
2. $p[(\phi\&{\sim}\psi) \vee (\phi\&\psi)] = p(\phi)$ From theorem 4 and step 1 (combining two steps into one).
3. $\phi\&{\sim}\psi$ and $\phi\&\psi$ are mutually exclusive. Pure logic.
4. $p[(\phi\&\psi) \vee (\phi\&{\sim}\psi)] = p(\phi\&\psi) + p(\phi\&{\sim}\psi)$. From step 3 and theorem 6 (combining two steps).
5. $p(\phi\&\psi) + p(\phi\&{\sim}\psi) = p(\phi)$. From steps 2 and 4.
6. $\phi\&{\sim}\psi$ entails ${\sim}\psi$. Pure logic.
7. $p(\phi\&{\sim}\psi) \leq p({\sim}\psi)$. From axiom K3 and step 6.
8. $p({\sim}\psi) = 1 - p(\psi)$. From theorem 2.
9. $p(\phi\&\psi) \geq p(\phi) + p(\psi) - 1$. From steps 6, 7, and 8 by algebra. QED

Theorem 8. If $\phi_1,\ldots,\phi_n$ entail ϕ and $p(\phi_1) = \ldots = p(\phi_n) = 1$ then $p(\phi) = 1$, and if they do not entail ϕ then there exists a probability function p such that $p(\phi_1) = \ldots = p(\phi_n) = 1$ but $p(\phi) = 0$.

Proof that if $\phi_1, \ldots, \phi_n$ entail ϕ and $p(\phi_1) = \ldots = p(\phi_n) = 1$ then $p(\phi) = 1$ in the case where $n = 2$.

1. Suppose that ϕ_1 and ϕ_2 entail ϕ and $p(\phi_1) = p(\phi_2) = 1$.
2. $p(\phi_1\&\phi_2) \geq p(\phi_1) + p(\phi_2) - 1$. From theorem 7.
3. $p(\phi_1\&\phi_2) \geq 1 + 1 - 1 = 1$. From steps 1 and 2 by algebra.
4. $\phi_1\&\phi_2$ entails ϕ. From step 1 by pure logic.
5. $p(\phi) \geq p(\phi_1\&\phi_2)$ From step 4 and K3.
6. $p(\phi) \geq 1$. From steps 3 and 5 by algebra.
7. $p(\phi) \leq 1$. From K1.
8. $p(\phi) = 1$. From steps 6 and 7. QED

Proof that if $\phi_1,\ldots,\phi_n$ do not entail ϕ then there exists p( ) such that $p(\phi_1) = \ldots = p(\phi_n) = 1$ but $p(\phi) = 0$.

1. Suppose that $\phi_1,\ldots,\phi_n$ do not entail ϕ.
2. There is a line in the truth table in which $\phi_1,\ldots,\phi_n$ are true but ϕ is false.[16] From 1, by pure logic.
3. The distribution that attaches probability 1 to that line and probability 0 to all other lines is a probability distribution. Let p( ) be the probability function generated by this distribution.
4. $p(\phi_1) = \ldots = p(\phi_n) = 1$ and $p(\phi) = 0$. From 3. QED

The most important thing that theorem 8 establishes is something that we have taken for granted all along: namely that so long as the premises of a valid inference are certainties the conclusion must also be a certainty. However, it also establishes something that may be less obvious: even if the premises are certainties the conclusion can have 0 probability if the inference is not logically valid. This means that so called *inductive reasoning* on which we most depend in our daily lives, and which will be discussed briefly in section 4.6, cannot logically guarantee any probability at all in its conclusions.

The last point is related to a technical matter in the proof of the second part of theorem 8. Step 3 in the second part of the argument only follows if axioms K1–K4 are interpreted as asserting that any function satisfying them is a probability function.[17] If some functions that satisfied the axioms were not probability functions, we would not be justified in claiming that the inductive reasoning to be commented on in section 4.5 cannot logically guarantee any probability in its conclusions.[18] In fact, the following section will discuss a special kind of assumption that is not justified by the axioms, but which is often made about probabilities, and which we will later see would justify certain kinds of inductive reasoning.

But let us end this section by considering *nontheorems*, which can be just as important to know about as theorems. An example of a nontheorem

[16]This step assumes that $\phi_1,\ldots,\phi_n$ and ϕ are formulas of the sentential calculus, to which truth-table methods apply. In fact, however, the theorem applies to formulas in any 'ordinary logic' such as the predicate and identity calculi discussed in appendices 7, 8, and 9.

[17]But note that the function p( ) considered here is a certainty distribution, and it is not regular in the sense of footnote 13.

[18]In fact, derivability from the purely logical Kolmogorov axioms can be regarded as the dividing line that separates deductive from inductive logic (though chapter 6's assumption that the probability of a conditional is a conditional probability can still be regarded as purely logical). That, logically, the conclusions of nondeductive inferences can have 0 probability when their premises are certainties reflects Hume's contention that future contingencies not only cannot be guaranteed with certainty, they cannot even be guaranteed with any probability.

is that p(E&L) should equal p(E) $\times$ p(L).[19] To prove that this is not a theorem, we give a *probability counterexample* to it, i.e., we describe a probability distribution that generates probabilities for p(E), p(L), and p(E&L) such that p(E&L) $\neq$ p(E) $\times$ p(L). In fact, distribution #2 in table 2.2 is a counterexample in this sense, since it generates probabilities such that p(E) = .3 and p(L) = .8, but p(E&L) = .2, which is not equal to .3 $\times$ .8. Using this method you can also prove that general propositions like the converse of K2, that if $p(\phi) = 1$ then ϕ is logically true, are not theorems. Substituting ~E&L for ϕ, distribution #3 in table 2.2 makes p(~E&L) equal to 1 although it is not logically true, which is a probability counterexample to the general claim that for any ϕ, if $p(\phi) = 1$ then ϕ must be logically true.

The exercises at the end of this section ask you to find probability counterexamples to other nontheorems, but before that we must stress that the fact that p(E&L) = p(E) $\times$ p(L) isn't 'logically true' in probability logic doesn't prove that it is false—just as the fact that the statement "All men are mortal" isn't logically true doesn't prove that it is false. That p(E&L) should equal p(E) $\times$ p(L) means that E and L are probabilistically independent in a sense already alluded to, which will be returned to in the following section, where we will see that it is assumed in many applications of probability. What our probability counterexample shows, however, is that this is a *factual* assumption, which cannot be justified by the pure laws of probability logic.[20]

Exercises

1. Prove the following theorems (while these are not stated in terms of *variables* like ϕ, ψ, etc., they are proved in the same way)

   a. $p(L) \leq p(E \rightarrow L)$.

   b. $p(A) + p(B) = p(A \vee B) + p(A\&B)$ (theorem 5).

   ⋆c. $p(A \vee B \vee C) = p(A) + p(B) + p(C) - p(A\&B) - p(A\&C) - p(B\&C) + p(A\&B\&C)$.[21]

   ⋆d. Prove theorem 6 in the case where n = 4.

---

[19] More generally, the probability of a conjunction may not equal the product of the probabilities of its conjuncts, which contrasts with the fact that the probability of a disjunction of mutually exclusive disjuncts is always equal to the *sum* of the probabilities of the disjuncts (the General Sum theorem).

[20] Section 9.9⋆⋆ will discuss the difficult question of what it is for a statement like p(E&L) = p(E) $\times$ p(L) to be 'factually true', or 'right' under certain circumstances.

[21] This is a *Poincaré Formula*, the general form of which gives the probability of a disjunction $\phi_1 \vee \ldots \vee \phi_n$ in terms of $p(\phi_1), \ldots, p(\phi_n)$, $p(\phi_1\&\phi_n), \ldots, p(\phi_{n-1}\&\phi_n)$, $p(\phi_1\&\phi_2\&\phi_3), \ldots$ and so on.

2. Give counterexamples to the following:
   a. If $p(E) = p(L)$ then $p(E\&B) = p(L\&B)$.
   b. $p(E) \neq p(\sim E)$.
   c. $p(L\&B) < p(L)$.

3. Which of the following are theorems? Give counterexamples to the nontheorems.
   a. $p(E \rightarrow L) = p(L) - p(E)$.
   b. $p(E \rightarrow L) = p(L) + p(E\&\sim L)$.
   c. $p(E\&L) = 1 - p(\sim E \vee \sim L)$.
   d. $p(L) + 1 \geq p(E \vee L) + p(\sim E)$.
   e. $p(E \leftrightarrow L) = p(E \rightarrow L) + p(L \rightarrow E)$.
   f. If $p(\phi) > p(\psi)$ then $p(\sim\phi) < p(\sim\psi)$.
   g. If $p(E\&L) = p(E) \times p(L)$ then $p(\sim E\&\sim L) = p(\sim E) \times p(\sim L)$.

4. a. Give an argument that for any formula ϕ, $p(\phi) = .5$ cannot be logically true, i.e., true for all probability functions p( ).
   ⋆b. Give an argument that for any given formulas ϕ and ψ, $p(\phi\&\psi) = p(\phi) \times p(\psi)$ can only be *logically* true, i.e., true for all probability functions p( ), if at least one of ϕ or ψ is logically true or logically false.

2.5⋆ Probabilistic Independence, Symmetry, and Randomness

Consider the following problem: You want to calculate the probability of a proposition K, that three persons, each of whom picks one of the numbers 1, 2, or 3 at random, will all pick different numbers. There are six mutually exclusive ways, A–F, in which this can happen, which can be represented in a table:

	person 1 picks	person 2 picks	person 3 picks
A	1	2	3
B	1	3	2
C	2	1	3
D	2	3	1
E	3	1	2
F	3	2	1

Since A,...,F are mutually exclusive, the probability that the three persons will all pick different numbers is the sum of the probabilities of these six 'cases'; i.e., $p(K) = p(A)+\ldots+p(F)$. But how do we calculate the probabilities of the cases? Let us concentrate on the case A, in which each person picks his own number.

In saying each person picks his number at random we are implicitly assuming that what one person does doesn't influence what the others do. That is, we are assuming *probabilistic independence*—the probability of person #1 picking 1 *and* person #2 picking 2 *and* person #3 picking 3 equals the probability of person #1 picking 1 *times* the probability of person #2 picking 2, *times* the probability of person # 3 picking 3.[22] Thus, we are assuming that

$$\text{p(A)} = \text{p(person \#1 picks 1)} \times \text{p(person \#2 picks 2)} \times \text{p(person \#3 picks 3)}.$$

Combining this with the assumption that persons #1, #2, and #3 are alike in their probabilities, hence p(person #1 picks 1) = p(person #2 picks 2) = p(person #3 picks 3), would imply that p(A) = p(person #1 picks $1)^3$. And, 'symmetries' of this kind lead to the conclusion that all of the probabilities $p(A),\ldots,p(F)$ actually equal 1/27, and therefore $p(K) = p(A)+\ldots+p(F) = 6/27 = 2/9 = .222$.

The assumption that the probability of a conjunction is the *product* of the probabilities of its parts is part of what is involved in randomness, which is very commonly assumed in mathematical statistics.[23] But it is easy to see that the assumption is not logically valid, since it is easy to imagine situations in which it would be false—for instance, person #1 might order the others to pick the same number that he picked, in which case there would be 0 chance that each person would pick his own number.[24] That independence and randomness are not logically valid is why we shall not generally assume them in this work, which concentrates on the consequences of the purely logical laws of probability.

But before concluding with these nonlogical assumptions we will note one more common application of them, which arises in reasoning about errors. Suppose that one person tells you E, that Jane will take ethics, and another persons tells you L, that she will take logic, but neither person is perfectly reliable. In these circumstances you might assume that while E and L aren't perfect certainties, they have equal probabilities, p, of being false, and these 'errors' are independent. Now, if you wanted to calculate

---

[22] *Probabilistic independence* is not the same as *causal independence*, where the conjuncts don't influence each other causally, and of course both of them differ from logical independence, as when the axioms of a theory are said to be independent. But the three kinds of independence are related, and probabilistic and causal independence are often mistaken for one another (cf. exercise ⋆4 at the end of this section).

[23] Especially in the *theory of sampling*, as when attempts are made to estimate general properties of large 'populations', e.g., the average (arithmetic mean) of the number of brothers and sisters that persons in a population have, from the average in a sample drawn at random from the population.

[24] Note too the 'result' that exercise 4b at the end of the previous section asked you to prove, namely that the probability of a conjunction can only be logically guaranteed to equal the product of the probabilities of its parts if one of the parts is either logically true or logically false.

the chance of Jane's taking either ethics or logic, p(E∨L), you might reason as follows. E∨L is equivalent to ∼ (∼E&∼L), hence according to theorems 2 and 4, p(E∨L) = p(∼ (∼E&∼L)) = 1−p(∼E&∼L). If p(∼E) = p(∼L) = p, and if errors are independent, the probability that both E and L are false should equal the probability of E's being false multiplied by the probability of L's being false, which would imply that p(∼E&∼L) = p(∼E) × p(∼L) = p^2, and therefore p(E ∨ L) = 1 − p^2. For instance, if there is 90% chance that Jane will take ethics and a 90% chance that she will take logic then $p = 1 - .9 = .1, p^2 = .01$, and p(E ∨ L) = 1 − .01 = .99; i.e., there is a 99% chance that Jane will take at least one of ethics and logic.

But of course you can't deduce the conclusion just arrived at from pure probability logic without assuming independence, which we have seen isn't a logically valid principle of probability. In fact if the probabilities were generated by distribution #2 in table 2.2 p(∼E) would be .7 and p(∼L) would be .2, hence p(∼E) × p(∼L) would be .14 but p(∼E&∼L) would be .1. Cases like this arise in real life, for example if Jane were certain either to take both ethics and logic or to take neither. As with symmetry, the independence assumption is seldom if ever logically valid, and we will not be much concerned with it here. This differs from standard courses in probability and statistics, where it is usually taken for granted.[25]

Exercises

1. a. Give a proof like those in section 2.3 that if A and B are probabilistically independent then so are A and ∼B.

   ⋆b. Give a proof like those in section 2.3 that if A, B, and C are probabilistically independent then so are A, B, and ∼C.

   ⋆c. Argue informally that if $A_1, \ldots, A_n$ are independent so that the probability of a conjunction of any of them is equal to the product of the probabilities of the members of the conjunction, then the same thing is true of any combination of $\pm A_1, \ldots, \pm A_n$, (where ±A stands either for A or for ∼A).

2. a. Propositions A, B, and C are *pairwise independent* if any two of them are independent. Find a distribution in which this is true, but all three of them are not independent; i.e., a distribution that generates probabilities such that p(A&B) = p(A) × p(B), p(A&C) = p(A) × p(C), and p(B&C) = p(B) × p(C), but p(A&B&C) ≠ p(A) × p(B) × p(C).

   ⋆b. Argue that there are probabilities such that any n − 1 of the propositions $A_1, \ldots, A_n$ are independent (the probability of the

[25]This is in part because these courses focus on random phenomena like tossing coins or rolling dice, where you are justified in assuming that one toss or roll doesn't influence another.

conjunction of any $n-1$ of them equals the product of the probabilities of these $n-1$ propositions), but the whole set of n propositions are not independent.

3. Prove that if any propositions ϕ and ψ are mutually exclusive and independent then at least one of them must be either logically true or logically false.

⋆4. The 3 marbles problem.[26] Probabilistic dependence, in which $p(A \& B) \neq p(A) \times p(B)$, should be distinguished from *causal dependence*, in which A causes or 'produces' B (cf. footnote 22). That is because probabilistic dependence is *symmetric*, i.e., if A is probabilistically dependent on B then B is probabilistically dependent on A, while causal dependence is *asymmetric*, i.e., if A produces B then B does not produce A. In spite of the difference, however, probabilistic reasoning is often influenced by causal considerations, and sometimes the result can be very confusing. Here is an example. Two persons, A and B, compete to see which of them can roll marbles closest to a 'target', but with the stipulation that although they are equally skilled, person A is allowed *one* roll while person B is allowed two. The problem is to figure out the probability that A will win if the rolls are independent, there is no chance of a tie, and A rolls first. One way to figure it is that A should have 1 chance in 3 of winning, since he has one roll and B has two, and they are equally skilled. But another way to figure it is that A should only have 1 chance in 4 of winning, since any one of B's rolls has a 50-50 chance of beating A's roll, and B has two independent chances of doing this. Which answer do you think is right, and why?

2.6 Uncertainty and Its Laws

The *uncertainty* of a proposition is the probability that it is false, which is 1 minus its probability. The uncertainty of ϕ will be written as $u(\phi)$, which is defined as $1-p(\phi)$, and we will call u( ) the uncertainty function *associated* with p( ).[27] We will see that using this symbolism makes it much easier to state important laws that connect logical deduction and probability. The simplest of these is an immediate consequence of axiom K3 and theorem 5:

[26] From Northrop (1944: 165). The solution given in the *Answers* follows the one given on p. 222 of this work.

[27] Three things should be noted about this definition. (1) The maximum possible uncertainty is 1, which is really a kind of certainty, namely certain falsity. (2) Whether or not this uncertainty equals 1 it differs from the *entropic*, distributional uncertainty commented on in footnote 8, which applies not to single propositions but rather to sets of mutually exclusive and exhaustive propositions. (3) This concept of uncertainty will be generalized in chapter 6, in application to so called *probability conditionals*, where it will still equal 1 minus probability, but it won't be the probability that something is false. How that is possible will be explained.

Theorem 9. If ϕ logically implies ψ then $u(\psi) \leq u(\phi)$, and if ϕ does not logically imply ψ then there is an uncertainty function u such that $u(\phi) = 0$ but $u(\psi) = 1$.

The proof of the first half of this is simple:

1. Suppose that ϕ logically implies ψ.
2. If ϕ logically implies ψ then $p(\phi) \leq p(\psi)$. By K3.
3. $p(\phi) \leq p(\psi)$. From 1 and 2 by *modus ponens.*
4. $u(\phi) = 1 - p(\phi)$ and $u(\psi) = 1 - p(\psi)$. By the definition of $u(\phi)$ and $u(\psi)$.
5. $p(\phi) = 1 - u(\phi)$ and $p(\psi) = 1 - u(\psi)$. From 4 by pure algebra.
6. $1 - u(\phi) \leq 1 - u(\psi)$. From 3 and 5 by pure algebra.
7. $u(\psi) \leq u(\phi)$. From 6 by pure algebra. QED

The second half of this theorem is negative, but it is also fairly simple and it will be left as an exercise.

Theorem 10. If ϕ is logically true then $u(\phi) = 0$, and if it is logically false then $u(\phi) = 1$.

Proof of the first part of the theorem.

1. Suppose that ϕ is logically true.
2. If ϕ is logically true then $p(\phi) = 1$. By K2.
3. $p(\phi) = 1$. From 1 and 2 by *modus ponens.*
4. $u(\phi) = 1 - p(\phi)$. By the definition of uncertainty.
5. $u(\phi) = 0$. From 3 and 4 by pure algebra. QED

Now we will state two theorems that are of very great importance in application to probabilistic aspects of deduction.

Theorem 11. Uncertainty Sum Theorem

$$u(\phi_1\&\ldots\&\phi_n) \leq u(\phi_1) + \ldots + u(\phi_n).$$

As with the probability sum theorem (theorem 6), this theorem has to be proved by mathematical induction, and we will confine ourselves here to proving the case where n=2, and ask the student to prove other cases in the exercises at the end of this section.

Theorem 11.2. $u(\phi_1\&\phi_2) \leq u(\phi_1) + u(\phi_n)$.

Proof

1. ϕ_1 is logically equivalent to $(\phi_1\&\phi_2) \vee (\phi_1\&{\sim}\phi_2)$. By pure logic.
2. If ϕ_1 is logically equivalent to $(\phi_1\&\phi_2) \vee (\phi_1\&{\sim}\phi_2)$ then $p(\phi_1) = p((\phi_1\&\phi_2) \vee (\phi_1\&{\sim}\phi_2))$. By theorem 4.
3. $p(\phi_1) = p((\phi_1\&\phi_2) \vee (\phi_1\&{\sim}\phi_2))$. From 1 and 2 by *modus ponens.*
4. $\phi_1\&\phi_2$ and $\phi_1\&{\sim}\phi_2$ are mutually exclusive. By pure logic.

5. If $\phi_1\&\phi_2$ and $\phi_1\&{\sim}\phi_2$ are mutually exclusive then $p((\phi_1\&\phi_2) \vee (\phi_1\&{\sim}\phi_2)) = p(\phi_1\&\phi_2) + p(\phi_1\&{\sim}\phi_2)$. By K4.
6. $p((\phi_1\&\phi_2) \vee (\phi_1\&{\sim}\phi_2)) = p(\phi_1\&\phi_2) + p(\phi_1\&{\sim}\phi_2)$. From 4 and 5 by *modus ponens.*
7. $p(\phi_1) = p(\phi_1\&\phi_2) + p(\phi_1\&{\sim}\phi_2)$. From 3 and 6 by algebra.
8. $p(\phi_1)+p(\phi_2) = p(\phi_1\&\phi_2)+p(\phi_1\&{\sim}\phi_2)+p(\phi_2)$. From 7 by algebra.
9. $(\phi_1\&{\sim}\phi_2) \vee \phi_2$ is logically equivalent to $\phi_1 \vee \phi_2$. By pure logic.
10. If $(\phi_1\&{\sim}\phi_2) \vee \phi_2$ is logically equivalent to $\phi_1 \vee \phi_2$ then $p((\phi_1\&{\sim}\phi_2) \vee \phi_2) = p(\phi_1 \vee \phi_2)$. By theorem 4.
11. $p((\phi_1\&{\sim}\phi_2) \vee \phi_2) = p(\phi_1 \vee \phi_2)$. From 9 and 10 by *modus ponens.*
12. $p(\phi_1) + p(\phi_2) = p(\phi_1\&\phi_2) + p(\phi_1 \vee \phi_2)$. From 8 and 11 by algebra.[28]
13. $u(\phi_1) = 1 - p(\phi_1)$, $u(\phi_2) = 1 - p(\phi_2)$, $u(\phi_1\&\phi_2) = 1 - p(\phi_1\&\phi_2)$, and $u(\phi_1 \vee \phi_2) = 1 - p(\phi_1 \vee \phi_2)$. By the definition of uncertainty.
14. $u(\phi_1)+u(\phi_2) = u(\phi_1\&\phi_2)+u(\phi_1 \vee \phi_2)$. From 12 and 13 by algebra.
15. $p(\phi_1 \vee \phi_2) \leq 1$. From K1.
16. $1 - u(\phi_1 \vee \phi_2) \leq 1$. From 13 and 15, by algebra.
17. $u(\phi_1 \vee \phi_2) \geq 0$. From 16, by algebra.
18. $u(\phi_1\&\phi_2) \leq u(\phi_1) + u(\phi_2)$. From 14 and 17, by algebra. QED

The following chapter will discuss in detail the consequences of the theorems just stated, which are the most important from the point of view of applications of the theory of logical consequence. We will end the present chapter with a theorem that relates these theorems to the theory of logical inconsistency.

Theorem 12. If $\phi_1, \ldots, \phi_n$ are logically inconsistent then $u(\phi_1) + \ldots + u(\phi_n) \geq 1$.

Proof

1. Supposc that $\phi_1, \ldots, \phi_n$ are logically inconsistent.
2. $\phi_1\& \ldots \&\phi_n$ is logically false. From 1.
3. $u(\phi_1\& \ldots \&\phi_n) = 1$. From 2 and theorem 10.
4. $u(\phi_1\& \ldots \&\phi_n) \leq u(\phi_1) + \ldots + u(\phi_n)$. Theorem 11.

[28]This is an immediate consequence of the theorem stated in exercise 1b at the end of section 2.4⋆.

5. $u(\phi_1)+\ldots+u(\phi_n) \geq 1$. From 3 and 4 by pure algebra.
QED

There is a superficial 'duality' between this theorem and the probability sum theorem because it states a relation between the uncertainty of a conjunction and the uncertainties of its conjuncts, and the probability sum theorem states a relation between the probability of a disjunction and the probabilities of its disjuncts. The duality isn't exact, however, since the probability sum theorem says that under some conditions the probability of the disjunction equals the sum of the probabilities of the disjuncts, while the uncertainty sum theorem says that the uncertainty of a conjunction is less than or equal to the sum of the uncertainties of its conjuncts. Note, however, that the theorem stated in the exercise below describes special conditions under which the uncertainty of the conjunction can actually equal the sum of the uncertainties of the conjuncts.

Exercise

1. Give an informal argument to prove the following:

Theorem If $\phi_1,\ldots,\phi_n$ are logically consistent and logically independent in the sense that no one of them is implied by the others, then for all nonnegative numbers $u_1,\ldots,u_n$ whose sum is no greater than 1 there is an uncertainty function u such that $u(\phi_1) = u_1, \ldots, u(\phi_n) = u_n$ and $u(\phi_1\&\ldots\&\phi_n) = u_1+\ldots+u_n$.

2.7 Glossary

Certainty Distribution A probability distribution that attaches probability 1 to one state-description and probability 0 to all others.

Kolmogorov Axioms Four axioms essentially due to A. N. Kolmogorov: $0 \leq p(\phi) \leq 1$ (K1); if ϕ is logically true then $p(\phi) = 1$ (K2); if ϕ logically implies ψ then $p(\phi) \leq p(\psi)$ (K3); if ϕ and ψ are mutually exclusive then $p(\phi \vee \psi) = p(\phi) + p(\psi)$ (K4). All of the laws of pure probability logic follow as theorems from these axioms.

Probabilistic Certainty; Practical Certainty Probability 1. Logical certainty (logical truth) is a sufficient but not a necessary condition for this.

Probability Counterexample A probability distribution that generates probabilities that do not satisfy a given condition; this shows that the condition cannot be a theorem of probability logic.

Probability Distribution A distribution of probabilities among the state-descriptions summing to 1.

Probability Function A function satisfying the Kolmogorov axioms that attaches probabilities to formulas.

Probabilistic Independence Propositions are probabilistically independent if the probability of their conjunction is equal to the product of their probabilities. This property, which is related to randomness, is commonly assumed in applications of probability theory, but it is a factual assumption that does not follow from the Kolmogorov axioms.

Probability Sum Theorem The probability of a disjunction equals the sum of the probabilities of its disjuncts, if they are mutually exclusive.

State-Description (SD) Logically consistent sentences to which probability distributions apply. Formally, a state-description is a logically consistent formula, α, such that for any formula at all, ϕ, of a language α logically implies either ϕ or $\sim\phi$.

Uncertainty Sum Theorem The uncertainty of a conjunction does not exceed the sum of the uncertainties of its conjuncts.

Uncertainty Probability of falsehood; equals 1 minus probability.

Uniform Probability Distribution A probability distribution that attaches equal probability to all state-descriptions.

3

Deduction and Probability Part I: Statics

3.1 Introduction

We have seen that if an inference is logically valid and its premises are certain its conclusion must be certain (theorem 8), and the lottery paradox shows that the premises can be nearly certain while the conclusion is certainly false. But we want to know whether that can happen in more 'ordinary' reasoning, such as in the inference $\{E \vee L, \sim E\} \therefore L$. If something like the lottery paradox were possible in this case too, then the fact that the inference was logically valid would be of little practical significance, since 'data' like "Jane will take either ethics or logic" and "Jane will not take ethics" are usually less than perfectly certain. We will now apply the general theorems about probability that were derived in the previous chapter to show that this isn't possible—at least assuming that the probabilities involved are 'static' and unchanging.

Anticipating, we will see that while we can expect the uncertainties of the premises of an inference to be to an extent 'passed on' to its conclusion, *the accumulated uncertainty of the conclusion can only be large if the sum of the uncertainties of the premises is large.* In fact, we have already seen this in the case of an inference with just one premise or whose conclusion is equivalent to the 'conjunction' of that premise. Thus, theorem 9 implies that if a valid inference has a single premise then its conclusion cannot be more uncertain than this premise, and theorem 11 says that the uncertainty of a conjunction cannot be greater than the sum of the uncertainties of its conjuncts. *Ipso facto*, if the conjuncts are the premises and their conjunction is the conclusion then the uncertainty of the conclusion cannot be greater than the sum of the uncertainties of the premises. We will now see that this holds in general, and this is of the utmost importance in justifying the application of deductive logic to reasoning with relatively few premises, each of which can be expected to have only a little uncertainty, like the logic and ethics example. The trouble with the lottery paradox was that even though each of its premises individually had very little uncertainty, it had so many of them that their uncertainties accumulated to the point of absurdity. Theorems in the next section make this clearer, and the section after that begins to consider the problem of limiting the uncertainties of the conclusions of inferences that have large numbers of premises like the lottery paradox, which can be important in certain kinds of practical reasoning. The final section of this chapter begins an inquiry into a subject that has come into increasing prominence in recent years: namely the nature and role of so called *default assumptions* and other practical rules that limit the uncertainties of conclusions in certain circumstances.

3.2 Two Hauptsätze of the Theory of Uncertainty Accumulation[1]

The following theorem of probability logic is the key to justifying the application of deductive logic to inferences with somewhat uncertain premises, so long as there are not too many of them:

Theorem 13. The uncertainty of the conclusion of a valid inference cannot exceed the sum of the uncertainties of its premises.[2]

Proof

1. Let ϕ be a logical consequence of premises $\phi_1, \ldots, \phi_n$.
2. $u(\phi_1 \& \ldots \& \phi_n) \leq u(\phi_1) + \ldots + u(\phi_n)$. theorem 11.
3. If ϕ is a logical consequence of premises $\phi_1, \ldots, \phi_n$ then it is a logical consequence of $\phi_1 \& \ldots \& \phi_n$. By pure logic.
4. ϕ is a logical consequence of $\phi_1 \& \ldots \& \phi_n$. From 1 and 3.
5. If ϕ is a logical consequence of $\phi_1 \& \ldots \& \phi_n$ then $u(\phi) \leq u(\phi_1 \& \ldots \& \phi_n)$. Theorem 9.
6. $u(\phi) \leq u(\phi_1 \& \ldots \& \phi_n)$. From 4 and 5.
7. $u(\phi) \leq u(\phi_1) + \ldots + u(\phi_n)$. From 2 and 6. QED

This result generalizes the positive part of theorem 9, that the conclusion must be certain when the premises are certain. When the premises are certain they have 0 uncertainty, and the conclusion's uncertainty cannot be greater the sum of these 0 uncertainties. The theorem also explains why, if the premises only have small uncertainties, there must be a lot of them for the conclusion to be very uncertain. In particular, if each premise has

[1]'Hauptsatz' (plural 'Hauptsätze') is the German term that is often used, even in writings in English, for mathematical statements of the highest importance.

[2]So far as the author knows, Suppes (1966) gives the first statement of a law akin to this *Hauptsatz.*

an uncertainty no greater v then there must be at least $1/v$ of them for the conclusion to have total uncertainty.[3]

The second Hauptsatz is a partial converse to the first:

Theorem 14. Let $\phi_1, \ldots, \phi_n$ be the premises and ϕ be the conclusion of an inference, and let $u_1, \ldots, u_n$ be nonnegative numbers whose sum is not greater than 1. If the premises are consistent and the inference is valid and all of its premises are essential in the sense that if any one of them were omitted the inference would be invalid, then there is an uncertainty function u() such that $u(\phi_i) = u_i$ for $i = 1, \ldots, n$, and $u(\phi) = u_1 + \ldots + u_n$.

Proof

1. Suppose that ϕ is a logical consequence of $\phi_1, \ldots, \phi_n$, but not of any proper subset of these premises, and $u_1 + \ldots + u_n \leq 1$.
2. Suppose first that $\phi_1, \ldots, \phi_n$ are logically consistent.
3. For each $i = 1, \ldots, n$ there is a line, ℓ_i, in the truth-table in which $\phi_1, \ldots, \phi_{i-1}, \phi_{i+1}, \ldots, \phi_n$ are true but ϕ is false, and there is a line, ℓ, in which all of $\phi_1, \ldots, \phi_n$ are true. From 1 and 2 because $\phi_1, \ldots, \phi_n$ are logically consistent, and ϕ is not a logical consequence of $\phi_1, \ldots, \phi_{i-1}, \phi_{i+1}, \ldots, \phi_n$.
4. Let the probability function p( ) be generated by the distribution that assigns probability u_i to line ℓ_i for $i = 1, \ldots, n$, and assigns probability $1 - u_i - \ldots - u_n$ to line ℓ.
5. If u( ) is the uncertainty function corresponding to the function p( ) defined in 4, then $u(\phi_i) = u_i$ for $i = 1, \ldots, n$. From 4, because ℓ_i is the only line that with positive probability in which ϕ_i is false.
6. $u(\phi) = u_1 + \ldots + u_n$. From 4, because the only lines with positive probability in which ϕ is false are $\ell_1, \ldots, \ell_n$. QED

We will end this section with comments on the contrast between a conclusion's maximum possible uncertainty, compatible with its premises having given uncertainties, and its actual uncertainty, which will generally be smaller than the maximum. To start with, note the application of theorem 14 to the curious inference with inconsistent premises "Jane will take ethics

[3]These were precisely the probabilities involved in the lottery paradox that was described in section 1.1. In that case each of the 1,000 tickets had 999 chances in 1,000 of losing, hence the uncertainty of losing on the ticket was $1/1{,}000 = .001$. On the other hand, the uncertainty of "No ticket will win," which is the logical consequence the 1,000 premises "Ticket #1 won't win," "Ticket #2 won't win," ..., "Ticket #1,000 won't win," equalled the sum of 1,000 of these uncertainties, which is 1.

and she won't take ethics; therefore she will take logic", which has the form $\{E, \sim E\} \therefore L$. Both of the premises are essential, and therefore theorem 13 implies that L's uncertainty can be as great as the sum of the uncertainties of E and ~E. But it is easy to show that $u(E)+u(\sim E) = 1$, so u(L) can be as great as 1, which is the same as 0 probability.[4] But that doesn't mean that L must have 0 probability, it is only that the premises don't give us any information about its probability.

Now turn to the more normal inference $\{E \vee L, \sim E\} \therefore L$, and suppose that $p(E \vee L) = .9$ and $p(\sim E) = .7$, which are the probabilities that distribution #2 in table 2.2 generates for these propositions. Given this, $u(E \vee L) = .1$ and $u(\sim E) = .3$, and since both of the premises are essential theorem 14 implies that the maximum uncertainty of the conclusion consistent with these uncertainties is $u(L) = .1+.3 = .4$. This is greater than the 'actual' uncertainty of L generated by Distribution #2, which is .2, and we have a case in which, while the maximum possible uncertainty gives some information about the actual uncertainty, it is still significantly less than the maximum. One way to think of theorem 13 is that it tells us what is the worst uncertainty that a conclusion can possibly have, but that is only an upper bound on its actual uncertainty, which is what a reasoner has a practical interest in. One of the two sections to follow will consider how the actual uncertainties may be limited to smaller values than their worst possible ones, and in certain cases these values can be estimated if certain 'default assumptions' are made about probabilities. Prior to that, though, we will consider how worst case uncertainties may be reduced by introducing redundancies into the premises of inferences, thereby making some of them not strictly, logically essential to the reasoning. This avoids the consequences of theorem 13, since that theorem only applies to inferences whose premises are all essential.

Exercises

1. If all of the uncertainties of the premises in the inferences below are equal, how small do these uncertainties have to be to guarantee that the conclusions of these inferences have probability at least .99?
   a. $\{A, B\} \therefore A \leftrightarrow B$
   b. $\{A, A \rightarrow B\} \therefore B$
   c. $\{A, A \rightarrow B, B \rightarrow C\} \therefore C$
   d. $\{A, B, \sim(A\&B)\} \therefore C$ (can the premises have the required uncertainties?)
2. Give probability counterexamples to the following:

[4]This generalizes a point made in section 2.3 that although anything follows from a contradiction, if all of its premises are essential it cannot guarantee any probability in its conclusion.

a. $u(L) \leq u(E) \times u(E \to L)$.[5]

b. If $u(\phi) \leq u(\psi)$ then ψ is a logical consequence of ϕ. (Here you should give specific formulas like E and L in place of variable formulas ϕ and ψ, *as well as* a specific uncertainty function u for which $u(\phi) \leq u(\psi)$ but ψ is a logical not consequence of ϕ.)

c. If $\phi_1 \vee \phi_2$ is a logical consequence of ϕ then there is an uncertainty function u for which $u(\phi) = u(\phi_1) + u(\phi_2)$.

d. If $\phi_1 \vee \phi_2$ is a logical consequence of ϕ then $u(\phi) \leq u(\phi_1)$ or $u(\phi) \leq u(\phi_2)$.

3. ⋆a. What is the maximum uncertainty of L, if E and L are probabilistically independent and the uncertainties of $E \vee L$ and $\sim E$ are .1 and .3, respectively?

b. Argue informally that if u(L) has the minimum value of .4 while $u(E \vee L) = .1$ and $u(\sim E) = .3$, then $p(E\&L) = 0$.

c. Prove that if $u_1 + u_2 > 1$ and ϕ_1 and ϕ_2 are independent in the sense that neither of them follows from the other, then there is an uncertainty function u( ) such that $u(\phi_1) = u_1$ and $u(\phi_2) = u_2$ and $u(\phi_1 \& \phi_2) = 1$.

d. It is possible for the uncertainties of all of the essential premises of a valid inference to sum to more than 1 while its conclusion is certain. Prove this by giving an example of a valid inference from two or more premises, all of which are essential to a conclusion, in which, when the premises are certainly false, the conclusion must certainly be true.

3.3⋆ Reinforcing Conclusions by Introducing Redundancy among Premises: Degrees of Essentialness

That introducing redundancy may reduce a conclusion's worst case uncertainty below the level that follows from theorem 13 is obvious in the inference "Jane will take ethics and she will take logic; therefore she will take *either* ethics or logic", which we can assume is of the form $\{E, L\} \therefore E \vee L$. In this case neither premise is essential, and it is clear that the uncertainty of the conclusion can't be greater than that of the least uncertain premise. In fact, it is clear that *given a valid inference with many premises, different subsets of which entail the conclusion, the conclusion's uncertainty cannot be greater than the total uncertainty of that subset with the smallest total uncertainty.* This suggests that you could treat these inferences as though their *only* premises were the ones in these 'sufficient subsets'. But there is

[5]Curiously, we will see in chapter 6 that $p(L) \geq p(E) \times p(E \to L)$ is valid when $\to$ is replaced by the *probability conditional*.

another possibility, and we will now see that premises outside these subsets can 'reinforce' a conclusion by reducing its possible uncertainty.

Consider the premises "Jane will take ethics", "Jane will take logic", and "Jane will take biology", and the conclusion "Jane will take at least two of ethics, logic, and biology", which may be symbolized as $\{E, L, B\}$ ∴ (E&L) ∨ (E&B) ∨ (L&B). Suppose also that each premise has probability .9 and uncertainty .1. In this case no single premise is essential, since the conclusion follows from the other two premises, and therefore its uncertainty cannot be greater than the sum of the uncertainties of any two of its premises, which is .2. But *can* the conclusion's uncertainty be that great if the premises' uncertainties all equal .1? Consider the distribution in diagram 3.1:

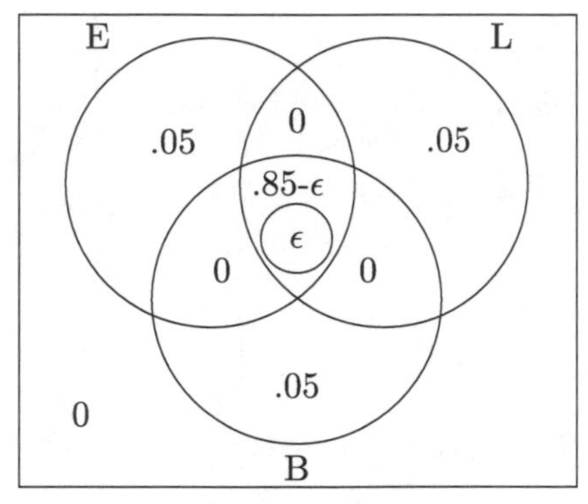

Diagram 3.1

You can easily verify in the diagram that each premise has probability .9 and uncertainty .1, while the probability that Jane will take at least two of ethics, logic, and biology is .85 and its uncertainty is .15, which is less than the combined uncertainties of any pair of premises that entail it. Note that as the diagram pictures it there is no chance of Jane's taking any two of ethics, logic, and biology, without actually taking all three; moreover the intersection of the E, L, and B circles, in which she takes all three, is divided into a larger part with probability $.85 - \epsilon$ and an 'infinitesimally small part' with probability ϵ. Now we want to argue intuitively that there is no way to redistribute the probabilities in the diagram in such a way as to lower the probability of Jane's taking any two of ethics, logic, and biology, without lowering the probabilities of one or more of the premises.

Suppose that you tried to lower the probability of the conclusion by transferring some part of the .85 probability in the intersection of the premises elsewhere—say the part with probability ϵ. This would reduce the conclusion's probability by ϵ, but since the same quantity is in each of the premise regions the transfer would also reduce each premise's probability, unless an equal quantity were put back into these regions. But if you think about it you will see that that could only be done by putting the quantity transferred out of the middle back into the middle, and that would mean that the probability of the conclusion wouldn't be lowered after all! This is rough reasoning, but assuming that it is right, it suggests that although none of the premises of the inference is essential to its validity, each of them contributes to reducing the possible uncertainty of the conclusion. This reduction is not nearly so great as that which follows on the assumption that the premises are independent, which implies that the conclusion's probability is .972 and its uncertainty is .028, which we will come back to. But now we want to generalize and consider inferences with more complex redundancies.

The key to this generalization is to measure the degree to which a premise may be essential to an inference, even when it is redundant in the way the premises in the ethics, logic, and biology example are. A crude but useful measure of a premise's essentialness is related to the sizes of the essential sets of premises that it belongs to. For instance in the inference $\{E, L, B\} \therefore (E\&L) \vee (E\&B) \vee (L\&B)$, E by itself isn't essential since the conclusion follows just from L and B, but the pair of E and L is essential since the conclusion doesn't follow if both of them are omitted. Thus, while E isn't 'completely essential', the two-member premise set {E,L} is, and since E is 1 out of 2 of its members, we can say that E's *degree of essentialness* is $\frac{1}{2}$. It is more than a coincidence that while the maximum uncertainty of the conclusion is less than the *sum* of the uncertainties of the premises, it is equal to the sum of these uncertainties weighted by the degrees of essentialness of the premises they apply to. That is, the worst case uncertainty in this case given by

$$\begin{aligned} u((E\&L) \vee (E\&B) \vee (L\&B)) &= e(E) \times u(E) + e(L) \times u(L) + e(B) \times u(B) \\ &= \frac{1}{2} \times .1 + \frac{1}{2} \times .1 + \frac{1}{2} \times .1 = .15, \end{aligned}$$

where e(E), e(L), and e(B) are the degrees of essentialness of E, L, and B, respectively. This doesn't hold in complete generality, but you can see that the argument that applied to diagram 3.1 also applies to the following variant:

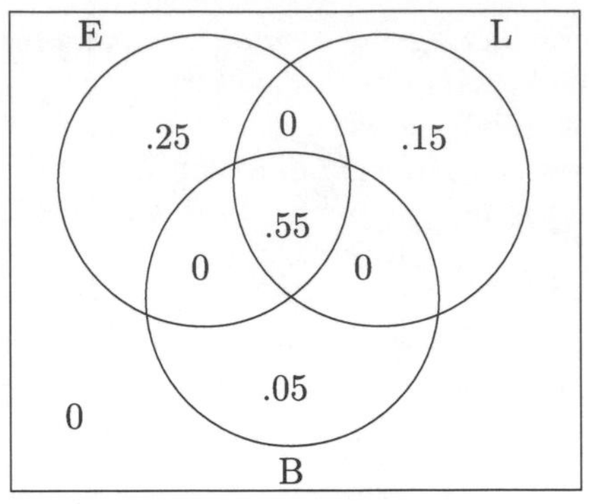

Diagram 3.2

Again, it is impossible to lower the probability of the conclusion without lowering that of at least one of the premises, hence the conclusion's probability is the lowest that is possible, compatible with the given premise probabilities. Moreover, the equation

$$u((E\&L) \vee (E\&B) \vee (L\&B)) = e(E) \times u(E) + e(L) \times u(L) + e(B) \times u(B)$$

still holds, although now u(E) = .2, u(L) = .3, and u(B) = .4, so

$$u((E\&L) \vee (E\&B) \vee (L\&B)) = \frac{1}{2} \times .2 + \frac{1}{2} \times .3 + \frac{1}{2} \times .4 = .45.$$

The foregoing generalizes as follows. We can say that in general *a set of premises is essential to an inference if its conclusion doesn't follow from the premises that are outside of the set.* Given this, we define:

Degrees of essentialness The degree of essentialness of a premise of a valid inference is the reciprocal of the size of the smallest essential set of premises to which it belongs, and if it doesn't belong to any minimal essential set then its degree of essentialness is 0.

Now we can state a generalization of the second Hauptsatz:

Theorem 15⋆. If ϕ is the conclusion of a valid inference with premises $\phi_1,\ldots,\phi_n$, whose degrees of essentialness are $e(\phi_1), \ldots, e(\phi_n)$, respectively, then $u(\phi) \leq e(\phi_1)\times u(\phi_1)+\ldots+e(\phi_n)\times u(\phi_n)$.[6]

[6]This result is proved using methods of Linear Programming in Adams and Levine (1975). Hailperin (1965), Nilsson (1986), and Pearl (1988, especially section 9.3) discuss more mathematically advanced aspects of this subject.

One thing worth noting about the result is that it shows that worst case uncertainties depend only on the 'entailment structures' of inferences, i.e., on what subsets of their premises entail their conclusions, and not on how strong or weak their premisses and conclusions are. For instance, the 'entailment functions' of the inferences $\{E, L\} \therefore E \leftrightarrow L$ and $\{E \vee L, {\sim}E\} \therefore L$ are the same, and since both premisses are essential to both inferences, their worst case conclusion uncertainties are the sums of their premises'

This applies to inferences in which not all premises have the same degree of essentialness, for example {A, B, C, D} ∴ A&(B ∨ C). In this case A is totally essential, since A&(B∨C) doesn't follow from {B, C, D}, B and C are 50% essential since the conclusion follows from the two-member sets {A, B} or {A, C} but from no smaller sets involving B and C, and D is totally inessential since it can be omitted from any set that A&(B ∨ C) follows from. Hence e(A) = 1, e(B) = e(C) = $\frac{1}{2}$, and e(D) = 0, and therefore according to theorem 15⋆, $u(A\&(B \vee C)) \leq u(A) + \frac{1}{2}u(B) + \frac{1}{2}u(C)$. The following diagram shows that the upper bound of $u(A) + \frac{1}{2}u(B) + \frac{1}{2}u(C)$ is actually attained:

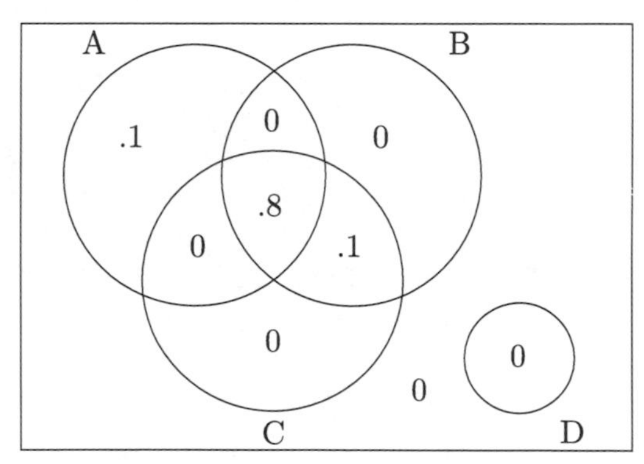

Diagram 3.3

Here u(A) = u(B) = u(C) = .1, and u(A&(B ∨ C)) = .2, which is equal to $u(A) + \frac{1}{2}u(B) + \frac{1}{2}u(C)$.[7]

The most interesting applications of theorem 15⋆ are to inferences with larger numbers of premises, and as a step in that direction we will briefly discuss two inferences with 10 premises, each of which concerns the political parties that voters will vote for in an upcoming election. The premises of the both inferences are the statements "Voter #1 will vote Democratic", "Voter #2 will vote Democratic", and so on up to "Voter #10 will vote

uncertainties. Significantly, the uncertainties that follow on the assumption of premise independence are not the same in these two inferences.

[7]It is significant that in this case u(A) + u(B) and u(A) + u(C) are equal, hence they are also equal to $u(A) + \frac{1}{2}u(B) + \frac{1}{2}u(C)$. If, for instance, u(A) + u(B) had been less than u(A) + u(C), then it would also have been less than $u(A) + \frac{1}{2}u(B) + \frac{1}{2}u(C)$, and it would have been the worst case uncertainty of the conclusion of {A, B, C, D} ∴ A&(B ∨ C). That would not show that theorem 15⋆ was wrong, only that $u(A) + \frac{1}{2}u(B) + \frac{1}{2}u(C)$ is not the worst case uncertainty of its conclusion in this situation.

A similar remark applies to the inference {E, L} ∴ E ∨ L. Here each premise's degree of essentialness is $\frac{1}{2}$ and theorem 15⋆ implies that $u(E \vee L) \leq \frac{1}{2}u(E) + \frac{1}{2}u(L)$. But $\frac{1}{2}u(E) + \frac{1}{2}u(L)$ is only the worst case uncertainty of E ∨ L when u(E) = u(L), and therefore both of them equal $\frac{1}{2}u(E) + \frac{1}{2}u(L)$.

Democratic", and we will assume that each prediction has a 90% probability of proving true.[8] If this is so it would not be reasonable to deduce "All 10 voters will vote Democratic", even though it is a logical consequence of the premises. In fact, even if errors were independent the conclusion's probability would only be $.9^{10} = .349$. But things could be significantly worse if the predictions were not independent. All of the 10 premises are essential to inferring "All 10 voters will vote Democratic", and since each of them has uncertainty .1, the sum of their uncertainties multiplied by their degrees of essentialness is $10 \times .1 = 1$. Therefore it follows from theorem 14 that "All 10 voters will vote Democratic" could have probability 0 even though each individual premise had probability .9. In fact it is not hard to imagine a situation in which this would be the case, even though it would be very far-fetched. If it were *a priori* certain that 9 of the 10 voters would vote Democratic and the 10th would vote Republican, but exactly which voter would vote Republican was entirely a matter of chance, then the probability of any particular voter voting Democratic would be 9 in 10, but there would be 0 probability that all 10 would vote Democratic.

Now consider the inference from the same premises as above to the conclusion that at least 8 of the voters will vote Democratic, which seems more plausible than the conclusion that all 10 of them would vote Democratic. In fact, assuming that each premise is 90% probable and errors are independent, the new conclusion must be 73.6% probable. But theorem 15$\star$ implies that even if errors are not independent, the conclusion must be at least 66.7% probable. This follows from the fact that each premise's degree of essentialness is $\frac{1}{3}$, and this itself follows from the fact that the smallest premise sets that are essential to inferring that at least 8 of the 10 voters will vote Democratic have 3 members. For instance, the three-member set of premises {Voter #1 will vote Democratic, Voter #2 will vote Democratic, Voter #3 will vote Democratic} is essential because the conclusion doesn't follow from the remaining 7 premises. On the other hand no smaller subset of this three-member set is essential, e.g., the two-member set {Voter #1 will vote Democratic, Voter #2 will vote Democratic} is not essential since the conclusion *does* follow from the 8 premises that remain when these two are omitted. So, since there are 10 premises, each of whose uncertainty is .1 and whose degree of essentialness is $\frac{1}{3}$, the sum of the uncertainties of the premises weighted by their degrees of essentialness is $10 \times .1 \times \frac{1}{3} = .333$. Given this, theorem 15$\star$ implies that the uncertainty of the conclusion cannot be greater than .333 and its probability cannot be less than .667.

Once again we can imagine a 'possible probabilistic state of affairs' in

[8]This is a little less than the usual reliability of voters' predictions as to how they will vote in American presidential elections. Information due to Professor James F. Adams.

which each premise of the inference above would be 90% probable while its conclusion would have a probability of .667, although it is considerably more complicated than the one described earlier. This involves two kinds of possibilities: (1) one in which all 10 voters vote Democratic, and (2) a set of possibilities in which exactly 7 of the 10 voters vote Democratic. 'Elementary combinatorics' shows that there are 120 possibilities of this kind, because there are 120 ways of selecting 7 things out of a set of 10.

Now suppose that the first possibility has probability .667, and the remaining probability of .333 is distributed equally among the 120 possibilities of type (2). A rather tricky argument as follows shows that in this situation there would a 90% probability that any one voter would vote Democratic and a .667 probability that at least 8 of them would.

If there is 1 chance in 3 that 7 of the 10 voters will vote Democratic, and there are 120 equally likely possibilities as to which 7 that would be, then the probability that any given set of 7 voters will vote Democratic and the rest will vote Republican should be $(1/3) \times (1/120) = 1/360$. Now, it is another combinatorial fact that any given individual voter belongs to 84 of the 120 sets of 7 voters, and this implies that the chance that exactly 7 of the 10 voters vote Democratic and that the given individual belongs to a set that votes Democratic is 84/360. Combining this with the fact that the voter is certain to vote Democratic if all 10 voters do, and that has a $\frac{2}{3}$ probability, it follows that her total probability of voting Democratic should be $(84/360) + \frac{2}{3} = .90$.

Students will not be asked to carry out the kinds of 'combinatorial' calculations that were involved in constructing the worst case scenarios just illustrated, but they should be able to determine the sets of premises that are essential to inferences, and the degrees of essentialness that depend on them, and thereby to calculate the worst case uncertainties that follow from them.

Exercises

1. Find the minimal essential subsets and the degrees of essentialness of the premises of the following inferences, and, using theorem 15$\star$ find lower bounds on the probabilities of their conclusions:
   a. $\{A, B, A \rightarrow B\}$ ∴ A&B. $p(A) = .9$, $p(B) = .8$, $p(A \rightarrow B) = .9$.
   b. $\{A \vee B, B \vee C, A \vee C\}$ ∴ $(A\&B) \vee (A\&C) \vee (B\&C)$. $p(A \vee B) = 1$, $p(B \vee C) = .9$, $p(A \vee C) = .8$.
   c. $\{A, B, C\}$ ∴ $A \vee (B\&C)$. $p(A) = .9$, $p(B) = .9$, and $p(C) = .8$.
   d. $\{A, B, C\}$ ∴ $(A \leftrightarrow B)\&(B \leftrightarrow C)$. $p(A) = .9$, $p(B) = .9$, and $p(C) = .8$.
   e. $\{A, B, C\}$ ∴ $(A \vee B \vee C) \rightarrow (A\&B\&C)$. $p(A) = .9$, $p(B) = .9$, and $p(C) = .8$.

f. $\{A \rightarrow B, B \rightarrow C, C \rightarrow A\} \therefore (A \vee B) \rightarrow C$. $p(A \rightarrow B) = .9$, $p(B \rightarrow C) = .8$, $p(C \rightarrow A) = .7$

2. Draw Venn diagrams in which the premises of inferences 1a–f in exercise #1 have the probabilities stated, and verify that their conclusion's probabilities are at least as high is the minimums calculated in your answers to exercise 1.
3. Given 100 premises $p(A_1) = p(A_2) = \ldots = p(A_{100}) = .99$ (i.e., there are 100 premises, each of which is 99% probable), and conclusions of the form "At least n of the premises are true", for $n = 0, 1, \ldots, 100$. Determine the following:
   a. If $n = 100$ (i.e., the conclusion is that *all* of the premises are true), what are the minimal essential sets of premises, and what is the degree of essentialness of each premise?
   b. What is the minimum probability of this conclusion?
   c. Describe a probability distribution in which all of the premises have probability .99, while the conclusion has probability 0.
4. ⋆a. Answer questions 3a and 3b if $n = 90$.
   b. What is the highest value of n for which the minimum probability of the conclusion is .75?

⋆5. Call a premise set *sufficient* if the conclusion of the inference in question follows from that set alone, when all of the other premises are omitted.
   ⋆a. Show that every sufficient premise set intersects every essential premise set.
   ⋆b. Show that every set that intersects every sufficient set is essential and every set that intersects every essential set is sufficient.

3.4⋆ Remarks on Defaults and Other Assumptions That Limit Conclusion Uncertainties

The maximum possible uncertainties of the conclusions of inferences that are compatible with given premise uncertainties are 'worst cases', but, as we have seen, these cases are often very far-fetched. Even in the ethics, logic, and biology inference we saw that the worst case uncertainty of the conclusion, assuming a 90% probability in each individual premise, was .15, but assuming that premises are independent entailed that the conclusion's uncertainty was only .028, which is less than one-fifth of the worst case uncertainty. This suggests that it would be unreasonable to be so skeptical as to think that the conclusion had 15 chances in 100 of being wrong, even though that might be logically compatible with the given probabilities of the premises. The question that we will consider in this section is: what makes a 'logically possible hypothesis', such as that there is a 15% chance of a conclusion being wrong, unreasonable? This is a subject of considerable recent interest in applied logical theory, which is not concerned just with extreme theoretical possibilities but with how it is 'reasonable to reason' in everyday life. However, the more deeply this subject is inquired into, the more complex it is seen to be, and what follows will only introduce some main themes.

Defaults are idealizing assumptions of a kind that is common in applied mathematical subjects, where the phenomena to which the subjects are applied are assumed to have particularly simple or 'elegant' mathematical properties. For instance, it is a common default in applied physics to assume that the only gravitational forces in the solar system act between the sun and the planets. Of course this is not exactly true, but, plausibly, it is 'close enough to the truth' that calculations that are made on the basis of it, such as that the planets travel in elliptical orbits around the sun, will be good first approximations. Similarly, *independence* is a practically ubiquitous default assumption of applied statistics,[9] especially in applying theories of error, where it is assumed that the chance of one 'observation' being subject to a given error is independent of another observation being subject to such an error.

Independence may also be a default in applied probability logic, on the assumption that the probabilities of errors in premises are independent. And, we have seen that these assumptions lead to uncertainty estimates that are much smaller than the worst cases that follow from theorems 13 and 15$\star$. But, with the increasing interest in applications of logical theory to reasoning in everyday life and science, independence is only one of many possible default assumptions that have recently begun to receive attention in logic. In fact, the assumption that this work is most concerned with, that the premises of inferences are certainties, is one of them.[10] However, this is not the place to enter into an extended discussion of this and similar defaults, and we will end this section with brief comments on another type of assumption that has the effect of limiting or otherwise influencing the uncertainties of conclusions in reasoning. These assumptions relate to the

[9]In view of the ubiquitousness of independence assumptions in applied statistics, and in view of the fact that statistical theory tends not to make a clear distinction between logical and factual assumptions, it is perhaps not unsurprising that independence assumptions are often treated as *a priori* truths in applied statistics.

[10]A perhaps equally important one is the idealization involved in treating concepts as though they were perfectly precise, which is presently a subject of widespread interest, and which the rapidly developing theory of fuzzy logic seeks to dispense with. But fuzzy logic has its own default assumptions, the central one being that 'fuzzy truth' is *componential*, e.g., that the 'fuzzy truth value' of a compound proposition like "Jones is old and wise" is uniquely determined by the fuzzy truth values of its 'components', "Jones is old" and "Jones is wise". This is a generalization of the standard truth-conditionality assumption of 'sharp', two-valued logic. Cf. Zadeh (1965).

source of the 'data' that are reasoned from, especially when they are verbal communications.

To illustrate this kind of assumption, which in this case does not necessarily imply reduced conclusion uncertainties, let us return to the ethics and logic case, i.e., to the inference $\{E \vee L, \sim E\} \therefore L$, in the case in which $u(E \vee L) = .1$ and $u(\sim E) = .3$. As we have the seen, the minimum value of u(L) consistent with these uncertainties is $u(L) = u(E \vee L) + u(\sim E) = .1 + .3 = .4$, and in fact diagram 3.4 depicts a situation in which $E \vee L$, $\sim E$, and L would have these uncertainties:

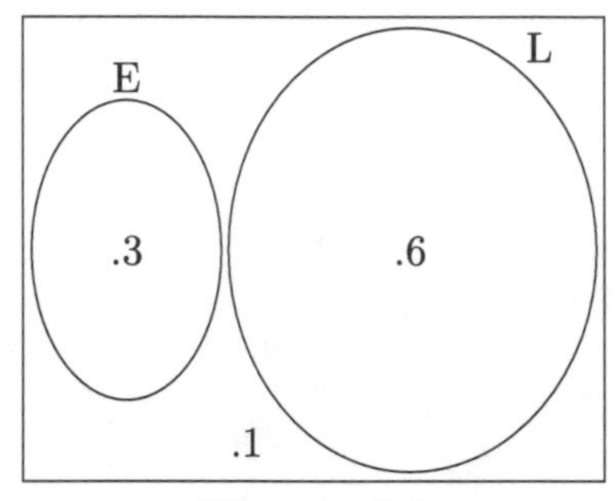

Diagram 3.4

The significant thing about this diagram is that it depicts E and L as being mutually exclusive. In fact p(E&L) must equal 0 if the value of u(L) is to be a minimum.[11] The question is: Is this a plausible state of affairs, and if not, what probabilistic states of affairs are plausible? As to the latter, it would not be plausible to assume that E and L are *independent*, which would entail that $p(L) = .857$ and $u(L) = .143$.[12] That is because if you were told "Jane will take either ethics or logic" you would conclude that this was an 'either... or situation' in which Jane would probably not take both.[13] This would be something like 'approximate mutual exclusiveness',

[11]Incidentally, this shows in what way the worst case uncertainty has to be extreme.

[12]This is not the same as assuming that the premisses, $E \vee L$ and $\sim E$, are independent, which would be inconsistent with the assumption that $p(E \vee L) = .9$ and $p(\sim E) = .7$. Given that $(E \vee L)\&\sim E$ is equivalent to $\sim E\&L$, if $E \vee L$ and $\sim E$ were independent it would follow that $p(\sim E\&L) = p((E \vee L)\&\sim E) = p(E \vee L) \times p(\sim E) = .9 \times .7 = .63$. Given that $p(\sim E) = .7$, this would imply that $p(\sim E\&\sim L) = .07$, hence, since $\sim E\&\sim L$ is equivalent to $\sim(L \vee E)$, it would follow that $p(\sim(L \vee E)) = .07$ and $p(L \vee E) = 1 - .07 = .93$, which is inconsistent with the assumption that $p(L \vee E) = .9$.

But then, even assuming that E and L are independent would be very implausible.

[13]That isn't to say that the "either... or" in "Jane will take either ethics *or* logic" has a mutually exclusive meaning. In fact, Jennings (1994) argues that the "either... or" of ordinary language never has this sense. But what we are suggesting is that it would be reasonable in ordinary circumstances for a person who heard someone say "Jane will take either Ethics or Logic" to conclude that she probably wouldn't take both, in the same way as it would be reasonable to conclude that she would probably take one of them—which will be returned to below. This would be a *conversational implicature* of

which would suggest that, while there could be an outside chance of Jane's taking both courses, the probability of Jane's taking logic was close to what it would be if she were certain not to take both; i.e., it would be close to .6.

The foregoing enters only a very little way into the 'conversational analysis' of the ethics and logic inference,[14] but this example cannot be pursued further here, and we will conclude with comments on two 'conversational defaults' that are central to the probabilistic logic that this work is concerned with. One is that, whatever are the sources of the 'data' from which we ordinarily reason, while they are usually probable, they are usually not perfectly certain. As to probability, the fact that persons making factual assertions convey or imply that what is asserted is *probable* is related to what Searle calls a 'preparatory condition for the speech act' of making them, namely that the persons are *justified* in making them.[15] As to uncertainty, it has already been stressed that very few factual propositions can be complete certainties, and it is generally accepted that Hume was right in claiming that future contingencies can never be. To restrict our statements to certainties would not only reduce us to silence in most practical matters, but prevent us from conveying the information whose high probability is nearly as useful as certainty from the practical point of view—as we will see in chapter 9.

The other apparent default of our probabilistic analysis is that the data from which we reason may have precise probabilities, and we know exactly what their probabilities are. Sometimes even the assumption that they have precise probabilities is an idealization. A person who said that Jane would take ethics or logic would certainly not assert or imply that he knew that her taking ethics or logic had a precise numerical probability, for example of .9, and a precise uncertainty of .1. To assume that would be no more plausible than to assume that $E \vee L$ was perfectly certain. A more plausible default would be a 'fuzzy' one; for instance that the probability of $E \vee L$ is about 90%. But our theory has not been concerned, it appears, with fuzzy probabilities and uncertainties. So, one would like to know what the connection is between the exact quantities that our theory is concerned with and the fuzzy uncertainties that are implicit in reasoning in ordinary life. There is no easy answer to this question, but two comments may be made about it.

---

a kind considered in some detail in Chapter 4 of Grice (1989). This would also explain why the inference "Jane will take either ethics or logic, and she will take ethics; therefore she won't take logic" has a kind of reasonableness in spite of being formally invalid.

[14]For instance, this leaves out of account both the consideration of how the reasoner learned the second premise, that Jane wouldn't take ethics, and how learning that might affect her confidence in the first premise. The latter in particular will be returned to in chapter 5.

[15]In this connection see Searle (1969: 63), also Adams (1986).

First, the exact probabilities and uncertainties that our theory deals with could be conceived of as being to fuzzy probabilities and uncertainties as exact heights and weights are to fuzzy heights and weights. Assuming this, there is a 'fuzzy sense' of "If Jane's taking either ethics or logic is 90% certain and her not taking ethics is 70% certain, then she is at least 60% certain to take logic" that is analogous to a fuzzy sense of "If John is 23 years old and Jane is 2 years younger than he is, then she is 21 years old", which 'means' something even if the ages referred to are not exact.[16]

The other point is that in spite of the fact that our theory is itself quantitative, certain aspects of it are qualitative, and these aspects have a significance that is independent of precise numerical assumptions. For example, that the conclusion of a valid inference from a single premise must be at least as certain as the premise is a qualitative inequality, and it gives us useful information whether or not we have an exact idea of the degrees of certainty involved.[17] Similarly, it is a qualitative fact that uncertainties in the premises of valid inferences can be passed on to conclusions, and accumulate to the point of absurdity when there are many premises and no redundancies among them.[18]

As said, these sketchy remarks do not do justice to the complex issues that they are concerned with, and we must now turn to other matters. However, we will end with pointing out a default assumption that we have made consistently in this chapter, and which will be the focus of the following one. That is that the probabilities of the propositions involved in the inferences we have considered all 'apply' to them at one time. But, to return again to the ethics and logic example, it is not usually the case when persons acquire information like "Jane will take either ethics or logic" and "Jane will not take ethics" that they learn these things at the same time. Moreover, by analogy with the fact that learning the second premise is likely to have a 'prospective effect' on the probability of the conclusion, that Jane will take logic, it is equally likely to have a 'retrospective effect' on the probability of the first premise. If that effect were to *lower* the probability of the first premise then, even though it had been 90% probable *before* the second premise became 70% probable, it wouldn't follow that "Jane will take logic" should be at least 60% probable afterwards. In fact,

[16]Only, whereas we have precise objective measures of height and weight, it is less clear what a precise objective measure of probability might be. This is the problem of the *foundations of probability*, and the problem of defining it will be commented on in section 9.9⋆⋆ and appendix 1.

[17]But, as pointed out in footnote 14 of chapter 2, the reasoner should know that the inference is valid.

[18]It might be objected that we don't require a theory to tell us these things because they are self-evident. But it is not self-evident how many premises there have to be with about 90% probability before a conclusion's probability can drop below 50%.

it was noted in chapter 1 that this sort of *nonmonotonicity* is a very real possibility in practical reasoning. And, it was also pointed out that apparent inconsistencies such as might be expressed by "Jane will take ethics" and "Jane will not take ethics" are most likely to be less-than-perfectly-certain pieces of 'information' acquired at different times, from which, far from anything following, nothing follows. The next two chapters begin a systematic study of possibilities of this sort.

3.5 Glossary

Default Assumption A (usually tacit) assumption that is plausible in the absence of evidence to the contrary, which affects the probabilities of conclusions. A common factual default assumption is that the premises of inferences are approximately independent, and a common mathematical default assumption is that the consequences of exact independence assumptions are close to those of the corresponding approximate independence assumptions.

Degree of Essentialness Given a valid inference with premises $\phi_1, \ldots, \phi_n$, for i = 1, ..., n, the degree of essentialness, $e(\phi_i)$, of ϕ_i is 1/k, where k is the size of the smallest essential set of premises to which ϕ_i belongs. If ϕ_i belongs to no minimal essential sets then $e(\phi_i) = 0$.

Essential Premise & Premise Set An essential set of premises of a valid inference is one whose omission would make the inference invalid; a single premise is essential if omitting it would make the inference invalid. Thus, essential premises are the members of singleton essential premise sets.

The Essentialness Theorem Given a valid inference $\{\phi_1, \ldots, \phi_n\}$ ∴ ϕ, $u(\phi) \leq e(\phi_1)u(\phi_1) + \ldots + e(\phi_n)u(\phi_n)$, where $e(\phi_1), \ldots, e(\phi_n)$ are the degrees of essentialness of $\phi_1, \ldots, \phi_n$, respectively.

Irrelevant Premise A premise that belongs to no minimal essential premise sets, hence whose degree of essentialness is 0.

Minimal Essential Premise Set An essential premise set that has no proper subsets that are essential.

Sufficient Premise Set A premise set that is sufficient in the sense that the inference is valid when all premises outside the set are omitted. A set is sufficient if and only if it intersects every essential premise set, and it is essential if and only if it intersects every sufficient premise set.

4

Conditional Probabilities and Conditionalization

4.1 Introduction: The Dependence of Probabilities on States of Knowledge

This chapter will turn away temporarily from applications of probability to deductive logic, and consider a 'dimension of probability' that has been left out of account in our discussions so far. This has to do with *conditional probabilities*, which make certain things that probabilities depend on explicit, and the consequences of the fact that what they depend on can change 'in the light of experience'. This sort of change will ultimately have consequences for applications to deductive reasoning because its premises are generally items of information, the learning of which can itself change probabilities. However, except for brief informal comments in this introduction, we will postpone considering the implications for deductive theory to chapter 5.

To illustrate the sort of dependency that we are concerned with, consider the probability of being dealt an ace *on the second card*, when the cards are dealt from a standard, shuffled pack of 52 cards. Assuming that there are 4 aces in the pack, the chance drawing an ace on the second card should be $4/52 = 1/13$. But what if we already know that the first card will be an ace? In these circumstances we would probably say that the chance of drawing an ace on the second card was $3/51 = 1/17$, because this card is 1 of the 51 remaining cards, 3 of which are aces. Thus, the chance of drawing an ace on the second card depends on what is known at the time it is drawn, and when this 'state of knowledge' changes the chances change. Conditional probabilities make this explicit,[1] and if we haven't done so previously it is

[1] Conditional probabilities are not usually characterized as quantities that show how 'ordinary', nonconditional probabilities depend on 'knowledge conditions', but this fits in more directly with the applications to deductive reasoning that will be considered in the following chapter. A drawback of this characterization is that it is apt to result in

because we have tacitly assumed that states of knowledge were unchanging, at least in the contexts that we have been considering.

The fact that conditional probabilities make explicit the changes in probabilities that result from acquiring knowledge makes them especially relevant to logic, and to the theory of knowledge in general, because these subjects are concerned with the acquisition and modification of knowledge by experience and reason. Moreover, there is a law of probability change that is particularly relevant in this connection, roughly as follows: *New experience becomes a part of the state of knowledge that determines probabilities.* The example of the aces illustrates this. Originally, before it is found out that the first card is an ace, the right odds to accept on a bet that the second card would be an ace is the proportion of aces in the whole deck, which is 4/52, but after it is found out that the first card is an ace the right odds to accept on this bet is the proportion of aces in the other 51 cards, which is 3/51. Moreover, the law that relates the *a posteriori* odds of 3/51 to *a priori* odds is simple: 3/51 is the ratio of the *a priori* odds on both the first and second cards being aces to the *a priori* odds on the first card alone being an ace. More generally, the rule is that the *a posteriori* odds on a proposition ϕ, after another proposition, ψ, is learned is the ratio of the *a priori* odds on the conjunction $\phi\&\psi$ to the odds on ψ alone.[2] This is sometimes called Bayes' principle,[3] and, though it is a

confusing what people think probabilities are with what they really are. Since 'real' and 'subjective' probabilities are assumed to satisfy the same formal laws, the distinction will not matter until we come to section 9.9★★, where we discuss the practical value of being 'right' about probability. Footnote 1 of the next chapter also comments briefly on this matter.

[2]In strictness, it should be assumed that the *a priori* probability of ψ is positive, since otherwise the ratio of the *a priori* odds on the conjunction $\phi\&\psi$ to the odds on ψ, which, since it is 0/0, isn't defined. The possibility of learning something that has 0 probability *a priori* cannot be ruled out *a priori*, but obviously it must be very remote, and we will not be much concerned with it. However, the problem of defining ratios of probabilities that are both equal to 0 has been extensively debated in the theoretical literature, and it will be returned to briefly in appendix 2, which concerns V. McGee's generalization of Bayes' principle to the case of new information whose *a priori* probability is 'infinitesimally small'.

Another comment has to do with the terms *a priori* and *a posteriori* (Latin for 'prior' and 'posterior'), which are used here in their standard probabilistic senses, rather than their philosophical senses. In traditional epistemology (theory of knowledge), *a priori* knowledge was supposed to be something that could be acquired prior to all experience, whereas in probability theory the *a priori* probability of a proposition is simply the probability it had prior to the acquisition of a specific piece of information, which is usually not prior to the acquisition of all information.

[3]Named for Reverend Thomas Bayes (1763) whose posthumously published paper "An Essay Towards Solving a Problem in the Doctrine of Chances" (cf. Bayes 1940) pioneered the ideas from which the present theory derives. It should be said, however, that the term 'Bayesian', referring to theories of probability and statistics, has no precise meaning. What we are here calling 'Bayes' Principle' is a default assumption that can

default assumption that will be examined in section 4.8⋆, we will see that it has enormous implications for logic and the theory of knowledge. In fact, its consequences for inductive reasoning will be discussed at some length in section 4.6 of this chapter, while its consequences for deductive logic will be the subject of the following chapter.

We will begin with a discussion of the formal concept of conditional probability and certain of its properties, before proceeding to consider the full implications of Bayes' principle.

4.2 The Formalism of Conditional Probability

Let A and B symbolize "The first card will be an ace" and "The second card will be an ace", respectively. Then the probability of B, given A, will be written as p(B given A), or more concisely p(B|A), where "|" stands for "given."[4] Thus, we assumed above that while $p(B) = 4/52$, $p(B|A) = 3/51$. Moreover it follows from Bayes' Principle that $p(B|A) = p(B\&A)/p(A)$. But this requires comment.

Although p(B|A) is often defined formally as the ratio p(B&A)/p(A), in practical applications p(B&A) would be more likely to be calculated by multiplying p(B|A) and p(A). That is because, while it is obvious intuitively that the chance of getting an ace on the second card given that the first card is an ace is 3 in 51, the chance of getting aces on the first two cards is much less obvious, and the simplest way to calculate it is to multiply p(B|A), which is already known, by p(A). In other words, while $p(B|A) = p(B\&A)/p(A)$ can be regarded as a definition from the formal point of view, in fact p(B&A) may have to be determined from p(B|A) and not the other way round. For our purposes it is better to think of the equation simply as a rule that allows any one of the three probabilities that it involves to be calculated from the other two.

Generalizing, the conditional probability of any proposition, ϕ, given the knowledge that ψ, will be written as $p(\phi|\psi)$, which is assumed to satisfy the equation

$$p(\phi|\psi) = \frac{p(\phi\&\psi)}{p(\psi)}$$

if $p(\psi) > 0$, and $p(\phi|\psi) = 1$ if $p(\psi) = 0$.[5] For instance, the probability of C,

fail in exceptional cases, some of which will be noted in section 4.8⋆.

[4]Sometimes the probability of B given A is written with a slanting stroke, p(B/A), but we want to avoid confusing B/A with a numerical ratio.

"Given" can be understood as meaning "given the knowledge", so that "the probability if getting an ace on the second card, given that an ace has been drawn on the first card" is short for "the probability if getting an ace on the second card, given the knowledge that an ace has been drawn on the first card." This sense of "given" is very close to the one it sometimes has in applied logic, where it might be said that "Jane will take logic" can be inferred, given that she will take either ethics or logic and she won't take ethics.

[5]That $p(\phi|\psi)$ should equal 1 when $p(\psi) = 0$ is a default assumption whose main func-

that an ace will be drawn on the third card, given the knowledge that both the first and second cards are aces, A and B, will be written p(C|A&B), and this is assumed to equal p(C&A&B)/p(A&B). How the unconditional probabilities p(C&A&B) and p(A&B) can themselves be calculated will be returned to in the following section.

Conditional probabilities can also be pictured in Venn diagrams, as *ratios of areas*. For instance, now going back to Jane's classes, the conditional probability of Jane's taking logic, given that she takes ethics, corresponds to the proportion of the 'ethics circle', E, that lies inside the logic circle, L:

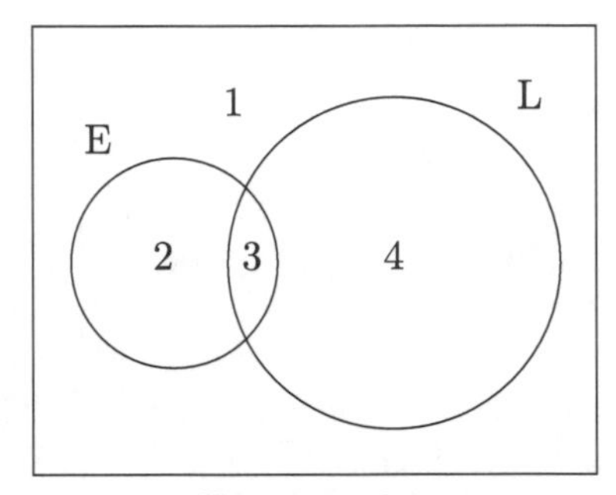

Diagram 4.1

The proportion of E that lies inside L in this diagram is quite small, hence we have represented the probability that Jane will take logic, given that she takes ethics, as being quite small.[6]

There is one more thing to note about diagram 4.1. If you fix attention on the interior of circle E, you see that conditional probabilities *given* E satisfy the same laws as unconditional probabilities. Thus, $p(\sim L|E) = 1 - p(L|E)$ and more generally, if we define the function $p_E(\phi)$ as $p(\phi|E)$ for any proposition ϕ, then p_E satisfies the Kolmogorov axioms.[7] But there

tion is to simplify the theory. However, not all theorists accept this default. Sometimes $p(\phi|\psi)$ is held to be *un*defined when $p(\psi) = 0$, and sometimes it is held to be the ratio of 'infinitesimal probabilities' in this case (cf. appendix 2). This matter is related to the question of whether conditional probabilities are really defined in terms of unconditional ones, since, as noted above, it is common for conditional probabilities to be known prior to the unconditional probabilities in terms of which they are supposedly defined. But we will ignore these questions here.

[6]This diagram might suggest that the probability of Jane's taking logic, given that she takes ethics, could be regarded as the proportion of 'cases' of Jane's taking ethics in which she also takes logic. This would be wrong, however, since Jane can only take these classes once. But there are close connections between conditional probabilities and frequencies of 'cases', some of which are pointed out in section 9.9★★.

[7]More exactly, it satisfies the following modification of the Kolmogorov axioms stated in section 2.3: for all ϕ and ψ,

K1. $0 \leq p_E(\phi) \leq 1$,

is another law of conditional probability that is particularly important in applications.

4.3 The Chain Rule: Probabilistic Dependence and Independence

Recall that we said that Bayes' principle, that $p(\phi|\psi) = p(\phi\&\psi)/p(\psi)$,[8] can be used to calculate any one of the quantities it involves from the other two, and in the case of the cards it is very natural to calculate the probability of getting two aces, p(A&B), as the product $p(A) \times p(B|A)$. This rule generalizes as follows:

Theorem 16. (*The chain rule*). If $p(\phi_1 \& \ldots \& \phi_{n-1}) > 0$ then $p(\phi_1 \& \ldots \& \phi_n) = p(\phi_1) \times p(\phi_2|\phi_1) \times \ldots \times p(\phi_n|\phi_1 \& \ldots \& \phi_{n-1})$.

In words: This rule says that the probability of a conjunction of arbitrarily many formulas is the probability of the first conjunct, times the probability of the second conjunct given the first, times the probability of the third conjunct given the first and second, and so on up to the probability of the last conjunct given all of the previous ones. As before, when a theorem involves arbitrarily many formulas it has to be proved by induction, and we will only prove it here in the case in which $n = 3$.

Theorem 16.3. $p(\phi_1 \& \phi_2 \& \phi_3) = p(\phi_1) \times p(\phi_2|\phi_1) \times p(\phi_3|\phi_1 \& \phi_2)$.

Proof

1. $p(\phi_2|\phi_1) = p(\phi_1\&\phi_2)/p(\phi_1)$. Definition of conditional probability.[9]
2. $p(\phi_1\&\phi_2) = p(\phi_1)\times p(\phi_2|\phi_1)$. From 1 by pure algebra.
3. $p(\phi_3|\phi_1\&\phi_2) = p(\phi_3\&\phi_1\&\phi_2)/p(\phi_1\&\phi_2)$. Definition of conditional probability.
4. $p(\phi_3\&\phi_1\&\phi_2) = p(\phi_1\&\phi_2) \times p(\phi_3|\phi_1\&\phi_2)$. From 3 by pure algebra.
5. $\phi_3\&\phi_1\&\phi_2$ is logically equivalent to $\phi_1\&\phi_2\&\phi_3$. By pure logic.
6. $p(\phi_3\&\phi_1\&\phi_2) = p(\phi_1\&\phi_2\&\phi_3)$. From 5, by theorem 4.
7. $p(\phi_1\&\phi_2\&\phi_3) = p(\phi_1\&\phi_2) \times p(\phi_3|\phi_1\&\phi_2)$. From 4 and 6.

K2. if ϕ is a logical consequence of E then $p_E(\phi) = 1$,

K3. if ψ is a logical consequence of ϕ&E then $p_E(\phi) \leq p_E(\psi)$,

K4. If $p(E) > 0$ and ϕ&E and ψ&E are logically inconsistent then $p_E(\phi \vee \psi) = p_E(\phi) + p_E(\psi)$.

[8] Assuming as before that $p(\psi) > 0$.

[9] This assumes that $p(\phi_1) > 0$, but that is a logical consequence of the assumption that $p(\phi_1\& \ldots \&\phi_{n-1}) > 0$.

8. $p(\phi_1\&\phi_2\&\phi_3) = p(\phi_1) \times p(\phi_2|\phi_1) \times p(\phi_3|\phi_1\&\phi_2)$.
From 2 and 7. QED

To see the usefulness of the chain rule, consider the problem of determining the chances of drawing four aces in a row from a shuffled deck. If A, B, C, and D are the events of drawing an ace on the first card, on the second card, on the third card, and on the fourth card, respectively, then the probability of getting aces on all four cards is p(A&B&C&D), and according to the chain rule this is:

$$p(A\&B\&C\&D) = p(A) \times p(B|A) \times p(C|A\&B) \times p(D|A\&B\&C).$$

Since there are 4 aces out of 52 to draw the first card from, 3 aces out of the 51 cards left if an ace has already been drawn, 2 out of the 50 cards left if two aces have already been drawn, and one ace out of the remaining 49 cards, it is clear that $p(A) = 4/52$, $p(B|A) = 3/51$, $p(C|A\&B) = 2/50$, and $p(D|A\&B\&C) = 1/49$, and therefore $p(A\&B\&C\&D) = (4/52) \times (3/51) \times (2/50) \times (1/49) = 24/6,497,400 = 1/270,725$; i.e., there is one chance in 270,725 of drawing 4 aces in a row from a shuffled deck of 52 cards.

The chain rule can also be used in combination with other rules to calculate more complex probabilities. For instance, suppose that you want to calculate the probability of getting exactly one ace in the first two cards. This is the disjunction of two conjunctions: namely A&~B, getting an ace on the first card but not on the second, and ~A&B, not getting an ace on the first card, but getting one on the second, and its probability is the sum of the probabilities of A&~B and ~A&B. The chain rule implies that $p(A\&{\sim}B) = p(A) \times p({\sim}B|A)$, and we already know that $p(A) = 4/52 = 1/13$. Since there are 48 cards left that are not aces after one ace has been drawn, $p({\sim}B|A)$ must equal 48/51, and therefore $p(A\&{\sim}B) = (4/52) \times (48/51) = 16/221$. The same reasoning shows that $p({\sim}A\&B) = p({\sim}A) \times p(B|{\sim}A) = (48/52) \times (4/51) = 16/221$, and therefore $p((A\&{\sim}B) \vee ({\sim}A\&B)) = (16/221) + (16/221) = 32/221 \cong .145$. Exercises at the end of this section will give you practice in calculating the probabilities of other simple compounds.

We conclude this section by returning briefly to the subject of probabilistic independence. Section 2.4⋆ defined propositions ϕ and ψ to be independent in this sense if the probability of their conjunction equals the product of their probabilities, i.e., if $p(\phi\&\psi) = p(\phi) \times p(\psi)$. However, an equivalent definition is that ϕ and ψ are independent if $p(\phi|\psi) = p(\phi)$, or, equivalently, if $p(\psi|\phi) = p(\psi)$.[10] In other words, ϕ and ψ are independent if the conditional probability of either given the other is equal to its uncon-

[10] Strictly, this only holds when $p(\phi)$ and $p(\psi)$ are positive. If either equals 0 then, trivially, $p(\phi\&\psi) = p(\phi) \times p(\psi)$, hence ϕ and ψ must be independent of each other. On the other hand, we have stipulated that if $p(\psi) = 0$ then $p(\phi|\psi) = 1$, and therefore it cannot equal $p(\phi)$ if the latter is less than 1. The qualification that all probabilities

ditional probability. This generalizes to probabilistic *dependence*, defined as follows.

ϕ and ψ are *positively dependent* if $p(\phi\&\psi) > p(\phi) \times p(\psi)$ and *negatively dependent* if $p(\phi\&\psi) < p(\phi) \times p(\psi)$. Again, these concepts can be defined in terms of conditional probabilities: ϕ and ψ are positively dependent if $p(\phi|\psi) > p(\phi)$, or, equivalently, if $p(\psi|\phi) > p(\psi)$, and they are negatively dependent if $p(\phi|\psi) < p(\phi)$, or equivalently, if $p(\psi|\phi) < p(\psi)$. In words: ϕ and ψ are positively dependent if the conditional probability of either given the other is greater than its unconditional probability, and they are negatively dependent if the conditional probability of each given the other is less than its unconditional probability.

Negative dependence is illustrated by *sampling without replacement* such as occurs when cards are drawn or 'sampled' from a deck one after another, without being replaced between drawings. If A and B are the events of getting aces on the first and second cards, respectively, the *un*conditional probability of B is $p(B) = 4/52 = 1/13$, while its conditional probability given A is $p(B|A) = 3/51 = 1/17$, which is less than p(B). Therefore A and B are negatively dependent, and the more aces that are sampled without being replaced, the less likely it is that succeeding cards will be aces. On the other hand, getting a king on the second card, K, is positively dependent on getting an ace on the first card, since $p(K) = 4/52$, while $p(K|A) = 4/51$, which is greater than p(K).

Instead of saying that getting an ace on the second card is negatively dependent on getting an ace on the first one, we may say that the first is negatively *relevant* to the second, in the sense that learning the first reduces the probability of the second. This looks forward to changes in probabilities, to be considered in section 4.5, which introduces the probability change law noted in section 4.1.

Exercises

1. Calculate the conditional probabilities of the following formulas, given the distribution below:
   a. $p(A|B)$
   b. $p(B|A)$
   c. $p(B|{\sim}A)$
   d. $p(A|A \vee B)$
   e. $p(A|A \vee {\sim}B)$
   f. $p(A|A \leftrightarrow B)$
   g. $p(A|B \rightarrow A)$
   h. $p(A \rightarrow B|B \rightarrow A)$
   i. $p(A \vee B|B \rightarrow A)$

should be positive also applies to the conditions stated below that are equivalent to positive and negative dependence.

j. $p(A \vee A|A \rightarrow \sim A)$

A	B	prob.
T	T	.1
T	F	.2
F	T	.3
F	F	.4

2. Calculate the following chances:
   a. That both of the first two cards are either kings or aces.
   b. That the first two cards are both diamonds.
   c. That the first two cards are of the same suit.
   d. That there will be one ace and one king in the first two cards.
   e. That there will be at least one ace in the first two cards.
3. a. Construct a probability distribution in which $p(A \vee B) = \frac{1}{2}$, $p(A|B) = \frac{1}{2}$, and $p(B|A) = \frac{2}{3}$.
   b. Construct a probability distribution in which A is positively relevant to B but $A \rightarrow B$ is negatively relevant to $B \rightarrow A$; i.e., in which $p(B|A) > p(B)$, but $p(B \rightarrow A|A \rightarrow B) < p(B \rightarrow A)$.
   c. Construct a probability distribution in which A is positively relevant to B and B is positively relevant to C, but A is negatively relevant to C.
   d. Describe real life possibilities A, B, and C, in which $p(B|A) > p(B|\sim A)$ and $p(C|B) > p(C|\sim B)$, but $p(C|A) < p(C|\sim A)$.
4. ⋆a. Give an informal argument that if A is positively relevant to B then it is negatively relevant to ∼B.
   b. Give an informal argument that if A is positively relevant to B then ∼A is also positively relevant to ∼B.

⋆5. *Simpson's Paradox* The following example of Simpson's paradox (Simpson, 1951) shows that some seemingly self-evident 'laws of conditional dependence' are invalid. During the 1985 baseball season, Steve Sax of the Los Angeles Dodgers outhit Ron Oester of the Cincinnati Reds both playing on natural grass (.257 for Sax and .225 for Oester), and playing on artificial turf (.333 for Sax and .323 for Oester), yet Sax's overall batting average was lower than Oester's (.279 for Sax and .295 for Oester). To prove that this isn't necessarily an arithmetical mistake, construct a probability distribution for the three atomic formulas S symbolizing 'Sax at bat', H symbolizing 'gets a hit', and G symbolizing 'playing on natural grass', (where we will assume that ∼S stands for 'Oester at bat' and ∼G stands for 'playing on artificial turf'), in which $p(H|S\&G) > p(H|\sim S\&G)$ (Sax has a better chance than Oester of getting a hit playing on natural grass), $p(H|S\&\sim G) > p(H|\sim S\&\sim G)$ (Sax has a greater chance than Oester of getting a hit playing on artificial turf), but p(H|S) < p(H|∼S) (Sax has less chance than Oester of getting a hit overall).

It is to be noted that this example applies conditional probabilities to collections of events like the class of all of Sax's hits, rather than to single events like Sax's getting a hit on a particular occasion. We will generally avoid doing this in this work, though the next exercise does do it.

⋆6. *Markov Processes* (cf. Doob 1953) Suppose that a cab driver driving city streets has different chances of turning left or right in different circumstances. In particular, suppose that following a left turn there is a 30% chance that his next turn will also be to the left, and a 70% chance that it will be to the right, while following a right turn there is a 40% chance that his next turn will be to the left and a 60% chance that it will be to the right. This can be viewed as a *Markov process* in which the probability of a turn in a given direction depends on the direction of the previous turn, and in which these dependencies can be represented in a so called *transition matrix*. Letting L symbolize "make a left turn" and R symbolize "Make a right turn", and supposing that R is equivalent to ∼L, the transition matrix can be written as follows:[11]

		transition probabilities		marginal probabilities
next turn	L	p(L|L)=.3	p(L|R)=.4	p(L)=p(L|R)/[p(L|R)+p(R|L)]=4/11
	R	p(R|L)=.7	p(R|R)=.6	p(R)=p(R|L)/[p(L|R)+p(R|L)]=7/11
		L	R	
		previous turn		

⋆a. Calculate the marginal probabilities that follow if p(L|L) = .3 and p(L|R) = .9.

⋆b. Prove that p(L|L) = p(L|R) implies that p(L) = p(L|L).

⋆c. Prove that if p(R|L) = p(L|R) then p(L) = p(R) = $\frac{1}{2}$.

4.4⋆ Ersatz Formulas and Truth Values

It is possible to calculate conditional probabilities directly from probability distributions, and tables representing these distributions have some interesting properties. Table 4.1 gives two examples:

[11] The symbolism used here for transition probabilities like p(L|L) mustn't be confused with ordinary conditional probabilities. E.g., here p(L|L) is the probability of a left turn being *followed* by a left turn, which the transition matrix assumes is .3, but this is obviously different from the conditional probability p(L&L)/p(L), which necessarily equals 1.

	distributions		formulas							
			atomic		compound			conditional		
regions ('cases')	arbitrary	general	E	L	~E	E∨L	E&L	L|E	L|~E	L|(E∨L)
3	.1	a	T	T	F	T	T	*T*	...	*T*
2	.4	b	T	F	F	T	F	*F*	...	*F*
4	.4	c	F	T	T	T	F	...	*T*	*T*
1	.1	d	F	F	T	F	F	...	*F*	.....
			a+b	a+c	c+d	a+b+c	a	$\frac{a}{a+b}$	$\frac{c}{c+d}$	$\frac{a+c}{a+b+c}$
			.5	.5	.5	.9	.1	.2	.8	.555

Table 4.1

(the 'cases' or state descriptions are numbered on the left above to correspond to the regions in diagram 4.1).

As in table 2.2 of chapter 2, the probabilities of ordinary formulas are computed as the sums of the probabilities of the cases or SDs in which they are true. This is extended to give a rule for computing the probability of a formula ϕ, given a condition ψ, as follows: p($\phi|\psi$) *is the sum of the probabilities of the cases in which* ϕ *and* ψ *are both true, divided by the sum of the probabilities of the cases in which* ψ *is true, and if there are no such cases then* p($\phi|\psi$) *is set equal to 1.* For instance, in the 'arbitrary' distribution in the third column of table 4.1, p(L|E) = .1/.5 = .2, because L and E are both true in case 3, whose probability is .1, while E is true in cases 2 and 3, whose total probability is .1 + .4 = .5.

The extended rule can be restated in terms of the *ersatz truth-values* such as are written in the top two rows in the column headed L|E, on the left in the 'conditional' part of table 4.1. These are in the rows in which the 'given', E, is true, and assuming this, an *ersatz* T is entered italicized into a row when L is true and *F* is written in when L is false, and when E is false (bottom two rows), nothing is written into the L|E column. Then p(L|E) is equal to the sum of the probabilities of the cases in which L|E has the value *T*, divided by the sum of the probabilities in which it has a value—either *T* or *F*. More generally, for any ϕ and ψ we can say: $\phi|\psi$ has a value when ψ is true, and it has the value *T* when ϕ is also true and the value *F* when ϕ is false; p($\phi|\psi$) is the sum of the probabilities of the cases in which $\phi|\psi$ has the value *T* divided by the sum of the probabilities of the cases in which it has a value.

There are two things to note about the ersatz formulas and truth values written into table 4.1. The first is that the fact L|E doesn't have a truth-value, even ersatz, when E is false shows that it isn't a formula in an ordinary sense. It is an expression that symbolizes "L given E", but this does not 'stand by itself', since it only forms part of larger statements like "the probability of L given E is .2." The second thing is that *T* and *F* are truth-values in name only, and there is no suggestion that this kind of 'truth' is something that should be aimed at in reasoning, or that it is better than falsehood. So far the only function of *T* and *F* is to simplify the calculation of conditional probabilities.[12] However, these will become very important in chapter 6, where they will be put to another use.

The following section focuses on the probability change aspect of Bayes' principle.

4.5 Probability Change and Bayes' Principle

Let us go back again to Jane and her chances of taking ethics and/or logic, whose probabilities are pictured in diagram 4.1. Recall what was pointed out in section 3.1, that these probabilities can change with circumstances, which implies that while the diagram may correctly depict the probabilities that pertain under certain circumstances, it may not picture them correctly under others. Assuming this, it should follow that when circumstances change probabilities may change with them, and therefore instead of one 'still picture' being needed to represent them, something more like a movie is needed. This section will be concerned with one kind of movie, which depicts the changes that take place in probabilities when a particular kind of 'epistemic force' is applied to them, namely the *acquisition of information.* For simplicity we will assume not that the new information is that Jane won't take ethics, but that she *will*, which would be expected to raise the probability of her taking ethics, and perhaps lower the probability of her taking logic.[13]

Diagram 4.1a below is like diagram 4.1 except for the addition of the two small regions A and B whose significance will be explained later, and diagram 4.1b depicts the altered probabilities that might result from learning that Jane will take ethics. Thus, the diagrams may be thought of as

---

[12]Various writers have proposed multivalued 'logics' for formulas like L|E, usually but not always assuming that they are 'most true' if L and E are both true, 'most untrue' when E is true but L is false, and they have 'intermediate' values when E is false (cf. Rosser and Turquette 1952). Recent work on conditional event logics is in this tradition, (cf. Goodman, Nguyen, and Walker 1991 and Calabrese 1994), though these authors depart from the tradition in their definitions of validity. But it is to be stressed that merely stipulating that L|E has the value *T* when L and E are both true and *F* when E is true and L is false doesn't give a complete many-valued truth-table for L|E, since it doesn't give its truth value in the cases in which E is false.

[13]This information is an *epistemic* force because it affects our beliefs about the classes that Jane will take, in contrast to a *physical* force, that affects the things that the beliefs are about.

successive frames in a 'logical' movie.[14] As already said, the 'probabilistic states of affairs' that they represent are usually called prior and posterior probabilities, or *a priori* and *a posteriori* probabilities.

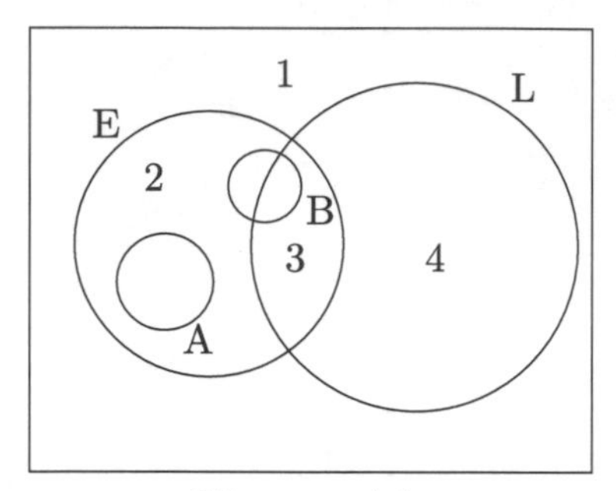

Diagram 4.1a

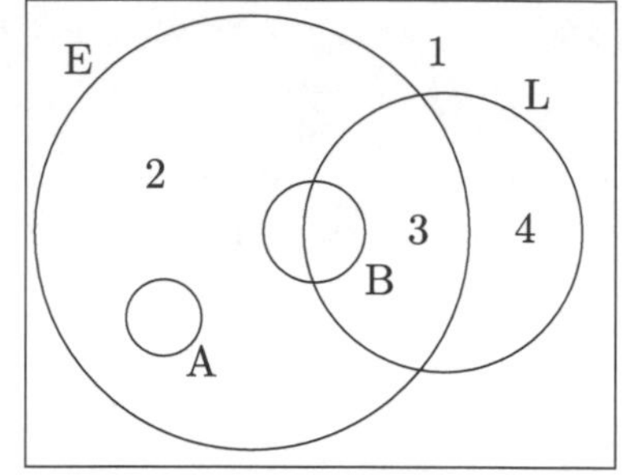

Diagram 4.1b

Diagram 4.1b shows that E's posterior probability is larger than its prior probability, while L's posterior probability is smaller than its prior probability, which we can easily imagine happening as a result of learning that Jane would take ethics. But now we are going to idealize this process in two ways, and change diagram 4.1b accordingly.

First note that diagram 4.1b pictures E as having increased its probability as a result of learning that Jane would take ethics, but it hasn't pictured it as becoming certain. That is actually fairly realistic, since a person would be unlikely to become 100% certain that Jane would take ethics just from being told that she would take it. But we will ignore this here, and assume that whatever information persons obtain becomes 100% certain once it is acquired. Assuming this, diagram 4.1b has to be changed to something like 4.1b′ in which 'E' is written into all of its regions:
Now E has been enlarged to fill the entire diagram, with the result that regions 1 and 4, which had been outside of E, no longer exist, though regions A and B remain, since they were inside of E originally. This is one idealization.

Now note a relation between the two diagrams. While A and B have both grown larger, as might have been expected, their relative *sizes* have changed, since A is larger than B in diagram 4.1a, but smaller than it in diagram 4.1b′. This seems to suggest that the new information told you more than just that Jane would take ethics, and you can see this if you think of A and B as special circumstances in which she might take it. Suppose, for instance, that A represents a situation in which Jane takes ethics from pro-

[14]The reader may find a curious analogy between the way that we have here 'pictured' thoughts, or systems of thoughts, and the famous 'picture theory of meaning' put forth in Ludwig Wittgenstein's *Tractatus Logico-Philosophicus* (Wittgenstein, 1922).

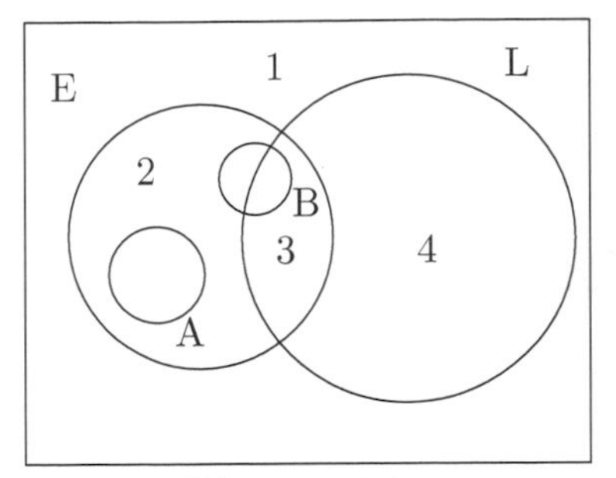

Diagram 4.1a

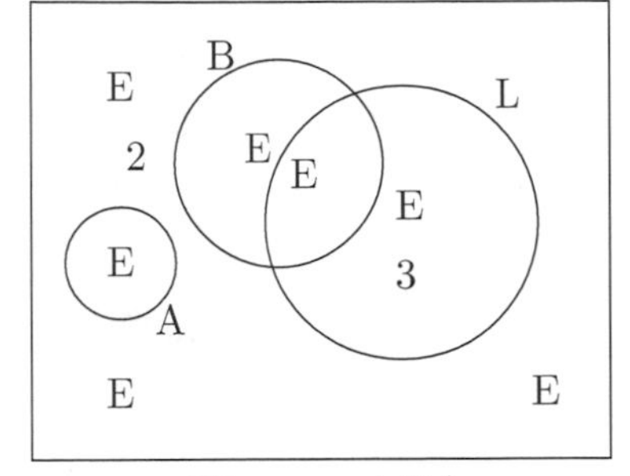

Diagram 4.1b′

fessor A while B is one in which she takes it from professor B. Then, that A was pictured as larger than B in diagram 4.1a would correspond to the fact that originally, before the new information was acquired, you thought that Jane was more likely to take ethics from professor A than from professor B. But the fact that B was pictured as larger than A in diagram 4.1b′ corresponds to the fact that after acquiring the new information it became more likely that Jane would take ethics from professor B. In other words, the new information told you not only that Jane would take ethics, but whom she was likely to take it from. Our second idealization excludes this, by stipulating, in effect, not only that the new information must be 100% certain after it is acquired, but that it is 'pure', and it tells you no more than what it states explicitly—in this case, that Jane would take ethics. Of course, real life only approximates this, just as it only approximates 100% certainty, but we will now see that if both certainty and purity are assumed then the new, *a posteriori* picture becomes very easy to draw. In fact, one way to draw it is a simple modification of diagram 4.1, which is reprinted on the left below with its modification on the right:

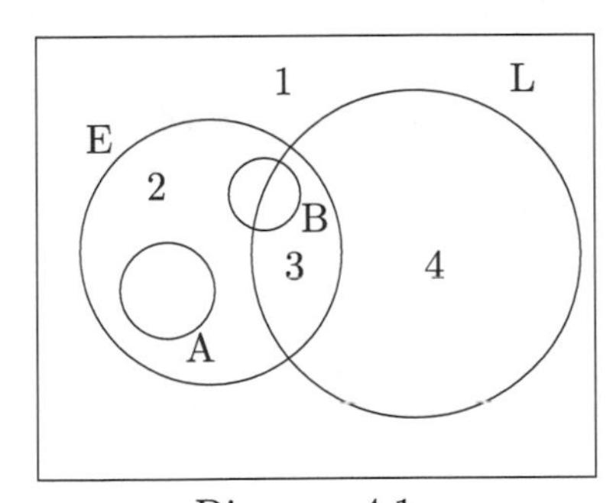

Diagram 4.1a

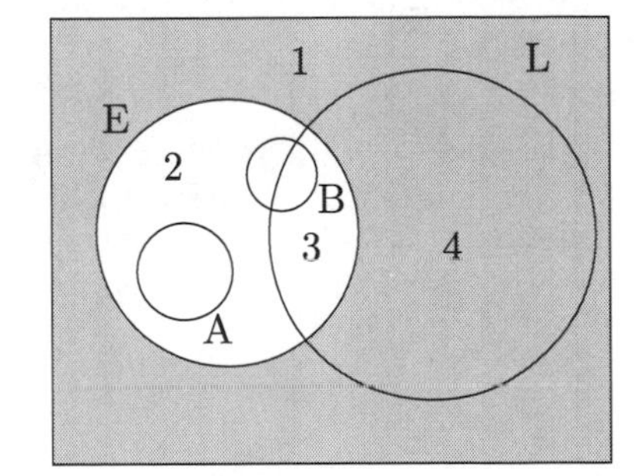

Diagram 4.1c

Going from diagram 4.1a to diagram 4.1c only involves shading out everything in diagram 4.1 outside of E, as though what was outside of E was no longer in the picture. In a sense, the camera has 'zoomed down' and 'psychologically enlarged' what is inside of E and cropped the rest of the picture out. Obviously, if this is the case then nothing inside E is affected, including the relative sizes of A and B. Moreover, assuming that $p_0(E)$, the *prior* probability of E, is positive and proportions are left unaltered in the way just described, there is a simple formula that describes the relation between the posterior probabilities, represented by the areas inside E, and the prior probabilities that apply to things outside of E as well. That is that if $p_1(\phi)$ is the posterior probability of a proposition, ϕ, and $p_0(\phi)$ is the corresponding prior probability, then

$$p_1(\phi) = p_0(\phi|E).$$ [15]

This is Bayes' principle again, but now making explicit the fact that $p_1(\phi)$ is the *a posteriori* probability of ϕ while $p_0(\phi|E)$ is the *a priori* probability of ϕ conditional on E. Thus, the principle now becomes: *The posterior unconditional probability of ϕ, following the acquisition of information E, is equal to the prior conditional probability of ϕ given E.* This is the Bayesian, or simple conditionalization rule of probability change, which we will apply to the probability changes that that are involved in deductive reasoning. However, it must be stressed that unlike the uncertainty sum theorem and other important rules that we have applied in the study of the 'static probabilities' associated with inferences, the 'dynamic' Bayesian principle is not a theorem of pure probability logic, among other things because it depends on the nonlogical certainty and pureness assumptions that were noted above. However, a brief evaluation of the principle will be undertaken in section 4.8⋆, where we will note a number of criticisms that have recently been directed against it. Nevertheless, this principle has been accepted for more than two centuries by some of the most eminent probability theorists who have ever lived,[16] and moreover we will see that even if it may be questioned in some of its applications, it is a very plausible default assumption.

The following exercises give practice in applying Bayes' principle, following which we will consider its application to inductive reasoning, and then comment on and criticize the 'epistemology' that is associated with it.

---

[15]But, remember the problem about defining $p_0(\phi|E)$ that arises when $p_0(E) = 0$, which is commented on in footnote 2 of this chapter. As said, we will largely ignore this problem, which means that we are assuming in effect that the new evidence, E, has nonzero *a priori* probability.

[16]For instance, Pierre Simon, Marquis de Laplace (1749–1827), whose great work, *Traité Analytique des Probabilités* (1812), did more than anything else to win the acceptance of probability as a serious scientific subject.

Exercises

1. Suppose that team A has a 50-50 chance of beating team B on every occasion that they meet, and they are scheduled to play a series to determine which of them wins the league championship.
   a. Calculate team A's *a priori* probability of winning the championship in a 3-game series, and its *a posteriori* probability of winning it if it wins the first game of the series.
   b. If this were a 5-game series, what would be team A's chances of winning it after winning the first game?
   c. After winning the first two gamcs?
2. a. Prove that if learning A would increase the probability of B (i.e., if $p(B|A) > p(B)$), then learning B couldn't decrease the probability of A.
   b. Prove that if A is a logical consequence of B then learning B would make A certain, and learning A couldn't decrease the probability of B.
   c. Prove that if A and B are logically inconsistent then learning either couldn't increase the probability of the other.
3. Assuming that the posterior probability of A given C is defined as $p_1(A|C) = p_1(A\&C)/p_1(C)$, what is the relation between $p_1(A|C)$ and prior conditional probabilities $p_0(A|B)$, $p_0(C|B)$, etc.? Assume that posterior probabilities arise from prior probabilities by 'conditioning on B' when the information that B is the case is acquired, i.e., that if B is the new information then for any proposition ϕ, $p_1(\phi) = p_0(\phi|B)$.

⋆4. If the posterior probability of any proposition ϕ arises from prior probabilities by conditioning on new information B, and the *posterior–posterior* probability of ϕ arises from posterior probabilities by conditioning on newer information C, state a direct relation between the posterior–posterior probability of ϕ and prior probabilities.

⋆5. *Universal Statements about Probabilities* If each team in the World Series has a 50-50 chance of winning any given game, does it follow that the team that wins the final game of the Series has a 50% chance of winning that game? If not, why doesn't this follow as a particular instance of the universal statement "each team in the World Series has a 50-50 chance of winning any given game"? And, if it doesn't follow, can you restate the universal statement so as to make clear exactly what follows from it?[17]

---

[17]This is related to the 'problem of referential opacity', which is commented on briefly in appendix 7. Cf. Quine (1953), chapter VIII.

4.6 On Induction[18]

We will start with some general comments on problems of inductive reasoning, before turning to applications of Bayesian principles. Sometimes the subject of inductive logic is stretched to include all nondeductive reasoning, such as the inference "Jane will take ethics or logic, and she will take ethics; therefore she won't take logic".[19] However, induction is more commonly regarded as a kind of inverse of deduction, that proceeds not from premises to conclusions but 'from conclusions to premises', and especially from the particular to the general, rather than from the general to the particular. For instance, while "All human beings are mortal, therefore Aristotle, Alexander the Great, Cleopatra, Joan of Arc, and Napoleon are mortal" would be a valid deductive inference, the inverse inference "Socrates, Alexander the Great, Cleopatra, Joan of Arc, and Napoleon are mortal, therefore all human beings are mortal" would clearly not be. But it might be 'inductively plausible'—a kind of learning from experience, just as coming to think that fire generally burns, from having been burnt by fire is learning from experience. Of course, it is vital to our survival that we can learn in this way, nondeductively; moreover, it is generally thought to be fundamental to scientific method. But while deductive logic has been studied intensively since the time of Aristotle, and there is widespread agreement on its fundamental principles, the principles of inductive logic have been controversial ever since David Hume drew attention to their importance.[20] We cannot enter into these controversies in detail, but because inductive reasoning is generally regarded as probable, and recent writing on the topic has made extensive use of mathematical probabilities, the theory of induction is relevant to our subject, and, as previously noted, it can be argued that Bayesian principles throw light on these matters. We will begin, however, with comments on the history of the problem of induction.

Hume argued that it is a purely psychological fact that, for instance, people who have been burned by fire come to expect that fire will burn them in the future, and no amount of 'evidence' can rationally justify forming this expectation. Of course Hume recognized that it is lucky that people have formed these kinds of expectations in the past, since the species would

[18]The subject of this section is very complex, and only the most superficial aspects of it will be sketched here. A few items cited in the footnotes in this and the following section are selected from the vast literature on this subject.

[19]Clearly this is not deductively valid, but H. P. Grice's principles of conversational quality (cf. footnote 13 of chapter 3) which suggest that a person shouldn't say "Jane will take either ethics or logic" solely on the ground that she will take ethics, also suggest that the person shouldn't assert the disjunction when he or she can assert "Jane will take both ethics and logic". Assuming this, it is reasonable to infer "Jane will not take Logic" after being 'given' that she will take one of the two and then learning that she will take ethics.

[20]See chapter XIV of Part III of Hume (1739).

quickly have become extinct otherwise, but this *is* a matter of luck, and we have no more reason to think that luck will continue than a person who has won ten times in a row betting that a coin will fall heads has to think that that will continue.

Very few philosophers have been able to accept Hume's total skepticism as to the rationality of inductive reasoning, and quite a few very prominent ones have sought to justify it—seeking to do something that Hume argued couldn't be done. For example, Immanuel Kant argued that reasoning about *causal connections*,[21] which is generally regarded as a kind of inductive reasoning, is justified in a sense by being *synthetic a priori*. And in his famous essay "On Induction", Bertrand Russell formulated principles of induction, thus:

(a) The greater the number of cases in which a thing of sort A has been found to be associated with a thing of sort B, the more probable it is (if no cases of failure of association are known) that A is always associated with B;

and

(b) Under the same circumstances, a sufficient number of cases of association of A with B will make it nearly certain that A is always associated with B... ("On Induction", p. 67).[22]

Read 'touching flame' for A and 'being burned' for B, and these principles would justify jumping to the conclusion that fire will continue to burn, on the basis of experiences of being burned by it. But it is significant that Russell ended his essay with almost Humian skepticism:

> Thus all knowledge which, on a basis of experience tells us something about what is not experienced, is based upon a belief which experience can neither confirm nor confute... The existence and nature of such beliefs... raises some of the most difficult and debated problems of philosophy ("On Induction", p. 69).

In other words, while Russell proposed principles that would justify inductive reasoning, he had to admit that he couldn't justify the principles themselves.

Russell's principles can be regarded as justifying a belief in the *uniformity of nature*, that whatever has been found true *in* our experience also holds true outside of it. And, since our experience is limited to what has

[21]For example, that the burn is caused by the flame. Kant's treatment of causality is set forth, among other places, in Book II, Chapter 3, Part C, of Kant (1781).

[22]Chapter VI of Russell (1912). Many writers both before and after Russell proposed similar principles, but this essay is exceptionally readable. This chapter also contains a famous example of a failure of induction: "...the man who has fed the chicken every day of its life at last wrings its neck..." (p. 63).

happened in the past, e.g., that fire has burned us, while what will happen in the future is unexperienced so far, our belief that that will be like what we have already experienced is a particular case of a belief in the uniformity of nature. Justifying this belief, which Russell held to be one of the most difficult and debated problems of philosophy, has come to be called the *old problem of induction.*

Nelson Goodman's *new riddle of induction* takes the old problem, or old riddle, as its point of departure, but goes beyond it by seeking to define the sense in which nature is uniform.[23] Goodman stresses what many writers including Hume and Bertrand Russell failed to recognize, or failed to give due weight to, namely that nature neither is nor is expected to be uniform in all respects, and it is by no means easy to specify in what respects we do expect it to be uniform. The famous Grue paradox illustrates this,[24] and it brings out the fundamentally important point that it is hopeless to try to do what Russell tried to do, namely to formulate purely formal principles of inductive logic, analogous to the formal principles of deductive logic.[25] Thus, it is hard to see what formal principles might justify jumping to the conclusion that all red blackberries are sour, just from having tasted one red blackberry, while it might take hundreds of cases of smokers getting lung cancer to convince us that there is a 'necessary connection' between the smoking and the cancer. Goodman's own theory is that what actually convinces us that certain apparent regularities are more than coincidences is deliberately 'trying them out', e.g., by carrying out scientific experiments. The details are too complicated and too controversial to be entered into here, but the fundamental idea seems very persuasive, and it fits rather neatly into still more recent probabilistic approaches to induction.

In some ways Bayes' principle for changing probabilities in the light of new information could be said to solve the old problem of induction, by specifying exactly what a person's posterior probabilities should be after

[23]Chapter III of Goodman (1955).

[24]Somewhat altering Goodman's own statement, something is *grue* at time t if it is either green at the time and that time is before the year 2000, or it is blue at the time, and that time is after the year 2000. The paradox consists in pointing out that all our experience that emeralds are always green is also experience that they are always grue, because our experiences to date have been prior to the year 2000, and to be green before then is also to be grue before then. But though we seem to have the same evidence for the claim that emeralds have always been and always will be green, even after the year 2000, as we do for the claim that they have always been and always will be grue, nevertheless we expect the first but not the second. As it were, we expect the future to resemble the past in greenness but not in grueness. It would be difficult to count the number of papers that have been written attempting to resolve this paradox.

[25]"The greater the number of cases in which a thing of sort A has been found to be associated with a thing of sort B, the more probable it is . . . that A is always associated with B" is a formal principle because it allows 'sort A' and 'sort B' to be defined by any properties that a thing might have—including being 'grue'

new information is acquired. For instance, assuming that the person already has *a priori* probabilities for touching fire, being burned, and being burned *after* touching fire, and he subsequently actually touches fire and is burned, he should measure the probability that he will be burned the next time he touches fire by the ratio of these prior probabilities; i.e., his posterior probability ought to be uniquely determined by his prior probabilities. Thus, if the old problem of induction was to specify how persons ought to learn from experience, then Bayes' rule solves it. On the other hand, the rule doesn't solve the old problem if it is conceived as the problem of justifying Russell's principles of induction or something like them. That is because Bayes' rule says nothing about the uniformity of nature, or about the future being like the past or any other regularities of the kind that Russell's principles apply to. It is even compatible with 'counterinductive' principles to the effect that what happens in the future will be *un*like what has happened in the past.[26] It is worth looking in some detail into how this might happen, and to do this we will consider an especially simple 'inductive problem'.

The problem is to predict the genders, female or male, of persons entering a room in which a meeting is about to take place. We will suppose that the persons come in in order, and as a matter of fact all of them are women, but this information is only acquired *a posteriori*, on the occasions when the women enter. The inductive problem is to describe the evolution of the probabilities of the next person being a woman, on successive occasions after it has been observed that all of the persons who have entered prior to that time have been women.

To simplify, suppose that only three persons are involved, namely persons #1, #2, and #3, and let the propositions that they are women be symbolized as W_1, W_2, and W_3, respectively. Then there are eight possible 'cases' or combinations of gender that these persons might have, which are entered in table 4.2. The table also lists four possible distributions of probabilities among these cases, namely D_1, D_2, D_3, and D_4. Table 4.3 then gives the unconditional and conditional probabilities of W_1, W_2, and W_3 that correspond to these distributions. The important thing is that all of these distributions and the probabilities they generate are compatible with Bayes' rule, but not all of them conform to Russell's principles of induction.

We are primarily concerned with the 'inductive sequences' $p(W_1)$, $p(W_2|W_1)$, and $p(W_3|W_1\&W_2)$ given in table 4.3, but let us look first at the distributions generating them, listed in table 4.2. Distribution D_1 gives variable values a–h summing to 1 to cases 1–8, and the values that

[26]This is not always as absurd as it seems. If by chance you keep drawing red cards—hearts and diamonds—from an ordinary deck of cards, you regard it as less and less likely that you will continue to draw red cards.

possible cases	'inductive sequences'						
	distributions				persons		
	D_1	D_2	D_3	D_4	#1	#2	#3
1	a	3/12	.125	.05	W	W	W
2	b	1/12	.125	.15	W	W	M
3	c	1/12	.125	.15	W	M	W
4	d	1/12	.125	.15	W	M	M
5	e	1/12	.125	.15	M	W	W
6	f	1/12	.125	.15	M	W	M
7	g	1/12	.125	.15	M	M	W
8	h	3/12	.125	.05	M	M	M

Table 4.2

properties of the distributions	$p(W_1)$	$p(W_2 \mid W_1)$	$p(W_3 \mid W_1 \& W_2)$
D_1 variable; not necessarily independent in any sense	a+b+c+d	$\frac{a+b}{a+b+c+d}$	$\frac{a}{a+b}$
D_2 gender and order independent; inductive, reflects experience positively	.5	.667	.75
D_3 uniform, gender and order independent; non-inductive, doesn't reflect experience	.5	.5	.5
D_4 gender and order independent; counter-inductive, reflects experience negatively	.5	.4	.25

Table 4.3

D_1 generates for the unconditional probability $p(W_1)$ and the conditional probabilities $p(W_2|W_1)$ and $p(W_3|W_1\&W_2)$ are listed as $a+b+c+d$, $(a+b)/(a+b+c+d)$, and $a/(a+b)$, respectively, in the top row of table 4.3. Skipping distribution D_2 for the moment, D_3 is the completely uniform distribution that gives all of the cases equal probabilities, which are .125 = 1/8, since there are eight cases. D_2 and D_4 are uniform except for the 'extreme cases' in which either all three persons entering are women (case 1) or all three are men (case 8). Both distributions give equal probability to the all women and all men cases, but D_2 makes these cases more probable than the nonextreme cases and D_4 makes them less probable.

Now consider the inductive series $p(W_1)$, $p(W_2|W_1)$, and $p(W_3|W_1\ \&\ W_2)$ listed in table 4.3 that are generated by the distributions in table 4.2. These are the probabilities of the next person entering the room being a woman: (1) *a priori*, before anyone has entered the room yet; (2) *a posteriori*, after one person has entered the room, who is a woman; and (3) '*a posteriori–a posteriori*', after two persons have entered the room, both of whom are women. This inductive series should conform to a principle very much like Russell's, namely "The greater the number of cases in which a thing of sort A has been found to be associated with a thing of sort B, the more probable it is ... that the next A will be associated with B", where in this case 'things' of sort A are women entering the room, and things of sort B are women following women who enter the room. This would imply that $p(W_2|W_1)$ should be greater than $p(W_1)$, and $p(W_3|W_1\&W_2)$ should be greater than $p(W_2|W_1)$—the chance of the next person entering the room being a woman should increase as more women keep entering without exception.

But now it is obvious that while all of the inductive series generated by the distributions are compatible with Bayes' rule, not all of them conform to Russell's principles, and therefore Bayes' rule does not solve the old problem of induction in the sense of justifying those principles or ones like them. In fact, Bayes' rule could even be regarded as supporting Humian skepticism about induction. To see this note that the values of $p(W_1)$, $p(W_2|W_1)$, and $p(W_3|W_1\&W_2)$ listed in table 4.3 that are generated by distributions D_2, D_3, and D_4 are all consistent with Bayes' principle. But notice how different the inductive sequences generated by D_1, D_2, and D_3 are. The first goes steadily up from .5, the second is steadily equal to .5, and the third goes steadily down from .5. The sequence corresponding to D_2 may seem to be the most 'natural' one, because it says in effect that the more women that are observed to enter the room, the more likely it is that the next person to enter it will be a woman. This fits in with the idea that we expect that the future will be like the past, or, in a sense, that nature will be uniform in this limited situation. Actually, this 'miniseries' conforms to Russell's inductive principles, hence it can be called an *inductive sequence.*

Some of its other properties will be returned to later, but before that, let us consider the sequences corresponding to D_3 and D_4.

That the probabilities in the series corresponding to D_3 are unchanging makes this series less like induction than like a blind *a priori* belief that everything that happens is pure chance, and past experience is no guide to what will happen in the future. The only thing to observe about this is that it is consistent with Bayesian principles, and therefore these principles cannot solve the problem of induction, if that means justifying things like Russell's principles. Actually, the series corresponding to D_4, which is equally compatible with Bayesianism, shows that things can be even worse.

The last line in table 4.3 shows that the values of $p(W_1)$, $p(W_2|W_1)$, and $p(W_3|W_1\&W_2)$ that correspond to distribution D_4 are .5, .4, and .25; i.e., the more people entering the room who are women, the less the probability is that the next person to enter will be a woman. This is *counterinduction*, which is directly contrary to inductive principles like Russell's. Again, the fact that this corresponds to a logically possible probability distribution proves that it is consistent with Bayesianism, hence Bayesianism not only doesn't entail principles of induction, but it is consistent with counterinduction. Nor can we rule out this possibility *a priori*, since there are many real life situations in which counterinduction is quite reasonable. In fact, we have already cited one: namely that the more aces that are drawn from a shuffled deck of cards, the less likely it becomes that the next card will be an ace. Even in the case of persons attending a meeting, if we know *a priori* that 50% of the persons to attend will be women and 50% will be men, then the more persons that enter it who are women, the less probable it is that the next person to enter will be one.

It may be objected that the kind of counterinduction just noted is only reasonable when there is *a priori* knowledge, e.g., that 50% of the persons attending a meeting will be women and 50% will be men, and that Russell-type inductive principles are only meant to apply in *tabula rasa* situations ('blank slate situations') in which there is no *a priori* knowledge. If so, this would be another difference between inductive and deductive principles, even probabilistic ones, since the latter are supposed to be valid in all situations whatsoever, whether there is prior knowledge or not. But a number of plausible '*tabula rasa* hypotheses' have been considered in the attempt to account for induction, which will be commented on briefly below.

4.7⋆⋆ Induction and Symmetry

Three interesting *tabula rasa* hypotheses are illustrated in distributions like D_2–D_4, and the kinds of independence properties that these distributions manifest: (1) They can be independent of the *order* in which persons enter the room, e.g., the probability that the persons enter in the order

possibilities	distributions			persons		
	D_3	D_5	D_6	#1	#2	#3
1	.125	.729	.427	W	W	W
2	.125	.081	.103	W	W	M
3	.125	.081	.103	W	M	W
4	.125	.009	.067	W	M	M
5	.125	.081	.103	M	W	W
6	.125	.009	.067	M	W	M
7	.125	.009	.067	M	M	W
8	.125	.001	.063	M	M	M

Table 4.4

woman-man-woman should equal the probability that they enter in the order woman-woman-man.[27] D_2, D_3, and D_4 are all independent in this sense, but obviously most Bayesian distributions do not have this kind of independence. (2) Distributions can be independent of *gender*. For instance, the probability of the series woman-man-woman is equal to the probability of the series man-woman-man, and, again, D_1, D_2, and D_3 all have this property, though most distributions do not. (3) Distributions can be *probabilistically independent*, for instance, so that the probability of the series woman-man-woman should equal the probability that the first person is a woman times the probability that the second person is a man times the probability that the third person is a woman. Of D_2, D_3, and D_4, only D_3 has this property, but all three properties have been thought to be important in inductive reasoning.[28] However, clearly these properties alone cannot justify induction since order and gender independence are consistent with counterinduction, while probabilistic independence is flatly inconsistent with it because sequences like $p(W_1)$, $p(W_2|W_1)$, $p(W_3|W_1\&W_2)$ that correspond to distributions that are independent in this sense are necessarily 'flat', and neither increase nor decrease. Hence, persons reasoning in this way could never learn from past experience. But that does not mean that probabilistic independence has nothing to do with inductive inference, and we will conclude this section by briefly describing a connection between them that has been much discussed recently.

Probabilistic independence obviously implies independence of order, but there is a less obvious inverse relation between these kinds of independence.

[27]This is commonly called *symmetry*, but it must be distinguished from other kind of symmetries such as gender-independence.

[28]For instance, the implications of 'symmetries' like independence of order and independence of gender are extensively studied in Carnap (1950). Order independence is essentially the same thing as what B. de Finetti calls *exchangeability* (de Finetti, 1975), generalizations of which have been studied extensively in recent publications (cf. Diaconis and Friedman 1980).

properties of the distributions	$p(W_1)$	$p(W_2\|W_1)$	$p(W_3\|W_1\&W_2)$
D_3 order, gender, and probability independent; non-inductive	.5	.5	.5
D_5 order and probability independent, not gender independent; non-inductive	.9	.9	.9
D_6 order independent, not gender or probability independent; inductive; 50-50 mixture of D_3 and D_5	.7	.757	.805

Table 4.5

Consider distributions D_3, D_5, and D_6 in tables 4.4 and 4.5 (note that D_3 appeared already in table 4.3).

Two things are to be noted immediately. One is that though all three of distributions D_3, D_5, and D_6 are order-independent, the first two are probabilistically independent but not inductive while the last is inductive but not probabilistically independent. The second is that the last is a 50-50 *mixture* of the first two, in the sense that the probability of each value in this distribution is the mean of the corresponding values in the other two distributions. E.g., the value of D_6 in the top line, which is .427, is the mean of the values D_3 and D_5 in that line, which are .125 and .729, respectively. These things illustrate three very important points.

First, mixtures of any number of distributions with any 'weights', are also distributions. For instance, distributions D_2–D_5 could have been mixed with equal weights of .25, or with unequal weights .1, .2, .3, and .4. If the individual distributions are thought of as defining 'possible probabilistic worlds', then a mixture of them is a weighted average of chances of being 'in' those worlds, e.g., having a 0.1 chance of being in D_2, having a 0.2 chance of being in D_3, and so on. Exercise $\star$4 at the end of this section is concerned with this conception, and with the idea that D_2, D_3, etc., can themselves be regarded as propositions with probabilities .2, .3, etc.

Second, mixtures of order-independent or inductive distributions are, respectively, order-independent or inductive, while mixtures of gender-independent or probabilistically independent distributions are not generally gender-independent or probabilistically independent.

Finally, so long as they don't assign any one distribution weight 1, mixtures of independent but not inductive distributions are inductive but not independent. However, *all inductive, order-independent distributions are mixtures of probabilistically independent distributions.* This is *de Finetti's Theorem* (de Finetti, 1937), and it has been given a great deal of attention in recent decades because it seems to be an important step towards solving the problem of induction.[29] That is because, in a sense, it derives induction from noninductive possibilities, assuming only that we don't know which of these possibilities is actual. But this also suggests limitations of the theorem.

The theorem seems to suggest that learning from experience can be regarded as learning 'which probabilistic world we are in'. But each of these worlds is itself one in which nature is totally random, and what has happened in the past is no guide to what will happen in the future. If we knew, for instance, that we were in the world corresponding to D_3, then no matter how many women we saw enter the room we would still give only a 50% probability that the next person to enter would be a woman. This is connected with the fact that in the simple form stated above the theorem does not account for learning 'nonrandom facts', e.g., that the people coming into the room might be married couples where the wife always enters before the husband, so that the actual series might go woman-man-woman-man-woman-man, and so on.[30] But then, not only doesn't the theorem explain how this nonrandom uniformity might be learned, it cannot solve the new problem of induction by explaining in what ways it is reasonable to expect nature to be uniform. But this is connected with more general limitations of the Bayesian approach, which will be commented on in the last section of this chapter.

Exercises

1. Calculate the values of $p(W_2|W_1)$ and $p(W_3|W_1\&W_2)$ according to the formulas in table 4.3 if $p(W_1) = x$, $p(W_1\&W_2) = x^2$, and $p(W_1\&W_2\&W_3) = x^3$, if $x = .3$.
2. Suppose that persons #1, #2, #3, and #4 are observed, the propositions that they are women are symbolized by W_1, W_2, W_3, and W_4, and we are interested in the series $p(W_1)$, $p(W_2|W_1)$, $p(W_3|W_1\&W_2)$, and $p(W_4|W_1\&W_2\&W_3)$. Then there are 16 possibilities represented in the table below.
   a. Give formulas for $p(W_1)$, $p(W_2|W_1)$, $p(W_3|W_1\&W_2)$, and $p(W_4|W_1\&W_2\&W_3)$ in terms of the variables a–p, like those in table 4.3.

[29]One interesting application is to justify Laplace's famous *Rule of Succession*, which will be discussed in exercise ⋆4 at the end of this section.

[30]Diaconis and Freedman (1980) generalize de Finetti's theorem so as to be able to account for this, but the theory is much more complicated.

b. How many different values of the variables a–p can there be if the distribution is order-independent?

c. How many if the distribution is gender and order-independent?

d. Give a particular order and gender-independent distribution which is 'inductive' in the sense that the series $p(W_1)$, $p(W_2|W_1)$, $p(W_3|W_1\&W_2)$, and $p(W_4|W_1\&W_2\&W_3)$ increases.

e. Give a particular order and gender-independent distribution which is 'noninductive' in the sense that the series $p(W_1)$, $p(W_2|W_1)$, $p(W_3|W_1\&W_2)$, and $p(W_4|W_1\&W_2\&W_3)$ stays constant.

f. Give a particular order and gender-independent distribution which is 'counterinductive' in the sense that the series $p(W_1)$, $p(W_2|W_1)$, $p(W_3|W_1\&W_2)$, and $p(W_4|W_1\&W_2\&W_3)$ decreases.

			#1	#2	#3	#4
1	a		W	W	W	W
2	b		W	W	W	M
3	c		W	W	M	W
4	d		W	W	M	M
5	e		W	M	W	W
6	f		W	M	W	M
7	g		W	M	M	W
8	h		W	M	M	M
9	i		M	W	W	W
10	j		M	W	W	M
11	k		M	W	M	W
12	l		M	W	M	M
13	m		M	M	W	W
14	n		M	M	W	M
15	o		M	M	M	W
16	p		M	M	M	M

⋆3. *Distributions as Causal Properties: Abduction* We can think of being 'in' distributions D_1, D_2, D_3, and D_4 in the example discussed in this section as themselves corresponding to mutually exclusive propositions, and then regard the probability of, say, W_1, that corresponds to D_4 as being the probability of W_1 *given* D_4. Assuming this, the value of $p(W_1)$ that corresponds to D_4 is $p(W_1|D_4)$, which equals .5. Similarly, the values of $p(W_1\&W_2)$ and $p(W_2|W_1)$ that correspond to D_4 are equal to $p(W_1\&W_2|D_4) = .2$ and $p(W_1\&W_2|p(W_1)) = p(W_1\&W_2|D_4)/p(W_1|D_4) = p(W_2|W_1\&D_4) = .2/.5$.

We can also think of conforming to a distribution as a property, perhaps a 'property of Nature'. For instance, conforming to distribution

D_4 can be thought of as the property of Nature being nonuniform in a particular way. This is a causal property, since nature's having this property 'causes' the chance of the next person to enter the meeting to decrease as the number of women entering it increases. And, whether the 'effects' are merely the genders of persons entering a meeting or more serious ones like the symptoms of a possibly serious illness, we would like to know something about the probabilities of their causes. This leads to a type of reasoning which C. S. Peirce called *abduction*, in which causes are 'inferred' from their effects, and which Thomas Bayes originally proposed his theory to account for.[31] In this sort of inquiry what we are not interested in are so called *direct inferences* 'from causes to effects', like $p(W_1\&W_2|D_4)$, but rather *inverse inferences* from effects to their causes, such as correspond to the conditional probability $p(D_4|W_1\&W_2)$. For example, given that the first two people to enter the room are women; we would like to know what the chances are of Distribution #4 being 'the right distribution'. To deal with this, however, we have to introduce new probabilities.

The following formula, which is actually a consequence of the Kolmogorov Axioms, states a key relation between the indirect and direct probabilities that we are concerned with:

Inverse Inference Law

$$\frac{p(D_i|W_1\&W_2)}{p(D_j|W_1\&W_2)} = \frac{p(D_i)}{p(D_j)} \times \frac{p((W_1\&W_2)|D_i)}{p((W_1\&W_2)|D_j)}$$

Here W_1 and W_2, that the first two persons to enter the room are women, are 'observed effects', and D_i and D_j are possible distributions that might 'produce' these effects. Assume, for example, that $i=2$ and $j=4$, so that D_i is the inductive distribution and D_j is the counterinductive one, we know the direct probabilities $p(W_1\&W_2|D_2)$ and $p(W_1\&W_2|D_4)$, namely $p(W_1\&W_2|D_2) = 4/12 = .333$ and $p(W_1\&W_2|D_4) = .20$, but we would like to know the inverse probabilities $p(D_1|W_1\&W_2)$ and $p(D_4|W_1\&W_2)$. Substituting the known values on the right in the inverse inference law gives:

$$\frac{p(D_2|W_1\&W_2)}{p(D_4|W_1\&W_2)} = \frac{p(D_2)}{p(D_4)} \times \frac{.333}{.20} = \frac{p(D_2)}{p(D_4)} \times 1.667$$

But, of course we don't know the values of $p(D_2)$ and $p(D_4)$, and these are what are usually called *prior probabilities* when Bayesian theory is applied to causal inferences.[32]

[31] Peirce (1903). This is sometimes called *inference to the best explanation*, and it has been written about very extensively in recent years.

[32] $p(W_1\&W_2|D_2)$ and $p(W_1\&W_2|D_4)$ are not usually called either *a priori* or *a pos-*

We cannot enter into the lively controversies that have raged over the past 60 years concerning 'Bayesian statistical applications' that involve unknown probabilities like $p(D_2)$ and $p(D_4)$, which lead many statisticians to reject this kind of reasoning,[33] but we can at least note that even without an exact knowledge of $p(D_2)$ and $p(D_4)$ or their ratio, the inverse inference formula provides qualitative information concerning the way that evidence can 'tilt the scales' in favor of one or another causal hypothesis. Thus, whatever the ratio of $p(D_2)$ was to $p(D_4)$ *a priori*, their ratio should be 1.667 times as great *a posteriori*. Since D_2 was the hypothesis that the 'future would resemble the past' and D_4 was the hypothesis that it would not, observing that the first two people to enter were women tilts the scales in favor of uniformity—that the future will be like the past.[34]

Calculate the posterior probability ratio $p(D_3)/p(D_2)$, given $W_1\&W_2\&W_3$.

$\star$4. *Rules of Succession* One interesting application of 'inference from effects to causes' is to justify some famous rules of inductive reasoning called 'rules of succession', which include a famous one due to Laplace. Applied to the persons entering the room in exercise $\star$3, the rule is that if n persons have already entered the room and k of them have been women, then the probability of the next person entering being a woman should equal $k+1/n+2$. Hence, if all persons entering have been women the probability of the next person being a woman should be $n+1/n+2$, which obviously gets closer and closer to 1 as n gets larger and larger.

A rather complicated argument shows that Laplace's rule actually follows from de Finetti's theorem on the assumption that all possible independent distributions are equally likely *a priori*, but a more complicated rule like Laplace's actually follows from the simpler assumption that the two independent distributions D_3 and D_5 in exercise $\star$3 are the only possible distributions, and they are equally likely *a*

---

teriori in applications of Bayesian statistics. Appendix 6 will point out that they have a curious relation to counterfactual conditionals like "If the actual distribution were D_2 then the probability of $W_1\&W_2$ would be .333."

[33]The most commonly accepted opposing theory is the so called Neyman-Pearson-Fisher theory (cf. Kendall 1948: 303), which rejects 'subjective priors'.

[34]This seems to be a kind of inductive reasoning about induction, namely that the fact that the future has been like the past in the past gives us good reason to think that it will be like the past in the future. Hume and others have argued that this assumes the very thing that it purports to justify, namely that Nature is uniform in the sense that what has been the case in the past, namely that it is like what came before it, will continue to be the case in the future.

priori. Then it can be shown that

$$p(W_{n+1}|W_1\&\ldots\&W_n) = \frac{.5^{n+1} + .9^{n+1}}{.5^n + .9^n}$$

and

$$p(W_{n+1}|k \text{ of the first } n \text{ persons have been women}) = \frac{.5^{n+1} + .9^{n-k+1} \times .1^k}{.5^n + .9^{n-k} \times .1^k}.$$

The interesting thing about this rule is that if $k = 0$ then $p(W_{n+1}|{\sim}W_1\&\ldots\&{\sim}W_n)$ (note that ${\sim}W_1\&\ldots\&{\sim}W_n$ is equivalent to none of the first n persons being women) approaches .5 as n approaches infinity, and if $k = n$ then $p(W_{n+1}|W_1\&\ldots\&W_n)$ (note that $W_1\& \ldots \&W_n$ is equivalent to all of the first n persons being women) approaches .9 as n approaches infinity. These are the values of the independent distributions D_3 and D_5 that enter into the probability mixture that determines the value of $p(W_{n+1}|k$ of the first n persons have been women) according to the rule above, and in general it can be shown that if more independent distributions are added to the mixture $p(W_{n+1}|k$ of the first n persons have been women) would have tended towards the ones whose values were close to the proportion, $(n-k)/n$, of women among the first n persons entering the room.

Supposing that the probabilities $p(W_{n+1}|k$ of the first n persons entering have been women) are generated by a 50-50 mixture of distributions D_5 and D_7 below:

independent distributions			mixtures	
D_3	D_5	D_7	$\frac{1}{2}(D_5+D_7)$	$\frac{1}{3}(D_3+D_5+D_7)$
.125	.729	.001		
.125	.081	.009		
.125	.081	.009		
.125	.009	.081		
.125	.081	.009		
.125	.009	.081		
.125	.009	.081		
.125	.001	.729		

a. Calculate the entries in the table above for the two mixtures $\frac{1}{2}(D_5+D_7)$ and $\frac{1}{3}(D_3+D_5+D_7)$.

b. Calculate the values of $p(W_1)$, $p(W_2|W_1)$, $p(W_3|W_1\&W_2)$, and $p(W_4|W_1\&W_2\&W_3)$ for probabilities generated by $\frac{1}{2}(D_5+D_7)$ and $\frac{1}{3}(D_3+D_5+D_7)$.

c. Determine the rules of succession of p(W_{n+1}|all of the first n persons entering the room have been women), corresponding to the mixtures $\frac{1}{2}(D_5+D_7)$ and $\frac{1}{3}(D_3+D_5+D_7)$.

4.8★ Critique of Bayesian Epistemology

Bayes' rule can be regarded as the fundamental principle of an extreme probabilistic empiricism, in which experience consists in acquiring information, and this in turn affects the mental state of the person who acquires it by changing his or her probabilities. As was seen in section 4.5, this rule pictures reasoners as 'travelling down the road of experience', by zooming in on and enlarging ever smaller parts of an *a priori* picture—after which they can ignore everything outside those parts, since the only thing required for 'updating' the picture is a knowledge of what was in the part most recently focused on prior to that.[35] But, of course, to make the whole system 'work' it is necessary to assume that in the beginning, reasoners have 'mental maps' of all possible experiences, or items of information that they might acquire at some time, and their probabilities, which they must be prepared to enlarge any part of because they cannot foresee what items of information they might acquire.

Obviously the 'Bayesian empiricism' characterized above is extreme even as an idealization,[36] since it assumes not only that new information is certain and 'pure', but that the persons who acquire it are ideal reasoners who never forget anything, and who are able to calculate logical relations among all of the propositions they consider. But it is worth noting similarities and differences between this epistemology and more traditional empiricisms.

Bayesian empiricism can be regarded as a variety of logical empiricism, or Logical Positivism (Ayer, 1959), differing from it principally in taking degrees of belief or confidence rather than 'all or nothing beliefs' to be the basic states of belief, but still assuming that these states do or should change with experience according the Bayes' rule. But it is important to note that in taking *beliefs* rather than sensations as 'sense data', both logical and Bayesian empiricism differ from classical empiricisms of the Lockian variety, in which sense experience is conceived to consist in acquiring 'ideas', which in some way 'represent things in the external world'.[37] By

[35]Thus, once the reasoner has learned that Jane will take ethics, it is as though what lies outside the E region no longer exists, and this part of the 'map' never needs to be consulted again. Another way to put it is to say that to calculate posterior probabilities Bayes' rule only has to be applied to the most recent prior probabilities, and the probabilities that existed prior to them never need to be taken into account.

[36]Vineberg (1996) examines and criticizes this idealization in detail, and especially 'Dutch Book' arguments for it of a kind that will be sketched in appendix 1. We will also see in section 6.7★ that the idealization must be abandoned if we are to account for how the acquisition of information affects the probabilities of conditional propositions.

[37]Cf. Locke (1689) especially Parts I–III of Book II.

taking beliefs, either all-or-none or of degrees, as basic, Bayesian and Logical Empiricisms bypass certain problems that confront more traditional empiricisms, while raising new problems of their own. A key problem that is sidestepped has to do with the way in which 'ideas' represent things in the external world, which was often supposed to be a kind of 'picturing', e.g., an idea of, say, the Golden Gate Bridge was supposed to be like a photograph of it. This involves many difficulties, including the facts that pictures are not very much like the things they are pictures of,[38] and that even if they did resemble 'things in themselves', the human mind, limited as it is to its own pictures, could never know this.

But, as said, in bypassing these problems, Logical Empiricisms are landed in a hornet's nest of problems of their own, two key ones of which should be noted. One is: What is a *proposition*, i.e., what is the content of a belief? The thinker previously didn't attach much confidence to the proposition that Jane would take logic, but later came to have a high degree of confidence in it. But what is the 'it' that the thinker came to have confidence in? Surely not a mental image! Then, even supposing that we could answer this question, we have to ask what relation these degrees of belief or confidence have to sense experience. In fact, it should be noted that there is nothing in the 'pure' Bayesian theory that requires 'information' to have any relation whatever to sensory observation. We do not suggest that these questions are unanswerable,[39] but we do suggest that answers to them can be expected to have profound consequences both for the 'philosophy' of the objects to which probabilities apply and for the theory of the probabilities that apply to them.

4.9 Glossary

Abduction C. S. Peirce's alternative to both induction and deduction, related to 'inference to the best explanation' and to inverse Bayesian inference.

Bayes' Principle Assuming that $p_0(\psi) > 0$, the posterior probability of a proposition, ϕ, after acquiring 'information' ι is the prior conditional probability of ϕ given ι: i.e., $p_1(\phi) = p_0(\phi|\iota)$. This idealizes by

---

[38] E.g., the picture is two-dimensional and it has a 'point of view', while the Golden Gate Bridge is three-dimensional, and it simply 'is'. Many, many epistemologists wrestled with this problem, and attempted to resolve it either by 'constructing' the external world out of subjective sense impressions (Bertrand Russell's 1914 essay, "The Relation of Sense Data to Physics," is a compact and easy to read example), or by denying that there is a world of real solid objects 'out there' (Berkeley and Hume were the foremost exponents of such views).

[39] Important work on these questions is now in progress (cf. Richard Jeffrey's theory of radical empiricism (Jeffrey, 1992; ch. 1)). Much recent writing in the field of cognitive science (cf. Fodor (1983), Churchland (1981), and Stich (1983)) advocates widely disparate views on these matters.

assuming both that ι is a certainty once it is acquired, and that it is all the information that is acquired.

Chain Rule A rule for calculating the probability of a conjunction of propositions, $\phi_1,\ldots,\phi_n$. The rule is:

$$p(\phi_1\&\ldots\&\phi_n) = p(\phi_1) \times p(\phi_2|\phi_1) \times p(\phi_3|\phi_1\&\phi_2) \times \ldots \times p(\phi_n|\phi_1\&\ldots\&\phi_{n-1}).$$

Conditional Probability The probability of a proposition given that some condition holds. If ϕ is the proposition and ψ is the condition, then the probability of ϕ given ψ, which is generally written as $p(\phi|\psi)$, is usually defined as the ratio $p(\phi\&\psi)/p(\psi)$.

Counterinduction That the more 'events' of a given kind occur in a sequence, the less likely it is that succeeding events in the sequence will be of that kind. Example: The more face cards are drawn from a deck of cards, the less likely it is that the next card will be a face card.

de Finetti's Theorem Due to B. de Finetti, *circa* 1937. The theorem is that any probability distribution over possible sequences of 'events', like the possible results of a sequence of coin-tosses, which is symmetric in the sense that any two sequences that are the same except for the order in which the events in them occur, is a mixture of independent probability distributions.

The Grue Paradox; Goodman's Paradox Due to Nelson Goodman who introduced the special predicate "Grue" and argued that while we have precisely the same evidence for "All emeralds are grue" as we do for "All emeralds are green", nevertheless we have every reason to think that the second will continue to hold true in the future but the first will not. This is a counterexample to the idea that we expect all uniformities that have held without exception in the past to continue to do so in the future. It also brings out the fact that, in contrast to the principles of deduction, the principles of induction cannot be formal.

Independence Propositions ϕ and ψ are probabilistically independent if $p(\phi\&\psi) = p(\phi)\times p(\psi)$, or equivalently, if $p(\psi|\phi) = p(\psi)$. Probabilistic independence differs from causal independence in that the latter isn't symmetric. Example: The growth of plants is causally dependent on rain and sun, but rain and sun aren't causally dependent on the growth of plants. On the other hand, each of these is probabilistically dependent on the other.

Inverse Bayesian Inference Sometimes thought of as reasoning from effects to causes. If any of a number of 'alternative hypotheses', $H_1,\ldots,H_n$ would explain or entail an observed 'event', E, then the ratio of the posterior probabilities of any two of them, H_i and H_j, is given by:

$$\frac{p_1(H_i)}{p_1(H_j)} = \frac{p_0(H_i)}{p_0(H_j)} \times \frac{p(E|H_i)}{p(E|H_j)}$$

Since the 'inverse probabilities' $p(E|H_i)$ and $p(E|H_j)$ of events given hypotheses are generally better known than the corresponding 'direct' probabilities $p(H_i|E)$ and $p(H_j|E)$, this is an especially useful formula.

Jeffrey Conditionalization A principle of probability change proposed by Richard Jeffrey that does not assume that new information becomes a perfect certainty after it is acquired. Jeffrey's rule reduces to the Bayes' rule when the new information is a certainty.

Markov Process A 'process,' like successively tossing a coin, in which the probability of the 'event' that happens at any stage of the process depends on what has happened at the previous stages.

New Problem of Induction—Goodman's Problem Nelson Goodman put forth the Grue Paradox in order to show that nature cannot be uniform in all conceivable respects, and he proposed the new problem, or riddle of induction as that of describing the respects to which Nature is regarded as uniform.

Old Problem of Induction—Hume's Problem The problem of justifying inductive reasoning, typically that regularities that have held in the past will continue to do so in the future. Hume argued that there is no 'rational' justification for any reasoning of this type, but many leading philosophers from Immanuel Kant to and after Bertrand Russell disagreed and tried to formulate inductive principles analogous to the principles of deductive inference that valid inductive reasoning should accord with.

Principles of Induction; Russell's Principles General principles of inductive reasoning, most of which are more precise statements of the intuitive idea that 'nature is uniform'. Many writers including Bertrand Russell formulated such principles, Russell's principles being, roughly, that the more cases found of 'associations' between things of two sorts, the more probable it is that they will be found to be associated in the future (provided there are no exceptions), and this probability approaches 1 as the number of cases increases without limit.

Prior and Posterior (*A Priori* and *A Posteriori*) Probabilities The probabilities that attach to a proposition before and after new information is acquired. Example: The probability that it will rain on a given day, before and after the state of the weather the previous evening has been observed.

Relevance Information is relevant to a proposition if acquiring it changes the proposition's probability; it is positively relevant if it increases the probability and negatively relevant if it decreases it.

Rules of Succession; Laplace's Rule Rules for updating probabilities of predictions of the form "The next thing will have property P" on the basis of successive 'observations' of the form "Object 1 has (does not have) property P", "Object 2 has property P",.... The most famous of these rules is attributed to Pierre Simon, Marquis de Laplace (1749–1827), one of the great pioneers in the theory of probability, which was that if n objects have been found to have property P and m objects have been found not to have it, then the probability that the next object will have it should equal

$$\frac{n+1}{n+m+2}$$

This rule entails but is not entailed by both of Russell's principles of induction, and authors such as Rudolf Carnap and Jaakko Hintikka have proposed more general rules of succession that also entail Russell's principles. Since the Bayes' assumption for updating probabilities does not entail Russell's principles, it does not entail any of the rules of succession that entail it. However, Laplace himself used inverse Bayesian reasoning to give an interesting 'proof' of his rule.

Uniformity of Nature, Principle of The idea that nature is uniform in the sense that whatever is true of the part of it that we are acquainted with holds true in general. This includes the idea that what has always held true in the past will continue to do so in the future. Russell's principles of induction can be regarded as special cases.

5

Deduction and Probability Part II: Dynamics

5.1 Introduction

When a person makes an inductive or deductive inference, her or his probabilities change, so that what was not regarded as probable prior to making the inference comes to be regarded as probable as result of it.[1] The last chapter briefly considered the changes in probability that arise in inductive reasoning, but neither classical logic nor the theory developed in chapter 3 has taken account of the analogous changes that take place in deductive reasoning. This chapter will make a start towards this—towards developing a 'dynamical' theory of deduction—by taking into account the same kind of effect on probability that was considered previously in the inductive context, namely that of acquiring new information. In the deductive case, acquiring a new piece of information can be regarded as learning a new premise, and so we will now begin to take seriously the fact that the premises from which we reason, even deductively, are generally learned in some order, and not all at once.[2]

[1]To speak of a change in what a person regards as probable as a change in probability is to make the confusion between 'subjective' and 'real' probability that was alluded to in footnote 1 of the previous chapter. In strictness, our concern here is with arriving at conclusions that are 'really' probable, and not with subjective 'feelings' about them. But, as noted previously, subjective feelings about probabilities are usually assumed to satisfy the same formal laws as real probabilities, and the distinction will only become important in chapter 9, especially section 9.9⋆⋆. So it will do no harm to conflate them at this stage.

[2]In a sense, inferences are 'timeless' in classical logic, which takes no account either of the fact that their conclusions are generally arrived at after their premises have been established, or of the fact that their premises are learned in a particular order. Abstracting from order in time is what makes it possible to reduce the 'dynamical' question of whether an inference is valid to the 'static' question of whether the negation of its conclusion is consistent with its premises. Taking time order into account makes this reduction questionable.

Of course, by restricting attention to the changes in probability that result from acquiring new premises, we idealize by ignoring perhaps the most important probability changes that accompany deductive reason, namely those that result from 'logical discoveries' such as that certain things follow from or are logically inconsistent with other things.[3] Even so, we will see that the 'dynamical' point of view throws interesting light on a number of important things. One is the phenomenon of *nonmonotonicity*, in which the acquisition of new information may not only *add* to what could previously be deduced, but may *subtract* from it as well. This too is something that neither classical logic nor the static probability logic that was developed in chapters 2 and 3 allows for, but which has become an important focus of interest in current work in the field of Artificial Intelligence and 'expert systems'.[4] For instance, as was noted in chapter 1, this throws further light on the principle that anything follows from a contradiction. This will be the subject of sections 5.2–5.6.

Section 5.7⋆ will be concerned with a problem that arises because the theory developed in sections 5.2–5.6, while allowing that 'old premises' may not be perfectly certain, still makes the Bayesian idealization that new information is certain *a posteriori*. A small step towards 'deidealizing' this takes into account the fact that information is generally acquired by fallible processes, especially by unreliable 'hearsay',[5] and in that case, while a reasoner may be sure of what someone said, e.g., that Jane would not take ethics, saying this may not make it a certainty. The gap between statement and fact has been mentioned in passing on more than one occasion previously, but this aspect of the 'probabilistic logic of conversation' will now become central to our investigation.

5.2 Deductive Dynamics, General

Continuing with the ethics and logic example, suppose that a person initially believes that Jane will take either ethics or logic, and then learns that she won't take ethics. Deducing that she *will* take logic would be a valid inference, hence according to the uncertainty sum theorem (section 3.2), the uncertainty of her taking logic cannot be greater than the uncertainty of her taking either ethics or logic plus the uncertainty of her not taking ethics. But if probabilities and uncertainties can change, even in the course of an inference, it is questionable how the theorem applies in such circum-

[3] So far as the author knows, the only serious attempt to deal systematically with these changes are due to J. Hintikka (cf. Hintikka 1970).

[4] cf. Pearl 1988.

[5] The same point can be made about the unreliable observations or recollections that often lead to unreliable hearsay, such as when someone thinks he or she has seen such and such a person at a gathering. For reasons that will be clear later, we will concentrate on hearsay.

stances. Does it follow from the fact that Jane's taking either ethics or logic was probable *a priori* and her not taking ethics became probable *a posteriori* that "Jane will take logic" should also become probable *a posteriori*? This is a complicated question.

Diagram 5.1a on the left below represents a situation like the one pictured in diagram 4.1c of section 4.5, in which Jane's taking ethics or logic is probable *a priori*, but now instead of learning that she will take ethics, it is learned that she *won't* take it. The shading in diagram 4.1c is reversed to represent this, and now the interior of E is shaded out since the new information excludes that (circles A and B are also omitted, since they play no part in the following discussion). Given this, to find out what the probability of Jane's taking logic ought to be *a posteriori*, after learning that she won't take ethics, we look at the proportion of the remaining unshaded region that lies inside the L region, which corresponds to her taking logic:

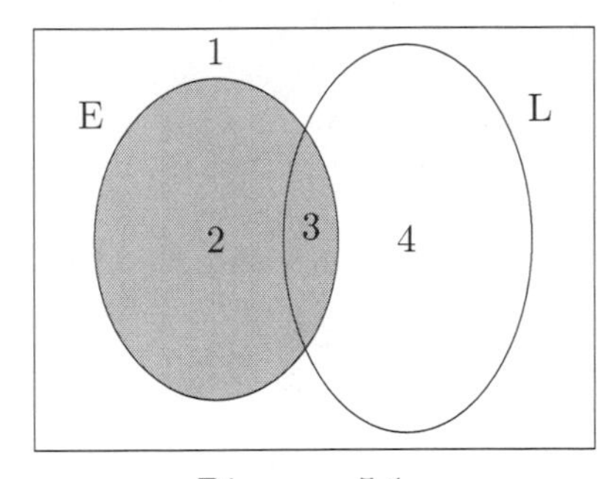

Diagram 5.1a

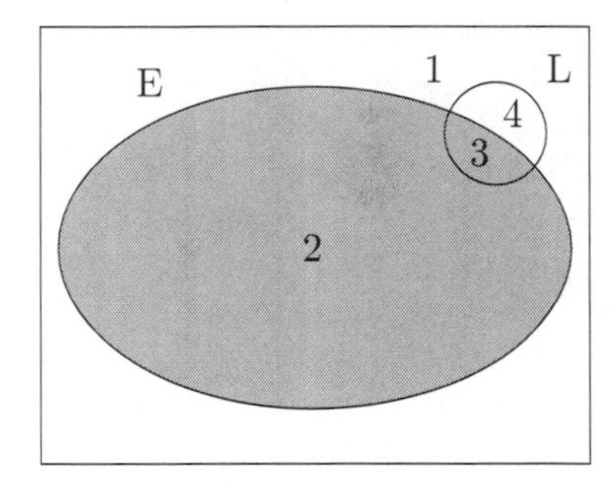

Diagram 5.1b

Diagram 5.1a shows most of this lying inside L, which suggests that after ~E is learned L ought to become probable. Thus, it seems 'logical' to deduce that Jane will take logic after learning that she won't take ethics. But is it? Diagram 5.1b also pictures $E \vee L$ as probable *a priori* and ~E as being learned *a posteriori*, but it doesn't show L as filling most of what is left after E has been shaded out. Table 5.1 confirms this numerically by describing a possible probabilistic state of affairs in which $E \vee L$ is probable *a priori*, ~E becomes certain *a posteriori*, but L does not become probable *a posteriori*. Prior probabilities and conditional probabilities are calculated in the table following the procedure described in section 4.4⋆.

Assuming that the new information is ~E and that Bayes' principle holds, and writing p_0 and p_1 for prior and posterior probabilities, respectively, the entries in table 5.1 show that $p_0(E \vee L) = .91$ but $p_1(L) = p_0(L|{\sim}E) = .1$. Hence, while $E \vee L$ was probable *a priori*, L was not probable after ~E was learned, even though it is a logical consequence of $E \vee L$ and ~E. So, perhaps it is not logical to deduce that Jane will take

	normal formulas and truth values			*ersatz* formulas and truth values	
distribution	E	L	E∨L	L|∼E	(E∨L)|∼E
.50	T	T	T	...	......
.40	T	F	T	...	......
.01	F	T	T	*T*	*T*
.09	F	F	F	*F*	*F*
→	.90	.51	.91	.10	.10

Table 5.1

logic after learning that she won't take ethics. But does this mean that the inference is invalid? That would be too simple; furthermore, we will see that to say that would be to condemn almost all inferences to invalidity. Rather than say that the inference {E ∨ L, ∼E} ∴ L is invalid because it is possible for E ∨ L to be highly probable and ∼E to be learned in circumstances in which it would be irrational to infer L, we will say that in these circumstances the second premise doesn't belong to the *scope* of the first. That is because, though E ∨ L was probable before ∼E was learned, it became improbable afterwards, and therefore, while L follows logically from E ∨ L and ∼E, both premises were not probable at the same time. This is nonmonotonicity, and, as noted in section 1.2, it is because classical logic tacitly assumes that premises are certainties that it doesn't allow for it. We now see that even just allowing that prior premises may not be absolute certainties opens the door to it. That is the kind of nonmonotonicity that is discussed in the next four sections,[6] the first two of which are restricted to the consideration of inferences with just two premises. However, it should be said that our discussion of the probability changes that take place in the course of deductive reasoning will be less formal and more intuitive than our discussion of the 'probabilistic statics' of deductive argument.

5.3 Nonmonotonicity and Scope: The Two-premise Case

In what follows we will focus on the ethics and logic example, calling "Jane will either take ethics or logic" the *prior premise*, and calling "Jane will not take ethics" the *posterior premise*, which is learned *after* the prior premise. We assume that the prior premise, E ∨ L, was probable *a priori*, but we want to consider how it might become improbable and have to be given up *a posteriori*, after the posterior premise, ∼E, is learned. To do this we must

[6]It should be said that there are currently other well known theories of nonmonotonicity (e.g., Gärdenfors (1988) and Lehman and Magidor (1988)). Most of these theories do not define nonmonotonicity in terms of probability in the way that we do here, and therefore they are not so much alternatives to the present theory as theories of other kinds of nonmonotonicity.

generalize the uncertainty sum theorem so as to make it applicable to prior and posterior uncertainties simultaneously. This can be done as follows.

Given that ~E and E ∨ L logically entail L, the 'static' uncertainty sum theorem implies

$$u_1(L) \leq u_1(\sim E) + u_1(E \vee L),$$[7]

i.e., the posterior uncertainty of L cannot be greater than the sum of the posterior uncertainties of E ∨ L and of ~E. Now, assuming that ~E is the new information, Bayes' principle applies to the posterior uncertainties on the right, i.e., it should be that

$$u_1(\sim E) = u_0(\sim E|\sim E)$$

and

$$u_1(E \vee L) = u_0(E \vee L|\sim E).$$

Moreover, obviously $u_0(\sim E|\sim E) = 0$,[8] and assuming this it follows immediately that

(I) $$u_1(L) \leq u_0(E \vee L|\sim E).$$

In words: *the posterior uncertainty of L, which is a logical consequence of the prior premise* E∨L *together with new information* ~E, *cannot be greater than the prior conditional uncertainty of* E∨L, *given the information.* This has important consequences, as follows.

Negatively, inequality (I) doesn't imply that learning ~E necessarily leads to giving up the prior premise E∨L, i.e., it applies whether or not nonmonotonicity is involved. Moreover, nonmonotonicity cannot be involved if the absolute difference $|u_0(E \vee L|\sim E) - u_0(E \vee L)|$ is small, i.e., if the prior uncertainty of E ∨ L, conditional on ~E, is close to its unconditional uncertainty. If this difference is small and E ∨ L was accepted *a priori*, hence $u_0(E \vee L)$ is small, then $u_0(E \vee L|\sim E)$ must be small, hence $u_1(L)$ should be small according to inequality (I). Thus, nonmonotonicity should only arise in our inference if, although the prior unconditional uncertainty of E ∨ L was small because it was accepted *a priori*, the corresponding conditional uncertainty $u_0(E \vee L|\sim E)$ was not small. The 'nonmonotonic values' given in table 5.1 illustrate this, since the prior unconditional uncertainty of E∨L equalled $1-.91 = .09$, which is low, but the conditional uncertainty of E∨L *given* ~E equalled $1 - .10 = .90$, which is high.[9]

[7]This can actually be strengthened to an equality, $u_1(L) = u_1(\sim E) + u_1(E \vee L)$, but it is left as an inequality here because it is an instance of the uncertainty sum inequalities to be discussed in chapter 7⋆, which define a precise notion of *probabilistic validity*. Thus, that $u(L) \leq u(\sim E) + u(E \vee L)$ should hold for any uncertainty function implies that the inference $\{\sim E, E \vee L\} \vdash L$ is not only classically valid, but it is probabilistically valid as well.

[8]The new information is certain *a posteriori*. This follows from the facts that $u_0(\sim E|\sim E) = 1 - p_0(\sim E|\sim E)$, and $p_0(\sim E|\sim E) = 1$.

[9]This is also borne out in diagram 5.1b, which depicts E ∨ L is probable *a priori*,

This can happen in real life, but, plausibly, it would be unlikely because it would mean that the new information, ~E, *contradicted* the original premise E ∨ L, not formally, but in the 'situational' sense that learning the one leads to giving up the other. This depends not just on what "Jane will not take ethics" and "Jane will take either ethics or logic" mean, but on features of the context in which they are uttered and accepted, including their order of acceptance, how firmly they are believed, and what other things are accepted along with them. These are 'situational', or 'pragmatic features',[10] and they enter into the idea of the *scope* of a proposition like "Jane will take ethics or logic" that is accepted by a person in a given situation.

For present purposes the scope of a belief held by a person in a situation can be roughly defined as *the class of propositions, the learning of which would not lead the person to give up the belief.* Hence, saying that learning that Jane wouldn't take ethics would lead someone to give up believing that she would take either ethics or logic is saying that her not taking ethics lies outside the scope of the belief in that situation. And, while we are saying that a person believing E ∨ L could give up this belief upon learning ~E, that would be unlikely. There are two reasons why this should be so, one peculiar to beliefs expressed as disjunctions, and the other general.

The special reason is that beliefs expressed as disjunctions should be 'robust' and not be given up just because one of the disjuncts proves to be false. Intuitively, we feel that a person who asserted "Jane will take ethics or logic" ought not to give it up simply as a result of learning that Jane wouldn't take ethics; i.e., Jane's not taking ethics ought to belong to the scope of "Jane will take ethics or logic". This 'logic of conversation' consideration will be returned to in section 5.6⋆, but let us put that off for now and turn to the other consideration, which has to do with why persons are unlikely to learn *anything* that falls outside the scopes of propositions that they accept.

Recall that nonmonotonicity arises in the ethics and logic case when, although the prior unconditional probability of E ∨ L is high, its prior conditional probability, *given* ~E, is low, or equivalently, when $u_0(E \vee L)$ is low but $u_0(E \vee L|{\sim}E)$ is high. The following inequality, which the student will be asked to prove as an exercise, relates these uncertainties to prior

but as improbable given ~E, because after region E has been shaded out very little of what is left lies inside region E ∨ L. But note that inequality (I) still applies, since $u_1(L) = u_0(E \vee L|{\sim}E) = .90$.

[10]The term 'pragmatic', referring to elements of the contexts in which statements are made, is used here in a sense that has become common in recent years, but this is not the same as the sense given to it in chapter 9, where it will be closer to the philosophical doctrine of Pragmatism.

probabilities:

$$p_0(\sim E) \leq \frac{u_0(E \vee L)}{u_0(E \vee L|\sim E)}$$

Now, when $u_0(E\vee L|\sim E)$ is high, the numerator of the ratio $u_0(E\vee L)/u_0(E\vee L|\sim E)$ is small and its denominator is large, hence according to the inequality, $p_0(\sim E)$ should be small. This means that if $E \vee L$ is accepted *a priori* and therefore $u_0(E \vee L)$ is small but $\sim E$ lies outside its scope because $u_0(E \vee L|\sim E)$ is high, $\sim E$ must be improbable *a priori*. This is really a general argument, and it implies that *anything falling outside the scope of a proposition that is accepted* a priori *must be improbable a priori*. This can be generalized: *the more probable an accepted proposition is* a priori, *the less probable any proposition falling outside its scope can be, and if it is certain, then any proposition falling outside its scope must be certainly false.*

Diagram 5.1a, which is repeated below, can be interpreted to picture the foregoing in the case of the ethics and logic inference:

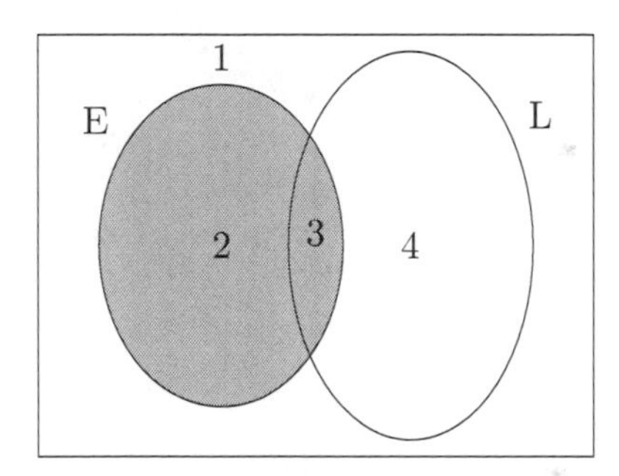

Diagram 5.1a

Here the *a priori* accepted proposition $E \vee L$ corresponds to the union of ovals E and L in the diagram, which nearly fills the 'universe' inside the outer rectangle. The more of the universe that this union fills, the smaller any region falling outside the union must be, and if it fills the entire universe the only conceivable region falling outside of it would have to be a point of zero size 'on the edge of the universe'.[11]

Finally, let us return to the question of what follows from a prior premise like $E \vee L$ together with a new premise or item of information like $\sim E$. The rough answer is that *in the likely case in which the information lies inside the scope of the prior premise, then any logical consequence of the two together follows from them in the sense that its posterior uncertainty must be small, and if the prior premise is a certainty then the new information is certain to belong to its scope*. Hence L can be inferred from the prior premise

[11]Bamber's proof (Bamber, 1997) that such cases must be 'statistically rare' is discussed in appendix 5.

$E \vee L$ and new information $\sim E$, unless that lies outside the scope of $E \vee L$. On the other hand, *in the unlikely case in which the new information falls outside the scope of the prior premise then nothing follows that doesn't follow from the new premise alone.*[12] Therefore, L cannot be inferred if $\sim E$ doesn't lie in the scope of $E \vee L$. But, *on the idealized assumption of classical logic, that all premises are certainties, anything that is a logical consequence also 'follows' in the sense of being a certainty.*

But the following section comments on what is warranted on the case of logically inconsistent premises.

Exercise

1. Prove informally that the inequality $p_0(\sim E) \leq u_0(E \vee L)/u_0(E \vee L|\sim E)$ is a theorem of pure conditional probability, i.e., ignore subscripts and simply show that $p(\sim E) \leq u(E \vee L)/u(E \vee L|\sim E)$.

5.4 What Follows from a Contradiction

Let us modify the ethics and logic example, and now suppose that the prior premise is simply that Jane *will* take ethics, not that she will take *either* ethics or logic, but the new information is, as before, that she won't take ethics. We would expect this information to lie outside the scope of the prior premise,[13] and a reasoner who accepted that premise would certainly not expect such news. But the scenario of old beliefs being contradicted by new information is familiar enough, and the question is: What should 'follow' in this situation? What would it be reasonable to deduce?

One point is that if the prior premise, E, isn't a certainty *a priori*, then the new information, $\sim E$, must lie outside its scope.[14] Therefore, as was said above, what 'follows' in this case is only what is logically entailed by $\sim E$ alone, and not 'anything' that follows from the contradictory combination of it with the prior premise, E.

But what if the prior premise is a certainty *a priori*, i.e., what if $p_0(E) = 1$? Then $p_0(\sim E) = 0$, Bayes' principle doesn't apply, and learning $\sim E$ would be 'learning the impossible'.[15] We will only make two observations

[12] This will be reconsidered in the following section and in section 5.7⋆. The former comments on the anomalous case in which the new information logically contradicts the prior premise, and the latter considers the possibility that posterior premises may not be certainties.

[13] That the denials of prior accepted propositions fall outside of their scopes, except possibly when they are certainties, is one of the formal laws of scope that will be discussed in section 5.5.

[14] That is because $p_0(E|\sim E)$ must equal 0 when $p_0(E) < 1$.

[15] However, we remind the reader of a point made earlier, that V. McGee has generalized Bayes' principle to the case of learning new information whose *a priori* probability is infinitesimally small (cf. appendix 2). If it were assumed that $p_1(\phi) = p_0(\phi|E) = 1$, even when $p_0(E) = 0$, this would contradict the Kolmogorov Axioms, since this would

concerning this possibility. One is that if E were a certainty *a priori*, it is hard to see how ~E could really be 'learned'. It is possible that someone might assert "Jane won't take ethics" *a posteriori*, but this would almost certainly be questioned in the circumstances—as we will see in section 5.7⋆. The other point is that however posterior probabilities are arrived at, it is plausible to suppose that they satisfy the Kolmogorov axioms, and therefore that $u_1(E) + u_1(\sim E) = 1$. But according to the uncertainty sum theorem this only implies that the uncertainty of a logical consequence of E and ~E cannot be greater than 1, which is no information at all. Hence, even if we choose to say that the contradiction between E and ~E logically entails anything, nothing follows from it in the sense that its uncertainty is guaranteed to be low.

Now we will generalize, and consider inferences with arbitrarily many premises, or, more exactly, with arbitrarily many *prior* premises.

5.5 More General Inferences

Consider a conclusion ϕ that is logically entailed by 'prior premises' $\phi_1, \ldots, \phi_n$ plus 'new information', ι. For instance, in the ethics and logic case $n = 1$, $\phi_1 = E \vee L$, $\iota = \sim E$, and $\phi = L$. Now Inequality (I) in section 5.3 generalizes to yield the following

Dynamical Uncertainty Sum Rule:

(Ig) $$u_1(\phi) \leq u_0(\phi_1|\iota) + \ldots + u_0(\phi_n|\iota).$$

In words: *The posterior uncertainty of a logical consequence of prior premises together with new information cannot be greater than the sum of the prior conditional uncertainties of the former, given the information.*

The argument for inequality (Ig) is a direct generalization of the argument for inequality (I). Since the 'static' uncertainty sum theorem applies to any uncertainties, and ϕ is entailed by ι and $\phi_1, \ldots, \phi_n$, it follows that it is satisfied by the posterior uncertainties, $u_1(\phi)$, $u_1(\iota)$, and $u_1(\phi_1), \ldots, u_1(\phi_n)$, i.e.:

$$u_1(\phi) \leq u_1(\iota) + u_1(\phi_1) + \ldots + u_1(\phi_n).$$

This can be combined with Bayes' principle, as follows. Given the fact that ι is the new information, we can replace the posterior unconditional uncertainties, $u_1(\iota)$, $u_1(\phi_1), \ldots, u_1(\phi_n)$, on the right of the above inequality by prior conditional uncertainties, conditional on ι. That is, $u_1(\iota)$ must equal $u_0(\iota|\iota)$, and $u_1(\phi_i)$ must equal $u_0(\phi_i|\iota)$ for $di = 1, \ldots, n$. Moreover, $u_0(\iota|\iota)$ is necessarily equal to 0. Substituting these values into the inequality above yields inequality (Ig).

Now let us consider the implications of the generalized rule.

---

entail that $p_1(\sim\phi) = p_1(\phi) = 1$.

First, as before, the rule is general, and it does not imply that acquiring new information necessarily leads to giving up prior premises; i.e., it applies whether or not nonmonotonicity is involved. In fact, nonmonotonicity should only arise in an inference when, although the unconditional uncertainties of its premises were small *a priori*, because they were accepted *a priori*, one or more of the corresponding conditional uncertainties $u_0(\phi_1|\iota),\ldots, u_0(\phi_n|\iota)$ was not small.[16]

This brings us back to the idea of the *scope* of a prior premise ϕ_i, roughly defined as the class of propositions, the learning of which would not lead a reasoner to give up the belief in ϕ_i, or, assuming Bayes' principle, as the class of possible items of information, ι, for which the uncertainty, $u_0(\phi_i|\iota)$, is low. As before, it is to be stressed that this is not a formal property of ϕ_i, since it depends on 'situational' factors related to the circumstances in which it is accepted.

Still generalizing from the ethics and logic example, we can also say, roughly, that *nonmonotonicity can only arise in an inference when its new premise does not belong to the scopes of all of its prior premises.* And, we can add that that is unlikely because prior probabilities and uncertainties have to satisfy the inequality:

$$p_0(\iota) \leq \frac{u_0(\phi)}{u_0(\phi|\iota)}.$$

This implies that if ϕ is probable *a priori*, but not conditionally *given* ι, then ι must be *im*probable *a priori.*

More generally still, we can state the following vague but useful 'rule of scope':[17]

Rule S of Scope Any item of possible information that lies outside the scope of a premise or other accepted belief must be improbable *a priori*, and the more certain that belief is *a priori*, the less probable such an item will be. If the belief has probability 1 (is absolutely certain) *a priori* then an item of possible information lying outside of its scope must be certainly false *a priori*,[18] but if the belief is not absolutely certain *a priori*, then there is nonzero probability that

[16]Of course, this ignores the kind of nonmonotonicity that can arise when there are too many premises, as in the lottery paradox. This will also be left out of account in the 'scope rules' to be formulated below.

[17]This and the rules to be formulated later are only first approximation 'rules of thumb'. In particular, most of them do not apply to cases of 'learning the impossible,' in which a reasoner acquires information that is inconsistent with prior premises that are themselves certainties, such as inferences from contradictions.

[18]Of course if premises are certain a priori, then there is zero probability that they will have to be given up *a posteriori*. But, as the previous footnote says, the rule does not apply in that case.

information lying outside of its scope may be found to be the case.[19]

In combination with our earlier comments this has the following consequences:

Corollary S1. The more certain the prior premises of an inference are the more likely it is that new information will belong to their scopes, and therefore that a conclusion that is logically implied by the prior premises together with the new information can be inferred after that information is acquired. If the prior premises are *a priori* certain, then any conclusion that is logically entailed by them together with the new information can be inferred after that information is acquired.[20] If the prior premises are not *a priori* certainties then it is possible that a conclusion that is logically implied by these premises together with items of new information may not become probable or be inferred after this information has been acquired.[21]

This is important practically. It warns us that in making inferences based on new information we must be sure that the information does not 'conflict' with whatever prior premises enter into the reasoning, and the only way we can be certain *a priori* that this won't happen is by being *a priori* certain of these prior premises. And, we can add that the *more* certain we are of the premises *a priori*, the less likely it is that new information will conflict with them in such a way as to force us to give them up.

The next corollary is a special case of Corollary S1 that bears on the rule that anything follows from a contradiction:

Corollary S2. The only conclusions that can be inferred when new information is acquired are ones that follow logically from this information together with the premises in whose scope it lies.

This generalizes what was said in the previous section about two-premise inferences, namely, that if the premises of such an inference are inconsistent, and they are pieces of information that are obtained in some order, then not only can't anything at all be inferred after the second premise is learned, but the only conclusions that can be inferred are ones that follow from the second premise alone.

Other corollaries of the rule S include some of the properties of an *algebra of scope*:

Corollaries S3–S9. Let ϕ, ϕ_1, and ϕ_2 be *a priori* probable propositions,

[19]This last follows because if ϕ is any proposition that is probable but not certain *a priori*, then $\sim\phi$ must have some positive probability *a priori*, and therefore it cannot lie inside the scope of ϕ unless ϕ is *a priori* certain.

[20]Though we must remember the caveat that the new information must be consistent with the prior premises.

[21]Again, this ignores problems that can arise when there are too many premises.

and let ψ be any proposition.

S3. If ψ is self-consistent but inconsistent with ϕ, and ϕ is not a certainty, then ψ does not belong to the scope of ϕ; hence $\sim\phi$ does not belong to the scope of ϕ unless it is an *a priori* certainty.

S4. If ψ logically implies or is logically implied by ϕ then it belongs to the scope of ϕ.

S5. Any *a priori* probable proposition belongs to the scope of any other *a priori* probable proposition.

S6. Either ψ or $\sim\psi$ belongs to the scope of ϕ.

S7. All propositions belong to the scope of a logically true proposition.

S8. ψ belongs to the scope of $\phi_1\&\phi_2$ if and only if it belongs to the scopes of both ϕ_1 and ϕ_2.

S9. If ψ_1 and ψ_2 belong to the scope of ϕ then so does $\psi_1 \vee \psi_2$, and if $\psi_1 \vee \psi_2$ belongs to the scope of ϕ then either ψ_1 or ψ_2 does.

Of these rules, S3 is the most practical. It tells us, in effect, that all propositions that are not *a priori* certainties are liable to be contradicted and have to be given up on receipt of new information—though we know too that the more certain they are, the less liable they are to have to be given up.

That anything entailed by ϕ belongs to its scope, according to rule S4, is hardly news; if ψ is entailed by an accepted proposition ϕ then learning ψ is only learning what is already deducible *a priori*, and it isn't news.[22] Similarly, if ψ is already probable *a priori* then learning it is not news either, so it should belong to the scope of ϕ—rule S5. Rules S6–S9 tell us something about the formal algebra of scope alluded to above.

S6 tells us that the scope of any accepted proposition must be 'large', since it must include at least half of all other propositions. However, the rule does *not* say that the scope is exactly half of these propositions, which would be false, since all propositions belong to the scope of a logically true proposition (rule S7). S8 and S9 give more detailed information about scopes, but the only thing we will note about them here is that the 'dual' of S8 is invalid: that ψ belongs to the scope of $\phi_1 \vee \phi_2$ if and only if it belongs to the scope of either ϕ_1 or ϕ_2. Giving an argument for this will be left as an exercise.

Exercises

1. a. Argue that if a proposition A neither logically entails nor is logically entailed by another proposition C, then there is a possible probability distribution in which $p(C) \geq .9$ but $p(C|A) = 0$, hence

[22] But, as before, this idealizes by ignoring the fact that the logically true proposition might not be known to be true.

A does not belong to the scope of C.

b. Argue that this cannot happen if A either entails or is entailed by C.

2. Describe a situation in which propositions A and B both belong to the scope of another proposition C, but A&B does not. State A, B, and C in English.
3. Give examples of propositions A, B, and C in English such that A belongs to the scope of B ∨ C but not to that of either B or C.
4. a. Can you give examples of propositions A and A → B, and a situation in which A → B is probable but A does not belong to its scope?

   b. Can ∼A lie outside the scope of A → B?

   c. Can you draw a diagram picturing A → B as probable, though it has a low conditional probability given A?
5. Can you give examples of propositions A, B, and C such that A belongs to the scope of B and B belongs to the scope of C, but A does not belong to the scope of C?
6. Describe propositions A and C, such that both A and ∼A belong to the scope of C.
7. Do you think it is possible for both A and B to belong to the scope of C while A ∨ B does not? Give reasons.
8. a. If B belongs to the scope of C and A logically implies B, does A necessarily belong to the scope of C? Give reasons.

   b. If A belongs to the scope of B and B entails C does A necessarily belong to the scope of C? Give reasons.
9. Give informal arguments for scope rule corollaries S3–S9.

5.6⋆ Defaults: The Logic of Conversation Again

We know from rule S3 in the previous section that anything that is not certain *a priori* may have to be given up as a result of gaining new information, but what information, aside from the direct denial of the proposition, could have this result has not been considered. This section will look a bit further into this matter, and in particular consider certain 'defaults' that we can expect would *not* require giving up an accepted proposition.

To begin with let us return to the 'logic of conversation' explanation given in section 5.3, as to why nonmonotonicity would not be expected in the ethics and logic inference. That was that it would be peculiar for a person to assert "Jane will take ethics or logic" in circumstances in which he or she would not be prepared to maintain it on learning that Jane wouldn't take ethics, hence Jane's not taking ethics ought to belong to the scope of the assertion "Jane will take ethics or logic". The explanation of this builds on the main argument of section 5.3, which consisted in pointing out that anything lying outside the scope of a probable proposition ought to be improbable, but it goes beyond that in an important way, since the conversational argument concerned the assertion "Jane will take ethics or logic" and not its probability. To assert something is to *say* something, which is why the argument is conversational, and that is neither a necessary nor a sufficient condition for what is asserted to be probable. But, as said, there is an important connection between the two arguments, which comes out in probabilistic aspects of H.P. Grice's logic of conversation.[23]

Grice formulated several 'maxims of conversational quality', one of which was to the effect, roughly, that *a speaker should not make a weaker, less informative, and less useful statement in circumstances in which it would be possible to make a stronger, more informative, and more useful one.*[24] This applies to "Jane will take ethics or logic" in a special way. If Jane's taking ethics or logic is probable enough to be asserted,[25] but her not taking ethics lies outside its scope, then it would have to be improbable, as we have seen. But then Jane's taking ethics would have to be probable, and if it were probable enough to be asserted the speaker should have asserted *that*, and not the weaker, less informative, and less useful statement that Jane would take either ethics or logic.

The foregoing suggests that, in general, in the absence of reasons to the contrary, one may expect that when a speaker makes an assertion of the 'Either E or L' form, neither E nor L by itself is probable, and therefore neither ~E nor ~L is improbable, hence both of the latter should belong to the scope of 'Either E or L'. Therefore if either ~E or ~L is learned, the disjunction should still be probable and the 'alternative disjunct' should be inferable.[26]

But 'absence of reasons to the contrary' is a default assumption, and

[23]Chapter 4 of Grice (1989) discusses the conversational logic of disjunctive assertions in detail.

[24]They can even be censured for doing so, in this case for not 'telling the whole truth'.

[25]Another conversational maxim establishes a very direct connection between conversation and probability, namely that speakers should only assert things that they have good grounds for, which is very close to saying that what they assert should be probable. See also Searle (1969).

[26]But there are two curious things to note in this connection. One is that it is not reasonable to assume that the disjuncts are not independently probable in the case of a 'Hobson's Choice' disjunction like "Jane will either take ethics or she will take ethics", which would have to be given up if it was learned that Jane wouldn't take ethics. Such disjunctions can be assertable under certain circumstances, which shows that the maxim "Don't assert a disjunction when one of its disjuncts can be asserted independently" has exceptions.

It is also curious that intuitionism, which is a special 'logic' that has been held to be 'the right logic' for mathematical arguments, where premises are usually supposed to be certainties (cf. Dummett 1977), holds that a disjunction shouldn't be asserted *unless* one of its disjuncts can be asserted. This is obviously inconsistent with the view that a disjunction shouldn't be asserted when one of its disjuncts can be asserted.

situations in which there *are* reasons to the contrary are easy to describe. Nothing in pure formal logic excludes the possibility that a speaker might 'opt out' of adhering to the maxim that he should not make a weaker, less informative, and less useful statement in circumstances in which it is possible to make a stronger, more informative, and more useful one. For instance, a speaker believing that Jane would take ethics could under certain circumstances say something like "I think I know what course Jane will take, but all I can tell you is that she will take either ethics or logic". This would be a case in which the new information "Jane will not take ethics" would lie *outside* the scope of "Jane will take either ethics or logic", and in which the person spoken to would not be entitled to infer "Jane will take logic" if she learned that Jane wouldn't take ethics.[27]

Now let us comment briefly on two other 'dynamical' patterns of inference, one of which is a formal variant of the disjunctive pattern that we have just been discussing. That is the 'negative conjunctive' pattern, as in "Jane won't take both ethics and logic, and she will take ethics; therefore she won't take logic". Of course, this inference is 'statically valid' and the uncertainty sum theorem implies that the uncertainty of its conclusion at any one time cannot be greater than the sum of the uncertainties of its premises at that time. Diagram 5.2a on the left below illustrates this, and suggests that the inference ought to be 'dynamically valid' in the sense that if its first premise is highly probable at one time and its second premise is learned, its conclusion should then become probable.

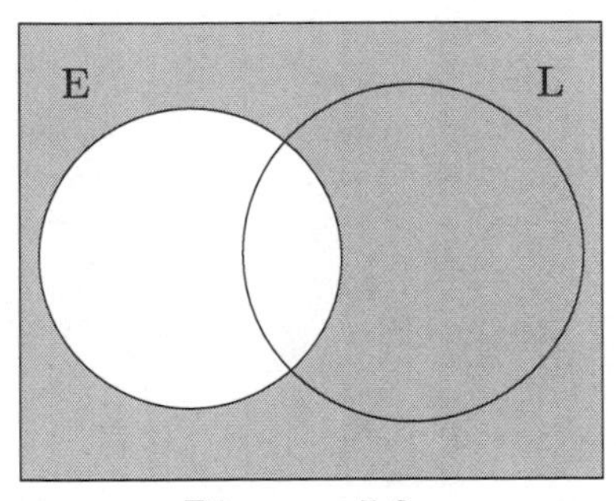

Diagram 5.2a

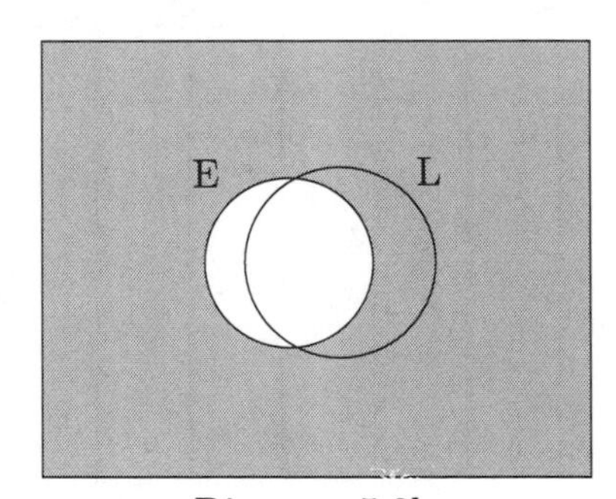

Diagram 5.2b

However, diagram 5.2b shows that no more than in the disjunctive case is the negative conjunctive inference always dynamically valid. In this case, $\sim$(E&L) is highly probable *a priori* but L occupies a large portion of what

[27]This should be qualified to allow for the fact that the hearer who heard that Jane would take either ethics or logic might have one set of probabilities while the person who said "I know what courses Jane will take, but all I am permitted to tell you is that she will either take ethics or logic" had another. Study of this aspect of the 'probabilistic dynamics of speech' is still very much in its early stages.

is left after everything outside of region E is shaded out, hence ~L is not pictured as probable *a posteriori*, after E is learned. And, again, this is supported numerically by a table analogous to table 5.1:

	normal formulas and truth values			*ersatz* formulas and truth values	
distribution	E	L	~(E&L)	~L|E	~(E&L)|E
.09	T	T	F	*F*	*F*
.01	T	F	T	*T*	*T*
.01	F	T	T	...	......
.89	F	F	T	...	......
→	.10	.10	.91	.10	.10

Table 5.2

Table 5.2 shows ~(E&L) as having the fairly high probability of .91 *a priori* but the probability of ~L *given* E is only .1, hence assuming Bayes' principle it should not become probable when E is learned. That is because the probability of the prior premise, ~(E&L), drops to .10 after E is learned, hence it and E are not both probable *a posteriori*. The new premise doesn't belong to the scope of ~(E&L), and ~L doesn't follow in the sense that it should become probable when the new information is acquired. We also see that the information was improbable *a priori*, i.e., $p_0(E) = .1$, which conforms to the principle that $p_0(\iota) \leq u_0(\phi)/u_0(\phi|\iota)$, where in this case $\iota =$ E, $\phi = \sim(E\&L)$, $u_0(\phi) = 1 - p_0(\phi) = .09$, and $u_0(\phi|\iota) = 1 - p_0(\phi|\iota) = .90$.

And, finally, the same 'default' conversational maxim that dictates that a person should not assert a disjunction when one of its disjuncts doesn't belong to its scope also dictates that a speaker shouldn't make an assertion of the ~(E&L) form when E doesn't belong to its scope. In such circumstances E must be improbable and therefore ~E must be probable, and it would be a stronger, more informative, and more useful statement than ~(E&L).[28]

Our last example is one that doesn't fall strictly within the formalism employed so far, but the principles we have been discussing can be applied to it nevertheless. In this case the prior premise is "Everyone who takes ethics will take logic", the new information is that Jane will take ethics, and the conclusion is that she will take logic, which is a logical consequence of the prior premise and the new information. But as before, the question is: given that "Everyone who takes ethics will take logic" was probable *a priori* and the information is acquired *a posteriori*, must the conclusion become probable *a posteriori*? The plausible answer is 'no'. Jane's taking

[28]Thus, nothing of the form ~(E&E) should be asserted. E.g., it would violate maxims of conversational quality to assert "Jane won't both take ethics and take ethics".

ethics can lie outside the scope of "Everyone who takes ethics will take logic", and even if the latter is accepted *a priori*, learning that Jane will take ethics *a posteriori* need not lead to the conclusion that she will take logic.

If we let E and L stand for "Jane will take ethics" and "Jane will take logic", as before, and let G stand for "Everyone who takes ethics will take logic", then a situation in which G is probable *a priori* but learning E should not lead to concluding L might be pictured as:

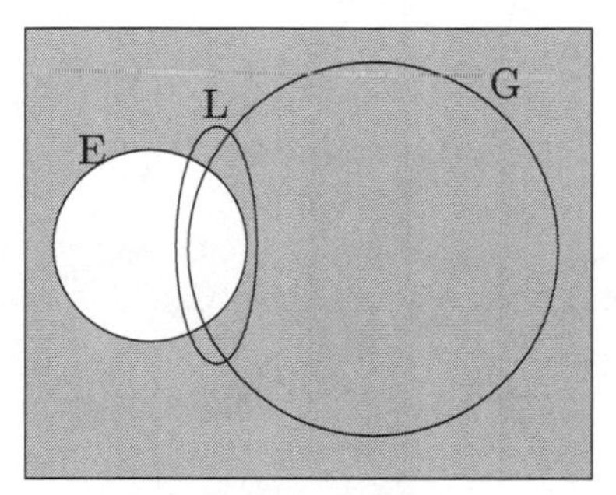

Diagram 5.3

Although the logical inappropriateness of symbolizing "Everyone who takes ethics will take logic" simply as G will be returned to immediately below, we can at least say that diagram 5.3 depicts $\{E, G\} \therefore L$ as logically valid by showing everything in the intersection of regions E and G lying inside L. But L isn't shown occupying most of what is unshaded after E is learned, so the diagram suggests that G could be probable *a priori* and E then learned, though L would not become probable *a posteriori*. It also shows E not belonging to the scope of G, since the region corresponding to G only fills a small part of the region corresponding to E.

However, we can't confirm the above by a probability distribution like the ones in tables 5.1 and 5.2. That is because a probability distribution is supposed to distribute probabilities 'over state descriptions' (section 2.2), and in languages that symbolize propositions like "Everyone who takes ethics will take logic" appropriately, 'state descriptions' are much more complicated than those considered here.[29] Nevertheless, we can describe a plausible 'real life possibility' in which Jane's taking ethics might lie outside the scope of "Everyone who takes ethics will take logic".

Suppose that it were thought *a priori* that Jane wouldn't take ethics,

[29] In fact, they are *models* or *interpretations* of languages of predicate logic, such as one in which "Jane will take Ethics", "Jane will take logic", and "Everyone who takes ethics will take logic" might be symbolized as Ej, Lj, and $(\forall x)(Ex \rightarrow Lx)$, respectively. Cf. appendix 7.

but that if she were to take it she wouldn't take logic. Moreover, assuming that Jane won't take ethics, it is believed that everyone who does take it will take logic. Now, in this 'scenario' learning that Jane would take ethics would cause a reasoner to give up believing that everyone who takes ethics would take logic, and not to conclude that Jane would take logic. More simply, Jane's taking ethics would fall outside the scope of "Everyone who takes ethics will take logic" and the only thing that could be concluded on learning that she would take ethics is what follows from that proposition alone. This leads to two observations.

One is to contrast the sort of scope under consideration here with what is traditionally called the scope of a universal generalization, namely the class of individuals that it is meant to 'apply' to. There is a clear formal difference between the two kinds of scope because all accepted propositions have scopes of the kind we have been considering, and these are classes of propositions, while traditionally only generalizations have scopes, and these are classes of individuals. However there could be a correspondence between the 'propositional scope' and the 'traditional scope' of a proposition that has both of them, namely a generalization. For instance, that Jane will take ethics falls outside the propositional scope of "Everyone who takes ethics will take logic" suggests that it isn't meant to apply to the person, Jane, and therefore she doesn't belong to the traditional scope of this generalization. The general rule that an individual who falls under the traditional scope of a generalization must be the subject of a proposition that falls under its propositional scope is perhaps a crude approximation, but it suggests that there are significant connections between them.

The second observation is that our examples illustrate a generalization of the point made earlier that anything not *a priori* certain is liable to have to be given up, which follows from rule S3. The generalization is that, given any proposition, ϕ, that is not certain *a priori*, and any other proposition, ψ, that is logically independent of ϕ, there are possible circumstances in which ψ would fall outside the propositional scope of ϕ and learning ψ would lead to giving ϕ up.[30] That is because, given that ϕ is not certain and $\psi \& {\sim}\phi$ is logically possible, it can have some positive probability, and learning it would necessarily lead to giving up ϕ. Our examples have demonstrated this in three cases that might seem surprising *a priori*, e.g., in which ϕ is E$\vee$L and ψ is ~E, which is logically independent of E$\vee$L. The point of our present discussion is examine details, with a view to formulating plausible default rules in which new items of information will belong to the scopes of previously accepted propositions, and 'standard logic' can be applied in reasoning from them.

But, though clearly the discussion in this section only scratches the

[30]This reinforces our earlier point that scope is not a formal property of a proposition.

surface of 'probabilistic conversational logic' that combines Gricean and probabilistic considerations, it is time to turn to other matters.[31] We will now consider another aspect of conversational logic, which throws light on the problem of accounting not only for prior premise uncertainties, but on posterior premise uncertainties as well.

Exercises

1. Describe a situation in which Jane's taking ethics would fall outside the scope of "Jane won't take ethics, logic, *and* history." Represent it in a probability diagram.
2. A phenomenon that Grice (1975) calls 'bracketing' is illustated by one person correcting another person's statement "Jane will take ethics or logic" by saying "No, she will take ethics or history." Supposing that the correction means Jane will take ethics or history, and not logic, what would be a situation in which it would fall outside the scope of "Jane will take ethics or logic"? Draw a probability diagram representing this.

5.7⋆ Posterior Premise Uncertainty

If all new information were certain *a posteriori*, and we assumed Bayes' principle as a default, it would follow that this new information would remain certain ever after, even when it in its turn became old information. That would imply that our only uncertain premises and beliefs would be 'totally *a priori*', and everything we learned afterwards would be completely certain. But of course that is absurd, as we argued in section 4.7⋆⋆. This section will briefly explore one way of avoiding this absurdity while retaining Bayes' principle, which takes into account the fact that the things we *say* we learn are themselves only highly probable inferences based on experience.

Still sticking with ethics and logic, we want to consider a 'Bayesian' way of representing learning that Jane won't take ethics, which doesn't make it a certainty *a posteriori*. The idea is to include not just E, L, and compounds like $E \vee L$ and $\sim$E in the analysis, but another atomic proposition $S_{\sim E}$, symbolizing the fact that a person, S, *says* that Jane won't take ethics. Then the new information is not really $\sim$E, that Jane won't take ethics, but rather $S_{\sim E}$, that S says that Jane won't take it. Then, assuming Bayes' principle, what posterior probabilities are really based on is $S_{\sim E}$, and not $\sim$E. How different is that from the result of basing posterior probabilities on $\sim$E?

[31] Also, this subject will be returned to in the following chapter, where we will consider rules of scope that apply to probability conditionals, and where the 'dynamic' scope rules commented on here 'translate' into static rules in the conditional logic.

The following complicated theorem of pure probability logic, which the student is asked to prove as an exercise, is a step towards answering the foregoing question:

(C) $\quad \|p_0(L|{\sim}E) - p_0(L|S_{\sim E})\| \leq u_0({\sim}E|S_{\sim E}) + u_0(S_{\sim E}|{\sim}E)$.

The quantity $\|p_0(L|{\sim}E) - p_0(L|S_{\sim E})\|$ on the left above is the absolute value of the difference between $p_0(L|{\sim}E)$ and $p_0(L|S_{\sim E})$,[32] which are the probabilities of the conclusion that Jane would take logic that would result from learning that she wouldn't take ethics, as against the result of learning that S said that Jane wouldn't take ethics. The whole inequality (C) says that the difference between these probabilities can't be greater than $u_0({\sim}E|S_{\sim E}) + u_0(S_{\sim E}|{\sim}E)$; i.e., it can't be greater than the uncertainty of Jane's not taking ethics given that S said that she wouldn't take it, plus the uncertainty of S saying this, given that Jane doesn't take ethics. If these uncertainties are small, then the difference between $p_0(L|{\sim}E)$ and $p_0(L|S_{\sim E})$ must be small, and we can reason 'approximately' as though what is learned when S says that Jane won't take ethics is that she won't take ethics. So the thing to do is to examine the two uncertainties.

There is good reason to think that $u_0({\sim}E|S_{\sim E})$ should be small. If S is a reliable person, and she says that Jane won't take ethics, then it is pretty certain that she won't take it. In fact, here we can think of $u_0({\sim}E|S_{\sim E})$ as a measure of S's 'unreliability'. Thus, the more reliable S is the smaller $u_0({\sim}E|S_{\sim E})$ will be.

However, probably $u_0(S_{\sim E}|{\sim}E)$ wouldn't be small, since even if Jane doesn't take ethics there is no reason for S to *say* so, even if she knows it. And, if $u_0(S_{\sim E}|{\sim}E)$ isn't small it appears that $\|p_0(L|{\sim}E) - p_0(L|S_{\sim E})\|$ will not be small. In fact, there could even be a large difference between $p_0(L|{\sim}E)$ and $p_0(L|S_{\sim E})$. For instance, if S would only say that Jane wouldn't take ethics if S knew that she wouldn't take any philosophy courses, then $p_0(L|S_{\sim E})$ would be low while $p_0(L|{\sim}E)$ could be quite high. This problem, concerning the relation between what is said and the fact that it is said, is a serious one, and dealing with it seriously would take us much further into the probabilistic logic of conversation than is practicable here. But we will close with a partial solution, which, as will be seen in the exercises, also applies to the relation between what is observed, perhaps unreliably, and the fact of observing it.

The partial solution is to suppose not only that S says that Jane won't take ethics, but that she does this in response to the question "Will Jane take ethics?" When the fact that this question was asked is added as another 'factor', Q, inequality (C) gets modified to:

[32] It is important that the absolute value of the difference appears here, rather than the 'signed' difference $p_0(L|{\sim}E) - p_0(L|S_{\sim E})$, which would be negative if $p_0(L|S_{\sim E})$ were larger than $p_0(L|{\sim}E)$.

$$(\mathrm{C}_Q)\ \|\mathrm{p}_0(\mathrm{L}|{\sim}\mathrm{E}\&\mathrm{Q})-\mathrm{p}_0(\mathrm{L}|\mathrm{S}_{\sim\mathrm{E}}\&\mathrm{Q})\| \leq \mathrm{u}_0({\sim}\mathrm{E}|\mathrm{S}_{\sim\mathrm{E}}\&\mathrm{Q})+\mathrm{u}_0(\mathrm{S}_{\sim\mathrm{E}}|{\sim}\mathrm{E}\&\mathrm{Q})$$

Now $\|\mathrm{p}_0(\mathrm{L}|\mathrm{Q}\&{\sim}\mathrm{E}) - \mathrm{p}_0(\mathrm{L}|\mathrm{Q}\&\mathrm{S}_{\sim\mathrm{E}})\|$ is the difference between the probabilities of the conclusion that Jane would take logic that would result from learning that she wouldn't take ethics, as against the result of learning that S said that Jane wouldn't take ethics, in both cases after the speaker has been asked whether Jane would take ethics. Also, as before, it is plausible that $\mathrm{u}_0({\sim}\mathrm{E}|\mathrm{Q}\&\mathrm{S}_{\sim\mathrm{E}})$ will be small, because it is probable that Jane won't take ethics, given that S said that she wouldn't take it after being asked whether she would. But now $\mathrm{u}_0(\mathrm{S}_{\sim\mathrm{E}}|{\sim}\mathrm{E}\&\mathrm{Q})$ can also be expected to be small, because, given that S is 'informed', it is probable, given that Jane won't take ethics, that S will say so when asked.

But again, what has just been said only scratches the surface of a subject that involves aspects of formal logic, probability theory, linguistics, Artificial Intelligence, and, the author believes, decision theory and psychology as well, and which is in a stage of rapid development.[33]

Exercises

⋆1. Prove the following theorem of pure probability logic:

$$\|\mathrm{p}(\mathrm{A}|\mathrm{B}) - \mathrm{p}(\mathrm{A}|\mathrm{C})\| \leq \mathrm{u}(\mathrm{B}|\mathrm{C}) + \mathrm{u}(\mathrm{C}|\mathrm{B})$$

⋆2. *The puzzle of the three job-seekers.*[34] The problem is as follows. Three job candidates, A, B, and C, have applied for the same position, and *a priori* their chances of getting it are equal, namely 1 in 3. Candidate A has learned that a decision has already been made, but when he approaches the manager to ask who got the job, the manager tells him that he can't say anything until the decision is announced publicly. Nevertheless, A persists and says "If you can't tell me whether I got the job, will you at least tell me the name of someone besides myself who *didn't* get it"? The manager then relents to the extent of replying "All right, I'll tell you in confidence that B didn't get it", which pleases A, because he now thinks his own chances of getting the job have risen from 1 in 3 to 1 in 2. Question: Was A right about this? After all, he knew *a priori* that the manager would say either that B didn't get the job or that C didn't get it, so why should learning that B in particular didn't get it affect his chances of getting it? Discuss this, and in particular whether it is plausible that the uncertainties on the

[33]Recent work of Brian Skyrms (1990) on the dynamics of deliberation, which concerns the approach to equilibrium of beliefs, should also be mentioned in this connection.

[34]Sometimes called the problem of the three prisoners, cf. Gardner (1961), pp. 226–9. Comments at the end of this short piece show that professional statisticians can disagree about the solution to the puzzle, but amateurs can still learn something from trying to work it out for themselves.

right in the inequality

$$\|p(A|Q\&S_{\sim B}) - p(A|Q\&{\sim}B)\| \leq u({\sim}B|Q\&S_{\sim B}) + u(S_{\sim B}|Q\&{\sim}B)$$

would both be small, where A = "A got the job", B = "B got the job", Q = "A tells the manager 'tell me the name of someone besides myself who didn't get the job'", and $S_{\sim B}$ = "The manager replied that B didn't get the job".[35]

5.8 Glossary

Dynamical Uncertainty Sum Rule If $\phi_1,\ldots,\phi_n$ are 'prior premises', ι is a 'new premise', and ϕ is a logical consequence of $\phi_1,\ldots,\phi_n$ and ι, then:

$$u_1(\phi) \leq u_0(\phi_1|\iota) + \ldots + u_0(\phi_n|\iota),$$

In words: *The posterior uncertainty of a logical consequence of prior premises together with new information cannot be greater than the sum of the prior conditional uncertainties of the former, given the information.* This depends on Bayes' principle, and therefore it isn't a theorem of pure probability logic.

Maxims of Conversational Quality Maxims due to H.P. Grice, roughly to the effect that *speakers should not make weaker, less informative, and less useful statements in circumstances in which it would be possible to make stronger, more informative, and more useful ones.*

Rules of Scope Rules involving the scopes of accepted beliefs, which are partly dependent on Bayes' principle. The most important is corollary S1: *The more certain the prior premises of an inference are the more likely it is that new information will belong to their scopes, hence the more likely it is that a conclusion that is logically implied by the prior premises together with the new information can be inferred after that information is acquired. If the prior premises are certain* a priori, *then any conclusion that is logically entailed by them together will new information can be inferred after that information is acquired. If*

[35]An alternative method for dealing with experiences that do not lead to perfectly certain conclusions is described in chapter 11 of Jeffrey (1983). An illustration involves someone looking at a piece of cloth by candlelight and concluding that it is probably green, though it could be some other color like blue. *A priori* the person might think that the probability of the cloth's being green, G, is .3, i.e., $p_0(G) = .3$, but observing it by candlelight changes this to some new value, $p_1(G)$, which is presumably greater than .3 but less than 1. But though the experience of looking at the cloth affects the person's probabilities, it may not even correspond to a proposition which would allow $p_1(G)$ to be represented as a prior conditional probability, $p_0(G|E)$, according to Bayes' principle, and Jeffrey's more general principle avoids this. Details are too complicated to enter into here, but the interested reader is urged to consult Jeffrey's work. Interestingly, however, the *radical probabilism* outlined in chapter 1 of Jeffrey (1992) seems to question even the Jeffrey Conditionalization principle.

the prior premises are not a priori *certainties then conclusions that are logically implied by these premises plus certain possible items of information may not be inferred after this information is acquired.*

Scope The class of possible items of information that are compatible with an accepted proposition in the sense that acquiring them would not lead to giving up the proposition. A rough rule of scope is that *the more certain the proposition is the broader its scope will be, and if it is perfectly certain its scope must be 'universal' in the sense that nothing could lead to giving it up.*

6

Probability Conditionals: Basics

6.1 Introduction

Recall expressions like L|E at the heads of the 'conditionals' columns on the right in table 4.1 of chapter 4, the nonquantitative part of which is reproduced below, with a new symbolism added underneath the *ersatz* formulas:

	formulas					conditionals		
regions	atomic		compound			L∣E	L∣~E	L∣(E∨L)
('cases')	E	L	~E	E∨L	E&L	E⇒L	~E⇒L	(E∨L)⇒L
3	T	T	F	T	T	*T*	...	*T*
2	T	F	F	T	F	*F*	...	*F*
4	F	T	T	T	F	...	*T*	*T*
1	F	F	T	F	F	...	*F*	......

Table 6.1

There were two reasons why an expression like L|~E above was regarded as being '*ersatz*': (1) although it forms a part of formulas like "p(L|~E) = .4/.5",[1] it doesn't symbolize or 'stand for' anything by itself, and (2) it doesn't have truth-values in all cases, and even when it does these values are *ersatz*. In spite of these reasons, however, we are now going to give these expressions a meaning, by using them to symbolize ordinary language conditional statements like "If Jane doesn't take ethics then she will take logic". We will also introduce a new symbolism, ~E ⇒ L, which is more like the symbolism of standard logic, to express this concept formally.[2] But this symbolization, which is written underneath the earlier symbolization, isn't only an alternative to it, since we will see that it leads to conclusions

[1] In this and most of the next chapter we will return to the 'static' point of view with respect to probabilities and uncertainties, and accordingly we will not write them with subscripts as in p_0(L|~E) or p_1(L|~E).

[2] Alternative symbolizations are sometimes used for probability conditionals. One of these is the large inequality sign, '>', which the student may find it easier to write.

about the reasoning it is used to symbolize that differ radically from the conclusions that follow when the reasoning is symbolized in the standard way, in terms of the *material conditional*, $\sim E \rightarrow L$.[3] This section will bring out certain fundamental differences, and the remainder of this chapter and all of the next will be devoted to working out their consequences.

The most basic difference between $\sim E \Rightarrow L$ and the standard symbolization, $\sim E \rightarrow L$, has to do with probabilities. If we use $\sim E \Rightarrow L$ as a substitute for $L|{\sim}E$ this means that we are assuming that $\sim E \Rightarrow L$ and $L|{\sim}E$ have equal probabilities, and, since $\sim E \Rightarrow L$ symbolizes "If Jane doesn't take ethics then she will take logic", it follows that we are assuming that the probability of the conditional statement should equal $p(L|{\sim}E)$. That is, we are hypothesizing that *the probabilities of conditional statements like "If Jane does not take ethics then she will take logic" are conditional probabilities.* This is the *probability conditional theory*,[4] which goes with the symbolism $\sim E \Rightarrow L$, and it will be fundamental to what follows in the next three chapters. However, it should be said that because this theory has consequences that differ radically from those of standard logic, it is not now widely accepted, although it is receiving increasing attention. The following comments may help to bring out the contrast between this and classical logic's theory of the conditional, as well as part of the reason why the probability conditional theory is controversial.

The two theories yield very different results when they are applied to certain conditional statements, and it can seem intuitively that the probability conditional theory is more 'right' than the material conditional theory is in these cases. One case is the statement "If the first card is an ace then the second one will be too", made about cards that are about to be dealt from a shuffled pack of 52 cards. If this is symbolized as a material conditional, $A \rightarrow B$, which is true in all cases except that in which A is true and B is false, its probability can be calculated as:

$$\begin{aligned} p(A \rightarrow B) &= 1 - p(A\&{\sim}B) \\ &= 1 - p(A)p({\sim}B|A) \\ &= 1 - p(A)(1 - p(B|A)) \\ &= 1 - (1/13) \times (1 - 3/51) \\ &= 205/221 \simeq .928 \end{aligned}$$

[3]Actually, it is only since Frege that the material conditional has become standard. The status of conditionals had long been a subject of controversy before that, as a remark attributed to the Hellenistic logician Callimachus attests: "Even the crows on the rooftops are cawing over the question as to which conditionals are true" (Mates, 1965: 203).

[4]This theory was advanced independently and almost simultaneously by Richard Jeffrey (1964), Brian Ellis (1973), and the author (1964).

On the other hand, according to the probability conditional theory, the probability of "If the first card is an ace then the second one will be too" should be

$$\begin{aligned} p(A \Rightarrow B) &= p(B|A) \\ &= 3/51 \simeq .059. \end{aligned}$$

This is very different from the .928 probability calculated according to the material conditional theory, and the value of 3/51 seems closer to our naive intuition that the probability of "If the first card is an ace then the second card will be too" ought to be 3 in 51. Of course, this does not settle the question of which theory is 'really right', which is a matter of considerable debate at this time, and we will return to this in chapter 8⋆, where it will be seen to lead to quite deep philosophical issues.[5] However, here we are concerned with the implications of the probability conditional theory, and we will end this section by noting another important difference between it and the material conditional theory.

We are used to 'embedding' material conditionals like $A \rightarrow B$ in larger formulas like $\sim(A \rightarrow B)$, but if we did this with probability conditionals and wrote, for instance, $\sim(A \Rightarrow B)$, this would seem to imply that its probability equalled $p(\sim(A \Rightarrow B))$. However, assuming that $A \Rightarrow B$ is merely another way of symbolizing $B|A$, it would follow that $p(\sim(A \Rightarrow B))$ should equal $p(\sim(B|A))$. But so far $p(\sim(B|A))$ has not been defined, hence it is formally meaningless. Therefore, while it might seem natural to use $\sim(A \Rightarrow B)$ to symbolize "It is not the case that if the first card is an ace then the second card will also be an ace", either we must define its probability or we must allow our formal language to contain a formula that not only doesn't have a truth value but doesn't have a probability. The second possibility is discussed briefly in appendix 4, but the fact that the probability conditional theory leads to this problem is one of the most important objections that many logicians have against it. As said, this is discussed in appendix 4, but we will sidestep the problem here by restricting ourselves to formulas like $A \Rightarrow B$, in which $\Rightarrow$ occurs only as the main connective, when it occurs at all. The next section is concerned with what many regard as the main advantage of the probability conditional theory, which is the way it resolves certain 'paradoxes' of material conditionals.

6.2 The Paradoxes of Material Implication

When the inferences

> The first card will not be an ace. Therefore, if the first card is an ace then the second card will be an ace.

[5] In any case, there are more than two alternative theories as to the proper analysis of the conditional, and one more of these will be described briefly in section 8.3.

and

> The first card will not be an ace. Therefore, if the first is an ace then it will not be an ace.

are symbolized using material conditionals, as $\sim$A $\therefore$ A $\rightarrow$ B and $\sim$A $\therefore$ A $\rightarrow$ $\sim$A, their validity is easily established by truth-tables. Thus, if $\sim$A is true then neither A $\rightarrow$ B nor A $\rightarrow$ $\sim$A can be false, since the truth of $\sim$A implies the truth of both A $\rightarrow$ B and A $\rightarrow$ $\sim$A. But this is paradoxical because these inferences seem absurd to most persons who haven't had courses in formal logic. Similarly, it is paradoxical that according to classical logic the statements "If the first card is an ace then the second card will be an ace" and "If the first card is an ace then the second card will not be an ace" are logically consistent, since they seem intuitively to be inconsistent,[6] and the single statement "If the first card is an ace then it will not be an ace" is classically consistent although it seems self-contradictory. Thus, when it is applied to conditional statements, classical logic has paradoxical consequences both of logical consequence and of logical consistency.

We will now see how the probability conditional theory can avoid these paradoxes, not by showing that the inferences are invalid in the sense that their premises can be true while their conclusions are false, or by showing that the intuitively contradictory statements cannot both be true, but rather by showing that the premises of the inferences can be highly probable while their conclusions are highly improbable, and by showing that the intuitively contradictory statements cannot both be probable.

First consider the two paradoxical inferences, which can be formalized in terms of probability conditionals as $\sim$A $\therefore$ A $\Rightarrow$ B and $\sim$A $\therefore$ A $\Rightarrow$ $\sim$A. Assuming the standard probabilities for cards drawn from a shuffled pack, we have:

$$\begin{aligned} p(\sim A) &= 12/13 \simeq .923, \\ p(A \Rightarrow B) &= p(B|A) = 3/51 \simeq .059, \text{ and} \\ p(A \Rightarrow \sim A) &= p(\sim A|A) = 0. \end{aligned}$$

The common premise of the inferences has the high probability of .923, but the first inference's conclusion only has probability .059 and the second inference's conclusion has probability 0. Each inference has a highly probable premise and a highly improbable conclusion, which would make it seem irrational for persons to reason in this way, at least if they hope to reach probable conclusions thereby.[7]

---

[6] Whether they are really inconsistent is another matter. This is part of what is at issue in Lewis Carroll's famous Barbershop paradox, which is described in exercise $\star$8 at the end of the following section.

[7] Since the inferences would be logically valid if the conditionals involved were material conditionals, it follows from the uncertainty sum theorem that their conclusions'

Similar considerations apply to the paradoxes of consistency. The statements "If the first card is an ace then the second card will be an ace" and "If the first card is an ace then the second card will not be an ace" are consistent when they are symbolized as material conditionals, $A \rightarrow B$ and $A \rightarrow \sim B$, but when they are symbolized as the probability conditionals $A \Rightarrow B$ and $A \Rightarrow \sim B$ their probabilities are $1/13$ and $12/13$, respectively. They are not both probable in this case, and it is hard to see how they could be in any case. Similarly, when "If the first card is an ace then it will not be an ace" is symbolized as $A \Rightarrow \sim A$, it has 0 probability, and therefore it would be totally absurd to maintain it.[8]

The following sections will explore other implications of the probability conditional theory, but we will conclude this one by contrasting the way in which this theory resolves the paradoxes with the way that standard logic approaches problems of inferential validity and contradiction. As said above, the probability conditional theory explains the 'fallacy' involved in the inferences not by showing that they can lead from true premises to false conclusions, but rather that they can lead from probable premises to improbable conclusions. And, the 'conflict' between "If the first card is an ace then the second card will be an ace" and "If the first card is an ace then the second card will not be an ace" is explained not by showing that they cannot both be true, but rather that they cannot both be probable. This is simple enough—the contrast is between the concepts of truth and falsehood on the one hand, and those of probability and improbability on the other.

However, another contrast is that standard logic's use of the 'all or none' concepts of truth and falsehood is perfectly precise, while our theory's use of the ideas of probability and improbability is imprecise, and therefore its 'resolution' of the paradoxes is vague. A precise definition of probabilistic validity will be given in section 6.4, but as it is more complicated than standard logic's concept of validity we will defer it and work for the present with the rough, intuitive ideas of probability and improbability that we have used to this point.

6.3 Intuitive Analysis of Other Inference Patterns

We will start by illustrating the general method that will be used to analyze inferences and inference patterns involving conditionals probabilistically,

---

probabilities must be at least as high as the probability of their common premise. Since the latter is .923, the conclusions' probabilities must be at least that high. In fact, we have seen that $p(A \rightarrow B) = .928$, and the same reasoning shows that $p(A \rightarrow \sim B) = .923$.

[8]This ignores the possibility that $p(A)$ could equal 0, in which case according to stipulation, all three of $p(A \Rightarrow B)$, $p(A \Rightarrow \sim B)$, and $p(A \Rightarrow \sim A)$ are equal to 1. This is the special 'limiting case' in which probabilistic consistency coincides with classical consistency, which will be returned to in section 7.8⋆⋆.

by applying the method to another kind of inference that is often regarded as paradoxical. These are ones of the form "B; therefore, if A then B", symbolized as B $\therefore$ A $\Rightarrow$ B.

The first step is by any means to construct a *diagrammatic counterexample* to the inference, in which its premise is represented as being probable while its conclusion is improbable. Here is a possibility:

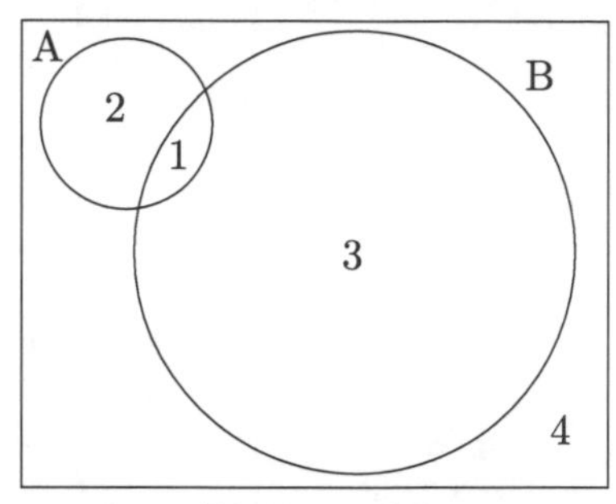

Diagram 6.1

This is a diagrammatic counterexample to the pattern B $\therefore$ A $\Rightarrow$ B because the premise, B, is shown as probable—the region corresponding to B fills most of the diagram—while the conclusion, A $\Rightarrow$ B, is shown as improbable. That is because its probability is assumed to equal the probability of B given A, and that is low because only a small part of the region corresponding to A lies in the one corresponding to B (see section 4.2).

The next step is to 'confirm' the diagram numerically, by constructing a probability distribution over the SDs involved, corresponding to regions 1–4, that generates probabilities for B and A $\Rightarrow$ B of a kind that the diagram would roughly represent. Here is a possibility of that kind:

		formulas		
regions	distribution	A	B	A⇒B
1	.01	T	T	*T*
2	.09	T	F	*F*
3	.89	F	T	...
4	.01	F	F	...
	→	.10	.90	.10

Table 6.2

Note that the conditional probability formula A $\Rightarrow$ B is entered at the top of the right hand column in table 6.2, but not the *ersatz* formula B$|$A, as in table 6.1. However, *ersatz* truth values are still written in in the lines in which the conditional's antecedent, A, is true, and the probability of the whole conditional is the ratio of the probability of the lines in which it has the *ersatz* value *T* to the sum of the probabilities of the lines in which it has either the value T or the value F. And, following this rule B gets the high probability of .9 while A $\Rightarrow$ B gets the low probability of .1, which 'confirms' the diagram by making B probable and A $\Rightarrow$ B improbable.

The last step in the analysis is to try to find 'real life counterexamples' involving statements of the forms B and A $\Rightarrow$ B, whose probabilities approximate those given in the diagram and table. Like constructing the diagram, this is a matter of ingenuity, but examining the diagram offers some clues as to what A and B might be. The diagram represents A as improbable and B as probable, but not 'perfectly probable' since there is a region outside of it that is largely occupied by A. So B is quite likely, except in the unlikely event of A's being the case. Trying to think of an A and a B satisfying these conditions, we can begin with B. Something that would be quite likely but not certain might be "The class will meet tomorrow", said on Monday of a class that normally meets on Tuesdays and Thursdays. Of course that wouldn't be a certainty, and to find A, we try to think of something that would exclude the class's meeting on Tuesday, or at least make it improbable. The professor might fall sick, for instance. So we try symbolizing: B = "The class will meet tomorrow" and A = "The professor will fall sick", which would make A $\Rightarrow$ B symbolize "If the professor falls sick the class will meet tomorrow." And, finally, we ask whether it would be 'reasonable' to infer this from "The class will meet tomorrow." If the inference seems unreasonable, we have 'real life confirmation' of the diagram and table, in the sense that we know that these correspond to something in ordinary life, and they are not just mathematical possibilities. In fact, we have just explained the intuitive paradoxicalness of the pattern B $\therefore$ A $\Rightarrow$ B by describing a case in which it would not be reasonable to reason in accord with it in real life.

Let us summarize the main steps in the method for analyzing an inference pattern illustrated above. Step 1 is to try to construct a diagrammatic counterexample like diagram 6.1, that represents the premises of the pattern as probable while the conclusion is improbable, following the rule that the probability of a proposition of the form $\phi \Rightarrow \psi$ is represented by the proportion of the region corresponding to ϕ that lies inside the region corresponding to ψ. This step requires intuition and ingenuity, since we do not yet have a systematic method for carrying it out. Step 2 is to 'confirm' the diagrammatic counterexample, by constructing a table like table 6.2, which gives a probability distribution that generates high probabilities for the premises of the inference and a low probability for its conclusion. This should be easier than step 1, because once a probability diagram has been constructed all that has to be done to construct the required distribution is to roughly estimate the areas of the minimal regions in the diagram (which correspond to SDs)—making sure that their total adds up to 1.

Step 3, the last step, is to try find real life propositions of the forms of the premises and conclusion of the inference which 'confirm the confirmation', because the premises would seem intuitively to be 'reasonable' while the conclusion seems to be unreasonable. Again, doing this requires intuition and ingenuity, but, as in our illustration, clues from the sizes and geometrical relations between the regions in the diagram itself can help with this.

Now that we know that certain inferences—the paradoxical ones—can lead from probable premises to improbable conclusions, even though they are valid in classical logic, we want to consider whether other classically valid inferences might also have probabilistic counterexamples and be questionable in real life for this reason. We will now examine eight such inferences, and find some rather surprising results. These are written below, with the premises above and the conclusions below a horizontal line:

Inference 1	Inference 2	Inference 3	Inference 4
$\dfrac{A\rightarrow B}{A\Rightarrow B}$	$\dfrac{A\vee B}{\sim A\Rightarrow B}$	$\dfrac{A\Rightarrow B}{\sim B\Rightarrow \sim A}$	$\dfrac{A\Rightarrow B}{A\rightarrow B}$

Inference 5	Inference 6	Inference 7	Inference 8
$\dfrac{A,\ A\Rightarrow B}{B}$	$\dfrac{B\Rightarrow C}{(A\&B)\Rightarrow C}$	$\dfrac{A\Rightarrow B,\ B\Rightarrow C}{A\Rightarrow C}$	$\dfrac{A\Rightarrow B,\ (A\&B)\Rightarrow C}{A\Rightarrow C}$

Note that four of these are of familiar forms, namely inferences 2, 3, 5 and 7. Inference 2 has a disjunctive premise and a conditional conclusion, and reasoning of this kind is very common in everyday life. Inference 3 is *contraposition*, which is less common than inferences of the form of inference 2, though it is important enough to have its own name. Inference 5 is *modus ponens*, which has been held to be the most fundamental of all deductive inference forms, and which, when probability dynamics come back into the picture, will be seen to have a special justification that doesn't apply to other inference patterns. And, inference 7 is the so called *hypothetical syllogism*, or *transitivity*, which is almost always involved in long chains of argument. Now we will use the method illustrated above to examine these inferences, to see whether it is possible to picture their premises as probable while their conclusions are improbable, i.e., we will try to construct diagrammatic and real life counterexamples to them.

Inference 1 looks trivial because its premise and conclusion are alternative symbolizations of the same thing, namely an ordinary language proposition of the form "If A then B". But even so, $A \rightarrow B$ and $A \Rightarrow B$ are different, and diagram 6.1 is a counterexample. Thus, $A \rightarrow B$ corresponds to the union of regions 1, 3, and 4, as we saw earlier, which nearly fills the diagram, while $A \Rightarrow B$ corresponds to a pair of regions, one corresponding to A and the other to B, and its probability is the proportion of the first that lies in the second, which is obviously not very large. The probabilities given in table 6.2 substantiate this by showing that $A \rightarrow B$ can have probability $p(A \rightarrow B) = .91$, while $A \Rightarrow B$ has probability $p(B|A) = .10$. Thus, inference 1 is not only not trivial, it is not even plausible, which is one more illustration of the difference between the material conditional and the probability conditional. We will see that the same thing is true of other inferences on our list, which would be valid if the probability conditional were the same as the material conditional. This is the case with inferences 2 and 3, but analyzing them brings out something more.

Diagram 6.2 below is a counterexample to both inferences:[9]

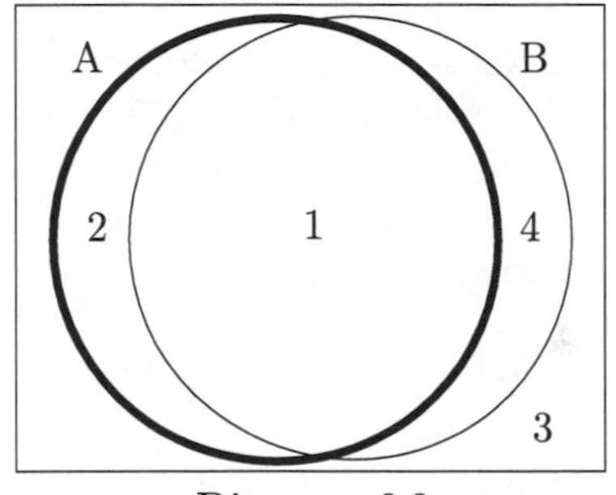

Diagram 6.2

(Note that the outline of region A is heavier than the outline of region B, in order to make it easier to follow).

Diagram 6.2 pictures $A \vee B$ as probable, since the union of regions A and B fills most of the diagram, while $\sim A \Rightarrow B$ is improbable because almost everything that lies outside of region A lies outside region B. Thus, we have a diagrammatic counterexample to inference 2. Similarly, the diagram is a counterexample to inference 3 because it pictures $A \Rightarrow B$ as probable, since most of what is inside A is inside B too, but $\sim B \Rightarrow \sim A$ is not probable since it is not the case that most of what lies outside of B also lies outside of A.

[9] Actually, from the probabilistic point of view inference 2 is only a variant of inference 1, and a diagrammatic counterexample of the latter can be turned into one of the former simply by replacing A by $\sim A$.

Table 6.3[10] confirms the diagram:

		formulas					
regions	distribution	A	B	A∨B	∼A⇒B	A⇒B	∼B⇒∼A
1	.900	T	T	T	...	*T*	....
2	.090	T	F	T	...	*F*	*F*
3	.009	F	F	F	*F*	...	*T*
4	.001	F	T	T	*T*	...	....
	→	.99	.901	.991	.10	.909	.091

Table 6.3

The distribution in the table confirms diagram 6.2 by generating probabilities p(A∨B) = .991 for the premise of inference 2 and p(∼A ⇒ B) = .10 for its conclusion, and probabilities p(A ⇒ B) = .909 for the premise of inference 3 and p(∼B ⇒ ∼A) = .091 for its conclusion. This proves that it is mathematically possible for A∨B to be highly probable while ∼A ⇒ B is improbable, and for A ⇒ B to be probable while ∼B ⇒ ∼A is improbable, which completes step 2 in the analysis of inferences 2 and 3.

The last thing to do is carry out step 3; i.e., to find real life counterexamples to the inferences. Diagram 6.2 provides hints for this in picturing A as highly probable and B as 'somewhat probable', but not as probable as A. Here are concrete 'interpretations' of A and B that satisfy these conditions:

A = the sun will rise tomorrow.
B = it will warm up during the day.

A is obviously very probable,[11] and under ordinary circumstances B should also be probable, though not as probable as A.

Given the above interpretations of A and B, A ∨ B, ∼A ⇒ B, A ⇒ B, and ∼B ⇒ ∼A get the following interpretations:

A∨B = either the sun will rise tomorrow or it will warm up during the day;[12]
∼A ⇒ B = if the sun doesn't rise tomorrow then it will warm up during the day;
A ⇒ B = if the sun rises tomorrow then it will warm up during the day;
∼B ⇒ ∼A = if it doesn't warm up during the day then the sun won't rise tomorrow.

[10]That the regions are ordered so that the one with A and B both false is third and not last is because we want their probabilities to be in descending order of magnitude.

[11]And obviously much more than 99% probable, but this isn't essential to the present argument. Of course neither this nor any of the other probabilities being considered are absolute certainties, but, as we have stressed repeatedly, absolute certainty is not to be had in matters of this kind.

[12]It might be held that it would be odd to say "Either the sun will rise tomorrow or it will warm up during the day tomorrow" when it would be more normal to assert

Assuming that it is intuitively doubtful whether it would be reasonable to deduce $\sim A \Rightarrow B$ from $A \vee B$ or $\sim B \Rightarrow \sim A$ from $A \Rightarrow B$ under these interpretations, we have real life counterexamples to inferences 2 and 3. But if that is doubtful we have a new problem: Why do these patterns of inference seem rational when they are considered in the abstract, in spite of the existence of counterexamples like the ones just given? This will be returned to in section 6.6⋆, but there are two things to note immediately about the counterexamples, the first of which is technical.

The technical point is that table 6.3 illustrates a correspondence between the *orders of magnitude* of the numerical probabilities in the distribution in the table and the areas of the regions they correspond to in diagram 6.3. Thus, region 1 is the largest in the diagram, and regions 2, 3, and 4 are in decreasing order of magnitude after that, and in table 6.3 the probabilities of the corresponding cases decrease by orders of magnitude from case 1 to case 4.[13] These orderings, which will be discussed in more detail in section 6.4, are theoretically important. That is because similar order-of-magnitude distributions show that the premises of 'probabilistically invalid' inferences, say of inferences 1–3, can have probabilities arbitrarily close to 1 while their conclusions have probabilities arbitrarily close to 0, though, interestingly, only if they are also classically invalid can their premises have probabilities equal to 1 while their conclusions have probability 0. This will be illustrated below, and discussed in more detail in the following section. But there is a very important informal remark to make on counterexamples to inference 2.

This is that the existence of these counterexamples doesn't just raise questions about the proper symbolization of conditional statements, about whether "if A then B" should be symbolized as a material or as a probability conditional. It also raises questions concerning the validity of informal reasoning involving conditionals, independently of the way it is symbolized. For instance, most people who were 'given' "Jane will either take ethics or logic" would infer "If Jane doesn't take ethics she will take logic" without hesitation, thus seemingly taking it for granted that the conditional can always validly be inferred from the disjunction. But now we are asking whether this assumption is justified.

The assumption is a very important one. If it were always rational to infer "If not A then B" from "Either A or B", then this, coupled with the

the conjunction, "The sun will rise tomorrow and it will warm up during the day". But that does not mean that the disjunction would not be probable enough to assert. The 'conversational logic of disjunctions', which was already commented on in section 5.6⋆, will be returned to in section 6.6⋆.

[13]But note that if the 'smallest probability' in case 4 were made equal to 0, the first conclusion, $\sim A \Rightarrow B$, would certainly be wrong, even though the area of the corresponding region in diagram 6.3 was positive.

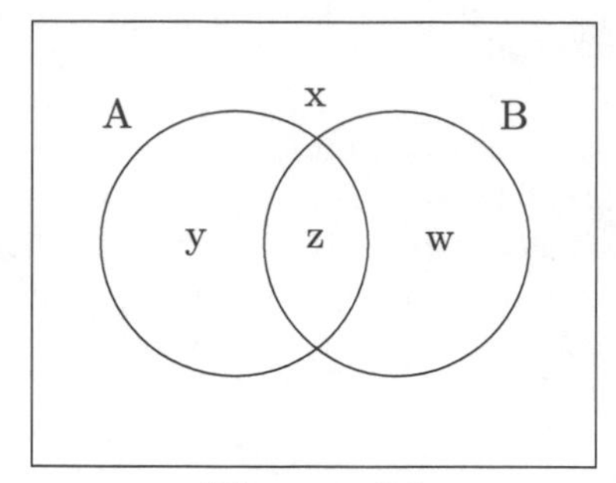

Diagram 6.3

fact that the inverse inference is always rational, as will be seen immediately below, would imply that "Either A or B" and "If not A then B" were logically equivalent. And by the same token this would imply that "If A then B" should be logically equivalent to "Either not A, or B", which is itself equivalent to a material conditional. Thus, it would follow that ordinary language conditionals were equivalent to material conditionals, and symbolizing them as such would not be arbitrary because it follows from the seemingly unquestionable assumption that they follow from disjunctions.[14] But we now see that there are cases in which it is not be rational to infer "If not A then B" from "Either A or B", and this shows that it is wrong to symbolize it as $\sim A \rightarrow B$ in those cases.

But now let us turn to other inferences.

Inference 4 is the inverse of inference 1, but the following argument shows that, unlike the latter, it has no counterexamples. Consider diagram 6.3, where, rather than numbering the regions, variable probabilities x, y, z, and w are entered into them.

This diagram gives the probability of $A \rightarrow B$ as $p(A \rightarrow B) = x+z+w$, while the probability of $A \Rightarrow B$ is $p(A \Rightarrow B) = z/(y+z)$. But it is easy to see that the first probability must be at least as high as the second. That follows from the fact that $1-(x+z+w) = y \leq y/(y+z)$ (because $y+z \leq 1$), and $y/(y+z) = 1-z/(y+z)$. Clearly, therefore, it is impossible to construct a diagrammatic counterexample representing $A \Rightarrow B$ as probable and $A \rightarrow B$ as improbable. And, it follows that steps 2 and 3 cannot be carried out, hence in particular no real life counterexample to inference 4 can be found.

The foregoing is significant for two reasons. One is that because material conditionals are equivalent to disjunctions, the fact that they can always be inferred from conditionals guarantees that disjunctions can always be inferred from conditionals—though not conversely, as we have seen.

---

[14]This is only one of several arguments that have been advanced to 'prove' that ordinary language conditionals are equivalent to material conditionals. Cf. Grice (1989: ch. 2).

The other point is that the fact that $p(A \rightarrow B) \geq p(A \Rightarrow B)$ guarantees that $u(A \rightarrow B) \leq u(A \Rightarrow B)$, which means that the inference $A \Rightarrow B \therefore A \rightarrow B$ satisfies the uncertainty sum condition in its application to inferences with a single premise. Recalling that satisfying this condition is a necessary and sufficient condition for inferences without conditionals to be classically valid (section 3.2), it now appears that a probabilistic condition that is equivalent to classical validity applies to inferences involving probability conditionals as well, even though they don't have classical truth values. We will see that this generalizes, and furnishes us with a *probabilistic criterion of validity* that applies to these inferences. Inference 5 furnishes another illustration.

Inference 5, *modus ponens*, is similar to inference 4 in one respect, but it is much more important, among other things because of its extremely common applications. This also satisfies the uncertainty sum condition,

$$u(B) \leq u(A) + u(A \Rightarrow B),^{15}$$

as can be verified by referring to diagram 6.3. The diagram pictures $u(B) = x + y$, $u(A) = x + w$, and $u(A \Rightarrow B) = y/(y + z)$, and therefore $u(B) \leq u(A) + u(A \Rightarrow B)$ follows if it can be shown that $x + y \leq x + w + [y/(y + z)]$. But this follows immediately because $y \leq y/(y + z)$.

Since *modus ponens* satisfies the uncertainty sum condition, it will turn out to be valid in the probabilistic sense to be defined in section 6.4. Section 7.6⋆⋆ will even show that it has another property that entitles it to a 'distinguished place' among inferences that are valid in this sense. For immediate practical purposes, however, this result shows that there are no diagrammatic or real life counterexamples to *modus ponens*, and therefore it is safe to reason in accord with it in everyday life.[16]

Now consider inferences 6 and 7, both of which are valid in classical logic. In spite of this, however, diagram 6.4 is a counterexample to both inferences. Here most of circle B lies inside circle C and therefore $B \Rightarrow C$ is represented as being probable, but none of region 3, which corresponds to A&B, lies inside region C, and therefore $(A\&B) \Rightarrow C$ is totally improbable. Hence this is a diagrammatic counterexample to inference 6. Similarly, most

---

[15] Actually, modus ponens satisfies the stronger condition that $p(B) \geq p(A)p(A \Rightarrow B)$. This and other conditions that are stronger than the uncertainty sum condition are discussed in Adams (1996).

[16] This must be qualified since McGee (1985) gives convincing ordinary language counterexamples to inferences of the form "A, and if A, then if B then C; therefore, if B then C". But this involves a conditional "if B then C" embedded in a larger conditional, which cannot be symbolized as $A \Rightarrow (B \Rightarrow C)$ in our formal language because this language only allows $\Rightarrow$ to occur as the main connective in any formula in which it occurs. Significantly, McGee's (1989) extension of the theory, which allows embedded conditionals, does not assume the validity of *modus ponens* in application to formulas of the form $A \Rightarrow (B \Rightarrow C)$. Appendix 4 comments briefly on this extension.

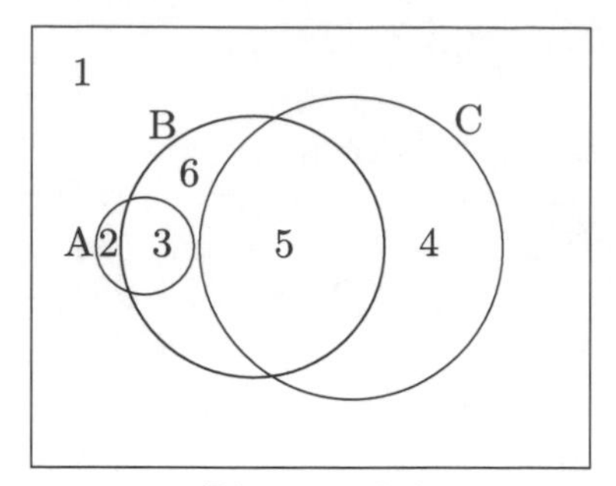

Diagram 6.4

of region A lies inside B and most of B lies inside C, but none of A lies inside C, and therefore A $\Rightarrow$ B and B$\Rightarrow$ C are pictured as probable while A $\Rightarrow$ C is totally improbable. Hence we have diagrammatic counterexamples to inferences 6 and 7.

Turning to step 2, table 6.4[17] below distributes probabilities among the regions of diagram 6.4 in a way that roughly corresponds to their areas.

regions	distribution	A	B	C	A$\Rightarrow$B	B$\Rightarrow$C	(A&B)$\Rightarrow$C	A$\Rightarrow$C
~~ ~~	~~ ~~	~~T~~	~~T~~	~~T~~	~~*T*~~	~~*T*~~	~~*T*~~	~~*T*~~
3	.09	T	T	F	*T*	*F*	*F*	*F*
~~ ~~	~~ ~~	~~T~~	~~F~~	~~T~~	~~*F*~~	~~...~~	~~.....~~	~~*T*~~
2	.00001	T	F	F	*F*	...	.....	*F*
5	.9	F	T	T	...	*T*	.....	...
6	.0009	F	F	T	...	*F*	.....	...
4	.00009	F	T	F	...	...	.....	...
1	.009	F	F	F	...	...	.....	...
	→	.09001	.9909	.90009	.9999	.9083	0	0

Table 6.4

Again, the distribution in the table generates high probabilities for the premises of inferences 6 and 7, and now it generates zero probability for their conclusions, which confirms the diagram and completes step 2 of the analysis. Note too the 'descending order-of-magnitude' of probabilities in the table, similar to those in table 6.3, although we have not bothered to

[17]Dashed lines are drawn through the SDs in which A, B, and C are all true, and in which A and C are true but B is false, since there are no regions in diagram 6.4 that correspond to them.

order them from top to bottom.

Turning to step 3, we go back to diagram 6.4 to find interpretations of A, B, and C in which $B \Rightarrow C$ is probable but $(A\&B) \Rightarrow C$ is improbable, and in which $A \Rightarrow B$ and $B \Rightarrow C$ are probable but $A \Rightarrow C$ is improbable. The diagram makes A look like an 'unlikely special case' of B, and C look probable given B, except in that case. The following interpretations fit these requirements.

Let A = The sun will rise exactly 5 minutes later tomorrow.
B = The sun will rise no more than 5 minutes later tomorrow.
C = The sun will rise less than 5 minutes later tomorrow.

Given the foregoing, $B \Rightarrow C$, $(A\&B) \Rightarrow C$, and $A \Rightarrow B$, and $A \Rightarrow C$ are interpreted as:

$B \Rightarrow C$ = If the sun rises no more than 5 minutes later tomorrow then it will rise less than 5 minutes later tomorrow.[18]

$(A\&B) \Rightarrow C$ = If the sun rises exactly 5 minutes and no more than 5 minutes later tomorrow, then it will rise less than 5 minutes later tomorrow.

$A \Rightarrow B$ = If the sun rises exactly 5 minutes later tomorrow it will rise no more than 5 minutes later tomorrow.

$A \Rightarrow C$ = If the sun rises exactly 5 minutes later tomorrow it will rise less than 5 minutes later tomorrow.

Hopefully, the reader will agree that with these interpretations $B \Rightarrow C$ would be probable but $(A\&B) \Rightarrow C$ would not be, and $A \Rightarrow B$ and $B \Rightarrow C$ would be probable but $A \Rightarrow C$ would not be, i.e., these would yield concrete 'real life' counterexamples to inferences 6 and 7. Of course, if there are such counterexamples to hypothetical syllogism, we have a problem similar to ones raised about contraposition and inferences of conditionals from disjunctions: does it call in question all of the uses of this principle in chains of reasoning? Again, however, this will be deferred to section 6.6$\star$.

Finally, let us note that inference 8 satisfies the uncertainty sum condition, hence it is implausible that either a diagrammatic or real life counterexample to it can be constructed. This is implied by the inequality on the left and the equation on the right below,

$$p(C|A) \geq p(B\&C|A) = p(B|A) \times p(C|A\&B),$$

which implies in turn that

[18]Note the change in the verb from "The sun *will* rise no more than 5 minutes later tomorrow" to "If the sun *rises* no more than 5 minutes later tomorrow then it will rise less than 5 minutes later tomorrow". This difference has led some to hold that "The sun will rise no more than 5 minutes later tomorrow" is not a 'logical part' of "If the sun rises no more than 5 minutes later tomorrow then it will rise less than 5 minutes later tomorrow".

$$u(A \Rightarrow C) \leq u(A \Rightarrow B) + u((A\&B) \Rightarrow C).$$[19]

We conclude this section by pointing out two other important generalizations that are illustrated by inferences 1–8. First and most important, the inferences divide into two classes: ones like inferences 4, 5, and 8 that satisfy the uncertainty sum condition, and ones like inferences 1, 2, 3, 6, and 7, whose premises can be arbitrarily highly probable while their conclusions have arbitrarily low or zero probability. Chapter 7⋆ will state a general theorem that implies that all inferences fall into one of these two classes. This motivates the definition of probabilistic validity, which involves the two classes, which will be stated and discussed in the following section.

The second point has to do with the way the class that an inference belongs to is determined. If an inference satisfies the uncertainty sum condition this is established by giving a general argument involving probabilities. On the other hand, showing that the premises of an inference can be arbitrarily probable while its conclusion has arbitrarily low or zero probability is established by constructing a diagrammatic counterexample and a corresponding order-of-magnitude probability distribution. In this case the diagram is also used to find a real life counterexample, in which the inference's premises are intuitively probable but its conclusion is intuitively improbable. But the method for doing this, and particularly that of constructing a diagrammatic counterexample, is itself intuitive, since nothing has been said about how these diagrams were constructed. Two systematic methods for doing this will be described in this book, the first and more cumbersome method in section 6.5⋆, and a second less cumbersome but also less informative method in chapter 7⋆. However, the exercises below ask you to use your ingenuity to assess the validity of other inferences, following the examples of inferences 1–8. In fact, you are likely to learn more from doing this than you will from applying the rote methods that will be described in section 6.5⋆ and in chapter 7⋆.

Exercises

1. For each of the inferences below, either construct a diagrammatic counterexample to it or argue that no counterexample can be con-

[19]The argument is as follows. The inequality on the left is a simple generalization of the law that $p(C) \geq p(B\&C)$, since C is a logical consequence of B&C, and the equation on the right is a simple generalization of the chain rule. Combining these two facts with the facts that $u(A \Rightarrow C) = u(C|A)$, $u(A \Rightarrow B) = u(B|A)$, and $u((A\&B) \Rightarrow C) = u(C|A\&B)$, it follows that:

$$\begin{aligned} 1 - u(A \Rightarrow C) &\geq (1 - u(A \Rightarrow B))(1 - u((A\&B) \Rightarrow C)) \\ &\geq 1 - u(A \Rightarrow B) - u((A\&B) \Rightarrow C) + u(A \Rightarrow B) \times u((A\&B) \Rightarrow C) \\ &\geq 1 - u(A \Rightarrow B) - u((A\&B) \Rightarrow C), \end{aligned}$$

and therefore $u(A \Rightarrow C) \leq u(A \Rightarrow B) + u((A\&B) \Rightarrow C)$.

structed.

a. $\dfrac{A \Rightarrow B}{B \Rightarrow A}$ b. $\dfrac{\sim(A\&B)}{A \Rightarrow \sim B}$ c. $\dfrac{A \vee B, A \Rightarrow B}{B}$

d. $\dfrac{A \Rightarrow B, \sim A \Rightarrow \sim B}{B \Rightarrow A}$ e. $\dfrac{(A \vee B) \Rightarrow C}{A \Rightarrow C}$

f. $\dfrac{A \Rightarrow C, B \Rightarrow C}{(A\&B) \Rightarrow C}$ g. $\dfrac{A \Rightarrow C, B \Rightarrow C}{(A \vee B) \Rightarrow C}$ h. $\dfrac{A \Rightarrow B, C \Rightarrow \sim B}{A \Rightarrow \sim C}$

i. $\dfrac{A \Rightarrow B, B \Rightarrow A, \sim A \Rightarrow \sim B}{\sim B \Rightarrow \sim A}$ j. $\dfrac{A \Rightarrow B, B \Rightarrow A, A \Rightarrow C}{B \Rightarrow C}$

2. For each inference in exercise 1 for which you can construct a counterexample, construct a distribution in which its premises have probability at least .9, while its conclusion has a probability no greater than .1.
3. For each inference in exercise 1 for which you can construct a diagrammatic counterexample, find ordinary language interpretations of its variables whose probabilities correspond to those in the diagram.
4. Symbolize the propositions involved in the following classically valid argument, and try to construct diagrammatic counterexamples to it. If you can, try to describe a 'scenario' in which it would be reasonable to accept its premises, but not its conclusion.

   If Ed wins first prize then either Fred wins second prize or George is disappointed. Fred does not win second Prize. Therefore if George is disappointed then Ed does not win first prize (this example was given in chapter 1).

5. Show that no truth values can be assigned to $A \Rightarrow B$ in the last two lines of table 6.2 in such a way that the sum of the probabilities of the lines in which $A \Rightarrow B$ is true is equal to $p(A \Rightarrow B)$ (reprinted below), i.e., in such a way that this sum equals 4/7.[20]

	formulas		
distribution	A	B	A⇒B
.4	T	T	*T*
.3	T	F	*F*
.2	F	T	...
.1	F	F	...
	.70	.60	4/7

6. a. Argue that if an inference is 'classically invalid' in the sense that it is invalid when '⇒' is replaced by '→', then there is a probability distribution that makes its premises absolutely certain and makes its conclusion certainly false.

[20] This is a special case of the 'triviality results' of A. Hájek (1994) which will be returned to in chapter 8⋆.

b. Argue that the foregoing implies that the classical validity of an inference is a necessary condition for its being probabilistically valid in the sense that it satisfies the uncertainty sum condition (note that the fact that inference 1 is classically valid but not probabilistically valid implies that classical validity may not be a sufficient condition for probabilistic validity).

c. Argue that if an inference is classically valid and its premises have probability 1, then its conclusion must have probability 1, whether or not it is probabilistically valid.

[Note: 6a–c imply that the so called *entailment function* which gives the maximum degrees of uncertainty that an inference's conclusion can have for any given degrees of uncertainty of its premises, must be *discontinuous* in the case of inferences like inference 1, which are classically valid but probabilistically invalid. That is because the maximum uncertainties of their conclusions must equal 1 for any premise uncertainties that are greater than 0, but they must equal 0 when their premises' uncertainties equal 0.]

⋆7. Consider the following 'proof' that A ⇒ B follows from ~A:

1. ~A given.
2. ~(A&~B) From step 1.
3. A ⇒ B From step 2. QED

Which step, if any, might not be probabilistically entailed by the steps it supposedly follows from, in the sense that its uncertainty may be greater than the sum of the uncertainties of the steps it follows from?

⋆8. *Lewis Carroll's Barbershop Paradox*

Lewis Carroll, who was a professional logician, wrote a short article in the journal *Mind*, in 1894, in which he invited his readers to 'diagnose' what is wrong with the following argument.[21]

Three barbers, A, B, and C, operate a barbershop under two rules: (1) at least one of them must be in the shop at all times during working hours, and (2) if B leaves the shop at any time then C is to go with him (to make sure he comes back). Then it is argued that A can never leave the shop, as follows:

1. A or B or C are in the shop. Given (rule 1)

[21] A more famous paper by Lewis Carroll entitled "What the Tortoise said to Achilles," *Mind*, 1895, was a kind of switch on Zeno's paradox of Achilles and the Tortoise. Zeno argued that if the tortoise got a head start in a race with Achilles then Achilles could never catch him, no matter how fast he ran. Lewis Carroll's 'story' had to do with what the tortoise said to Achilles after the race was over. The tortoise 'proved' that even if Achilles won he couldn't prove that he won, because to do that he would have to prove that *modus ponens* was valid, and he couldn't prove that without assuming what he was trying to prove. Interestingly, McGee (1985) argues not only that you can't prove that *modus ponens* is valid, but in certain cases it is invalid! This is connected with our limitation of conditionals to formulas in which ⇒ occurs only as the main connective.

2. If B is not in then C is not in. Given (rule 2)
3. Suppose that A is not in.
4. If B is not in then C is in. From 1 and 3.
5. Step 4 contradicts step 2.
6. Since the supposition that A leaves the shop leads to a contradiction, A can't leave the shop.

a. Symbolize steps 1–6 above, using A, B, and C to symbolize "A is in the shop", "B is in the shop", and "C is in the shop".
b. Discuss the question of whether Lewis Carroll's argument is valid, and if it is invalid, which of its steps is mistaken, if any.

⋆9. Some rules of *indirect* inference that are valid in classical logic are not valid in probability logic. For example, in classical logic if you want to prove an 'if... then... ' proposition you add the 'if' part to the premises and show that the 'then' part follows from it (this is the rule of *conditional proof*, sometimes called the *deduction theorem*). However, it is not the case that an inference with premises $\phi_1 \Rightarrow \psi_1,\ldots,\phi_n \Rightarrow \psi_n$ and conclusion $\phi \Rightarrow \psi$ is valid if and only if the inference with premises $\phi_1 \Rightarrow \psi_1,\ldots,\phi_n \Rightarrow \psi_n$ plus ϕ, and conclusion ψ, is valid. Can you give a concrete example?

6.4 Probabilistic Validity

As said at the end of the previous section, and as will be proved in the next section, inferences in our formal language, including probability conditionals, are 'partitioned' into two classes: (1) those that satisfy the uncertainty sum condition, and (2) those whose premises can be arbitrarily highly probable while their conclusions have arbitrarily low probabilities. Given that to this point we haven't had a criterion of validity for inferences in our language, this result suggests defining one in terms of the partition; i.e., it suggests defining such an inference to be *probabilistically valid (abbreviated p-valid) if and only if the uncertainty of its conclusion cannot exceed the sum of the uncertainties of its premises.* Thus, inferences 4, 5, and 8 are p-valid, while inferences 1, 2, 3, 6, and 7 as well as the paradoxes of material implication are probabilistically invalid (p-invalid). This idea will be very important in what follows, and a few comments may help to fix it in the reader's mind.

The first point is that we have not defined the validity of an inference in the ordinary 'truth-conditional' way, i.e., as it not being possible for its premises to be true while its conclusion is false. But this is something that can't be done in the case of inferences involving probability conditionals, since the ordinary concepts of truth and falsity don't apply to them.

The second point is that while we haven't defined p-validity in the ordinary way, we have defined it in such a way as, in a sense, to extend the ordinary definition. Specifically, 'purely factual inferences' that don't involve probability conditionals, like $\{E \vee L, \sim E\} \therefore L$, are valid in the ordinary truth-conditional sense if and only if they satisfy the uncertainty sum condition, and therefore we can say that they are valid in this sense if and only if they are valid in the probabilistic sense that we have just defined. But the ordinary truth-conditional concept of validity doesn't apply to inferences involving probability conditionals like $A \vee B \therefore \sim A \Rightarrow B$, and only the probabilistic one does.

The third point is that unlike probability itself, which is a matter of degree, p-validity is an all-or-none property of an inference. *Modus ponens* is 'simply' p-valid while $A \vee B \therefore \sim A \Rightarrow B$ is simply p-invalid, and there are no gradations in between. *However*, when we come to defaults and enthymemes in section 6.6⋆ we will reconsider inferences of the form $A \vee B \therefore \sim A \Rightarrow B$, and see that in a certain sense they are 'more rational', or at least more often rational than, say, ones of the form $A \therefore \sim A \Rightarrow B$.

Finally, a number of other properties of p-validity will be described in section 6.5⋆ and in chapter 7⋆, which provide further justification for defining the validity of inferences involving probability conditionals in this way, and one of these is worth mentioning at this point. That is that p-validity is a *deduction relation* in the following formal sense.[22] (1) Any premise in a set of premises is p-entailed by those premises; i.e., all inferences whose conclusions are among the premises are p-valid. (2) If the conclusion of one inference is p-entailed by the conclusions of a set of inferences, then it is p-entailed by the union of the premises of these other inferences. In short, consequences of consequences of premises are consequences of them.

That p-validity is a deduction relation in the sense of satisfying conditions (1) and (2) above is presupposed in sections 7.1–7.4 of chapter 7⋆, which briefly develop a formal deductive theory of p-entailed consequences of premises. The following exercise involves the derivation of a property that that theory will itself presuppose: namely that a p-valid consequence of two premises, each of which is p-entailed by a subset of a three-premise inference is p-entailed by these premises.

Exercise

⋆1. You can argue as follows that the fact that $\{A, A \Rightarrow B\} \therefore B$ and $\{B, B \Rightarrow C\} \therefore C$ are both p-valid inferences guarantees that $\{A, A \Rightarrow B, B \Rightarrow C\} \therefore C$ is p-valid:

(1) $\{A, A \Rightarrow B, B \Rightarrow C\} \therefore A$ is p-valid; set of premises p-entails member.

(2) $\{A, A \Rightarrow B, B \Rightarrow C\} \therefore A \Rightarrow B$ is p-valid; set of premises p-entails member.

[22] cf. Montague and Henkin 1956.

(3) $\{A, A \Rightarrow B\} \therefore B$ is p-valid. given (*modus ponens*).

(4) $\{A, A \Rightarrow B, B \Rightarrow C\} \therefore B$ is p-valid; because B p-entailed by A and $A \Rightarrow B$, and A and $A \Rightarrow B$ are both p-entailed by $\{A, A \Rightarrow B, B \Rightarrow C\}$

(5) $\{A, A \Rightarrow B, B \Rightarrow C\} \therefore B \Rightarrow C$ is p-valid; set of premises p-entails member.

(6) $\{B, B \Rightarrow C\} \therefore C$ is p-valid. given (*modus ponens*)

(7) $\{A, A \Rightarrow B, B \Rightarrow C\} \therefore C$ is p-valid; because C p-entailed by B and $B \Rightarrow C$, and B and $B \Rightarrow C$ are both p-entailed by $\{A, A \Rightarrow B, B \Rightarrow C\}$.

Give a similar argument that the fact that $\{A, B\} \therefore A\&B$ is p-valid guarantees that $\{A, B, C\} \therefore A\&B\&C$ is p-valid.

6.5⋆ Order-of-Magnitude Orderings and Distributions

This section will be concerned with the first of the two systematic methods to be considered in this book, for assessing inferences involving conditionals to determine the classes to which they belong. Essentially, the method simply generalizes the descending order of magnitude probability distributions illustrated in tables 6.3 and 6.4, which in turn order the SDs to which these probabilities apply and determine whether the conditionals involved in the inferences have high or low probabilities.[23] The present method will be somewhat sketchily described, because the method to be discussed in chapter 7⋆ is simpler, although it is of somewhat more limited usefulness.

To get the basic idea of an order-of-magnitude distribution (OMP-distribution) let us return to table 6.4. It attaches the highest probability of .9 to region 5, which corresponds to SD #5. For brevity, let us write its probability as p_5. The next highest probability of .09 attaches to SD #3 whose probability may be written as p_3. The next highest is .009, attaching to SD #1, whose probability can be written p_1, and so on. These probabilities conform to a simple rule:

$$p_5 = .9 = .1^0 \times .9,\ p_3 = .1^1 \times .9,\ p_1 = .1^2 \times .9,\ p_6 = .1^3 \times .9,\ p_4 = .1^4 \times .9, \text{ and } p_2 = .1^5$$

And, we get a more general rule by choosing any small quantity υ, and setting:

$$p_5 = 1-\upsilon = \upsilon^0(1-\upsilon),\ p_3 = \upsilon^1(1-\upsilon),\ p_1 = \upsilon^2(1-\upsilon),\ p_6 = \upsilon^3(1-\upsilon),\ p_4 = \upsilon^4(1-\upsilon),\ p_2 = \upsilon^5.$$

[23]The method was first presented in Adams (1966) and the order-of-magnitude probability orderings that were introduced there are the converses of what are called possibility orderings in David Lewis's (1973) book *Counterfactuals*. Recent work in fuzzy logic makes extensive use of these orderings.

It is not hard to prove that this series of values sums to 1,[24] and therefore they constitute a probability distribution, which we will call a *υ-distribution.* The significant thing about υ-distributions is that they always generate either high or low probabilities for factual or conditional formulas that are formed from A, B, and C. This is illustrated by the values in table 6.4, all of which are either greater than .9 or less than .1. More generally, we can say that for any formula $\phi \Rightarrow \psi$ or simply ψ, the probability generated by the OMPυ-distribution for $\phi \Rightarrow \psi$ or ψ is either greater than or equal to $1 - \upsilon$ or less than or equal to υ. Assuming that υ is close to 0, this implies that all formulas either have probabilities close to 1, which is almost like truth, or close to 0, which is almost like falsity. Moreover, there is an easy rule for determining whether a formula's probability is close to 1 or to 0, and we can use this as a kind of 'probabilistic truth-table test' for determining the class to which an inference belongs. Two more bits of terminology will help us to do this.

All of the υ-distributions above order the probabilities of the SDs to which they apply in the same way, and this ordering determines one on the SDs themselves, called an *order of magnitude probability ordering* (abbreviated *OMP-ordering*). This can be written as $5 \succ 3 \succ 1 \succ 6 \succ 4 \succ 2$, which means that the probability of SD #5 is an order of magnitude greater than the probability of SD #3, the probability of SD #3 is an order of magnitude greater than the probability of SD #1, and so on.

Now we introduce the important idea of a factual formula ψ or conditional formula $\phi \Rightarrow \psi$ *holding* in an OMP-ordering like $5 \succ 3 \succ 1 \succ 6 \succ 4 \succ 2$. ψ will be said to hold in the ordering if it is consistent with the SD that is highest in the ordering, and $\phi \Rightarrow \psi$ holds in the ordering if the highest SD that is consistent with $\phi\&\psi$ is higher than the highest SD that is consistent with $\phi\&{\sim}\psi$ (and, if $\phi\&{\sim}\psi$ is not consistent with any SDs then $\phi \Rightarrow \psi$ holds). For example, ~A&B holds but A&~B doesn't hold in the ordering $5 \succ 3 \succ 1 \succ 6 \succ 4 \succ 2$ because ~A&B is consistent but A&~B is inconsistent with SD #5, which makes A false and B and C true (see table 6.4). And, A ⇒ B holds in the ordering because the highest SD in it that is consistent with A&B is SD #3, which is higher than the highest SD consistent with A&~B, which is SD #2. Conversely, A ⇒ ~B doesn't hold in this ordering because the highest SD that is consistent with A&~B is SD #2, which is not higher than the highest SD consistent with A&~~B, which is SD #3.

Given the idea of holding, we can give a rule for determining whether

[24]This is most easily proved by adding up the terms in reverse order: $\upsilon^5 + \upsilon^4 \times (1 - \upsilon) + \upsilon^3 \times (1 - \upsilon) + \ldots$. Then

$$\upsilon^5 + \upsilon^4 \times (1-\upsilon) = \upsilon^4,\ \upsilon^5 + \upsilon^4 \times (1-\upsilon) + \upsilon^3 \times (1-\upsilon) = \upsilon^4 + \upsilon^3 \times (1-\upsilon) = \upsilon^3,$$

and so on down to $\upsilon^0 = 1$.

$\phi \Rightarrow \psi$ has probability $\geq 1 - \upsilon$ or $\leq \upsilon$ in an OMPυ-ordering: *If* p() *is the probability function generated by the* υ*-ordering then* $p(\phi \Rightarrow \psi) \geq 1 - \upsilon$ *holds if and only if* $\phi \Rightarrow \psi$ *holds in the ordering, and* $p(\phi \Rightarrow \psi) \leq \upsilon$ *if and only if* $\phi \Rightarrow \psi$ *does not hold in it.* Similarly, $p(\psi) \geq 1 - \upsilon$ *holds if and only if* ψ *holds in the ordering, and* $p(\psi) \leq \upsilon$ *if and only if* ψ *does not hold in it.*

The distribution in table 6.4 and the OMP-ordering that it generates provide some examples. The ordering was $5 \succ 3 \succ 1 \succ 6 \succ 4 \succ 2$, and we saw that A $\Rightarrow$ B holds in it, essentially because the only SD that is consistent with A&~B is SD #2, which is lower in the ordering than SD #3, which is consistent with A&B. Similarly, B $\Rightarrow$ C holds because the only SDs that are consistent with B&~C are SDs #3 and #6, which are lower than SD #5, which is consistent with B&C. On the other hand, A $\Rightarrow$ C does not hold because A&~C is consistent with SDs #2 and #3, and there are no SDs that are consistent with A&C.[25] The distribution is an OMPυ-distribution with $\upsilon = .9$, and, as the table shows, A $\Rightarrow$ B and B $\Rightarrow$ C both have probabilities greater than .9 while A $\Rightarrow$ C has a probability less than .1. Therefore, we can call our ordering an *OMP-counterexample* to the inference $\{A \Rightarrow B, B \Rightarrow C\} \therefore A \Rightarrow C$, because in the corresponding OMPυ-distribution $p(A \Rightarrow B)$ and $p(B \Rightarrow C)$ are both greater than or equal to $1 - \upsilon$, while $p(A \Rightarrow C)$ is less than or equal to υ.

Now, the fact that an OMP-ordering isn't a counterexample doesn't show that an inference satisfies the uncertainty sum condition and belongs to Class I, any more than the fact that an ordinary model or 'interpretation' isn't a counterxample doesn't show that an inference is valid in the classical sense. However, we can show that if there are no OMP-counterexamples to an inference then it must satisfy the generalized uncertainty sum condition. But to do this we must consider more general OMP-orderings.

The ordering $5 \succ 3 \succ 1 \succ 6 \succ 4 \succ 2$ can be turned into another OMP-ordering simply by reordering its members, for example as $2 \succ 4 \succ 6 \succ 1 \succ 3 \succ 5$. However, both this and the previous ordering leave out the two lines in table 6.4 that are 'strike-outs', with lines through them, because they don't correspond to regions in diagram 6.4. But, they do correspond to regions in diagrams like diagram 6.5.

Now, forgetting about the actual sizes of the regions, the following table might represent a distribution over possible sizes.

---

[25]Since the SDs correspond to lines in table 6.4, it follows that an OMP-ordering corresponds to an ordering of the lines. Given this, an equivalent condition for a formula to hold in the ordering is that the highest line on which the formula has the real or ersatz value T must be higher in the ordering than some line on which it has the value F. Then whether a formula holds in an ordering can be determined by simple inspection of the table. For instance, A $\Rightarrow$ B holds because it has the value F only on line 2 in table 6.4, but that is lower in the ordering than line 3, on which A $\Rightarrow$ B has the value T. On the other hand, A$\Rightarrow$ C does not hold because it has the value F on lines 2 and 3, and it doesn't have the value T on any line.

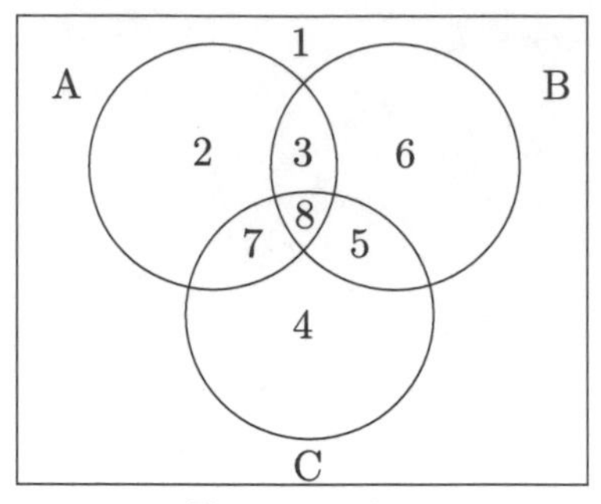

Diagram 6.5

region	distribution	A	B	C	$A \Rightarrow B$	$B \Rightarrow C$	$(A\&B) \Rightarrow C$	$A \Rightarrow C$
8	0	T	T	T	*T*	*T*	*T*	*T*
3	.09	T	T	F	*T*	*F*	*F*	*F*
7	0	T	F	T	*F*	...	.....	*T*
2	.00001	T	F	F	*F*	...	.....	*F*
5	.9	F	T	T	...	*T*	.....	.....
6	.0009	F	T	F	...	*F*	.....	.....
4	.00009	F	F	T	...	...	.....	.....
1	.009	F	F	F	...	...	.....	.....
	→	.09001	.9909	.90009	.9999	.9083	0	0

Table 6.5

Note that table 6.5 is like table 6.4 except for filling in the lines with strikeouts, and giving them 0 probability. But now the OMP-ordering includes lines 7 and 8, and these can be written into the ordering by putting a 'tail' on the end of it, as follows:

$$5 \succ 3 \succ 1 \succ 6 \succ 4 \succ 2 \succ 7 \approx 8 \approx \emptyset.$$

What the tail, $7 \approx 8 \approx \emptyset$,[26] at the end means is that the last SD in the original ordering, SD #2, is an order of magnitude more probable than

[26]The $7 \approx 8 \approx \emptyset$ in the tail of $5 \succ 3 \succ 1 \succ 6 \succ 4 \succ 2 \succ 7 \approx 8 \approx \emptyset$ isn't the only possible 'tie' that can occur in an OMP-ordering. For instance, $5 \approx 3 \approx 1 \succ 6 \succ 4 \succ 2 \succ 7 \approx 8 \approx \emptyset$ would be an ordering in which atoms 5, 3, and 1, had probabilities of roughly the same order of magnitude, which is higher than that of atoms 4, 6, etc. Such orderings don't need to be considered in the theory being developed here, but they become important in the logic of high probability developed in Adams (1986). These orderings correspond more strictly to the *possibility orderings* of David Lewis mentioned in footnote 23, while orderings with ties only in their tails correspond the *nearness orderings* of R. Stalnaker's (1968) theory of conditionals (cf. section 8.3)

SDs #7 and #8, both of which have 0 probability. Thus, adding the tail means simply adding $p_7 = p_8 = 0$ at the end of the corresponding series of probabilities:

$$p_5 = .9 = .1^0 \times .9,\ p_3 = .1^1 \times .9,\ p_1 = .1^2 \times .9,\ p_6 = .1^3 \times .9,\ p_4 = .1^4 \times .9,\ p_2 = .1^5, \text{ and } p_7 = p_8 = 0.$$

More generally, for any small quantity υ

$$p_5 = 1 - \upsilon = \upsilon^0 \times (1-\upsilon),\ p_3 = \upsilon^1 \times (1-\upsilon),\ p_1 = \upsilon^2 \times (1-\upsilon),\ p_6 = \upsilon^3 \times (1-\upsilon),\ p_4 = \upsilon^4 \times (1-\upsilon),\ p_2 = \upsilon^5, \text{ and } p_7 = p_8 = 0.$$

Adding the tail also requires a slight modification of the definition of holding in the case of conditionals $\phi \Rightarrow \psi$, to say that $\phi \Rightarrow \psi$ holds in an OMP-ordering if either (1) every SD that is consistent with $\phi\&{\sim}\psi$ is lower in the ordering than some SD that is consistent with $\phi\&\psi$, or (2) all SDs that are consistent with $\phi\&{\sim}\psi$ are equivalent to $\emptyset$ in the ordering. Then we can still say that any formula that holds in this ordering has a probability at least $1 - \upsilon$ and any one that does not hold has a probability less than or equal to υ. But we can also state the following:

OMP-counterexample theorem. If there is an OMP-ordering in which all of an inference's premises hold but its conclusion does not hold then its premises can have probabilities arbitrarily close to 1 while its conclusion has a probability arbitrarily close to 0, and it belongs to Class II. If there is no such ordering then it satisfies the generalized uncertainty sum condition and it belongs to Class I.[27]

Although this theorem is only stated and will not be proved here (it is not as easy to prove as the theorems in earlier chapters), it is a theorem of probability logic, that follows from the Kolmogorov axioms together with the 'meaning postulate' that $p(\phi \Rightarrow \psi)$ is a conditional probability. The practical importance of the theorem lies in the fact that a given set of SDs only have a limited number of OMP-orderings, and since we have a mechanical method for determining whether formulas hold in them, we can run through the orderings one by one to find out whether any of them is an OMP-counterexample to the inference, in the sense that all of the inference's premises hold in the ordering but its conclusion doesn't hold. In fact, this method works for any inference for which it is possible to construct a truth-table, no matter what atomic formulas it involves, since each OMP-ordering corresponds to an ordering of lines in the table, and there are only a finite number of them.[28]

[27]Though we won't need to use this fact immediately, it can be shown that if the inference belongs to Class II then its premisses can have probabilities arbitrarily close to 1 while its conclusion's probability equals 0.

[28]But the method doesn't work for inferences with infinitely many premisses, such as are considered in more advanced branches of logic, since truth-tables cannot be applied

But the method is often a very impractical one. Even when there are only three atomic formulas the number of OMP-orderings is very large and the method is impracticable. That is because the number of orderings of an 8-line truth table is 1+2!+3!+...+8! = 46,233,[29] and even with only four lines there are 1+2!+3!+ 4! = 33 possible OMP-orderings. We will see at the end of this section that there are practical shortcuts, and the following chapter will describe a far shorter shortcut, but first let us illustrate the method in application to another inference which shows the need to consider OMP-orderings with tails.

Consider the following:

$$\frac{(\text{A\&B}) \Rightarrow \sim\text{A}}{\text{A} \Rightarrow \text{C}}$$

The premise holds in the OMP-ordering

$$5 \succ 3 \succ 1 \succ 6 \succ 4 \succ 2 \succ 7 \approx 8 \approx \emptyset,$$

because the only SDs that are consistent with $\phi\&\sim\psi$ = (A&B)&~~A are 7 and 8, which are equivalent to $\emptyset$. But the conclusion A $\Rightarrow$ C does not hold in this ordering because SD #2 is consistent with A&~C, and that is higher than all of the SDs that are consistent with A&C, which are #7 and #8. It follows that p((A&B) $\Rightarrow$ ~A) $\geq 1 - \upsilon$ and p(A $\Rightarrow$ B) $\leq \upsilon$ in the OMPυ-distribution that corresponds to this ordering, and therefore the inference isn't valid in the probabilistic sense.

But there is still a problem of finding OMP-counterexamples like $5 \succ 3 \succ 1 \succ 6 \succ 4 \succ 2 \succ 7 \approx 8 \approx \emptyset$ out of all of the 46,233 OMP-orderings of the 8 SDs, and we conclude this section with comments on a way of doing this. This is most easily described in application to inference 7, namely {A $\Rightarrow$ B, B $\Rightarrow$ C} $\therefore$ A $\Rightarrow$ C.

Returning to diagram 6.5, suppose that we want to order SDs #1–#8 in it in such a way that the premises of inference 7, A $\Rightarrow$ B and B $\Rightarrow$ C, come out to be probable, while the conclusion A $\Rightarrow$ C comes out to be improbable. It is easy to see that in order to make A $\Rightarrow$ B probable we must make A&B much more probable than A&~B, and therefore much more probability must go into the union of SDs #3 and #8, which are consistent with A&B, than into the union of SDs #2 and #7, which are consistent with A&~B. In order to make this true, the SD that is highest in the ordering between SDs #3 and #8 must be higher than the one that is highest between SDs #2 and #7. We will express this requirement

to them. Special problems arise in applying probability logic to them because p-validity is not a *compact* relation. I.e., it is possible for an inference with infinitely many premisses to be p-valid even though its conclusion isn't p-entailed by any finite subset of these premisses.

[29] 8! is the factorial of 8, i.e., 8! = 1×2 × 3 × 4 × 5 × 6 × 7×8 = 40,320.

by writing '$\{3,8\} \succ \{2,7\}$'.[30] Similarly, making B $\Rightarrow$ C probable requires putting much more probability into the union of SDs #5 and #8, which are consistent with B&C, than into the union of SDs #3 and #6, which are consistent with B&~C. Therefore the highest of SDs #5 and #8 must be higher in the ordering than the highest of SDs #3 and #6, which can be expressed by writing $\{5,8\} \succ \{3,6\}$. Finally, to make A $\Rightarrow$ C improbable we must put much less probability into the union of #7 and #8, which are consistent with A&C, than into the union of #2 and #3, which are consistent with A&~C, and therefore the highest among #2 and #3 must be higher than the highest among #7 and #8, i.e., we must have $\{2,3\} \succ \{7,8\}$. The total ordering of the SDs must therefore satisfy three 'inequalities':

(a) $\{3,8\} \succ \{2,7\}$
(b) $\{5,8\} \succ \{3,6\}$
(c) $\{2,3\} \succ \{7,8\}$

Now we want to ask whether there is some OMP-ordering of SDs #1–#8 that is consistent with the inequalities above, i.e., whether there is one in which the highest among SDs #3 and #8 is higher than the highest among #2 and #7 (consistency with inequality (a)), the highest among #5 and #8 is higher than the highest among #3 and #6 (consistency with inequality (b)), and the highest among #2 and #3 is higher than the highest among #7 and #8 (consistency with inequality (c)). In fact, the ordering

$$5 \succ 3 \succ 1 \succ 6 \succ 4 \succ 2 \succ 7 \approx 8 \approx \emptyset$$

actually satisfies this condition, but now let us see how we could have found such an ordering without searching all of the 46,233 possible OMP-orderings.

If inequalities (a)–(c) are all to hold at the same time in some ordering then the highest of the SDs involved in these inequalities cannot occur on the right hand side of any of them, because if it did then some other SD would be higher than it. SDs #2, #3, #6, #5, #7, and #8 are the only ones involved in these inequalities, and the only one of these SDs that doesn't occur on the right of any of the inequalities is SD #5. Therefore putting SD #5 highest in the ordering would guarantee that inequality (b) was satisfied, i.e., the ordering we are trying to construct should rank SD #5 at the top.

But the remaining SDs must still be ordered in a way that guarantees that inequalities (a) and (c) are satisfied. SDs #2, #3, #7, and #8 are the only ones involved in these two inequalities, and ordering them in a way that satisfies both inequalities requires that the highest among these SDs must not occur on the right of either of these inequalities. SD #3 is the

[30]This is equivalent to at least one of the SDs in $\{3,8\}$ being higher in the ordering than both of the SDs in $\{2,7\}$.

only one that is not on the right in either inequality (a) or inequality (c), and ordering it highest among SDs #2, #3, #7, and #8 would guarantee that these two inequalities were satisfied. Furthermore, if SD #3 is put below SD #5 but above SDs #2, #3, #7, and #8 this would guarantee that inequality (b) was also satisfied, hence all three inequalities would be satisfied. In sum, if SD #5 is put at the top, SD #3 is next below it, and SDs #2, #7, and #8 are below that in the ordering, all of the inequalities will be satisfied. One way to do this is to order the SDs in the order

$$5 \succ 3 \succ 1 \succ 6 \succ 4 \succ 2 \succ 7 \approx 8 \approx \emptyset,$$

as we have already seen. Of course there are many other ways to order the SDs with SD #5 at the top, SD #3 next, and all other SDs below that, and the variety of these ways is a crude indication of the range of 'possible probabilistic worlds' in which the premises of inference 7 would be probable while its conclusion was improbable.

The final thing to note is that there is no ordering of SDs that would make the premises of inference 8, namely A ⇒ B and (A&B) ⇒ C, probable while its conclusion, A ⇒ C, was improbable. We have already seen that making A ⇒ B probable and A ⇒ C improbable requires satisfying the inequalities $\{3,8\} \succ \{2,7\}$ and $\{2,3\} \succ \{7,8\}$. To make (A&B) ⇒ C probable as well would mean making (A&B)&C much more probable than (A&B)&~C, i.e., the inequality $\{8\} \succ \{3\}$ would have to be satisfied. Combining this with the other two inequalities we would have:

$$\{3,8\} \succ \{2,7\}$$
$$\{2,3\} \succ \{7,8\}$$
$$\{8\} \succ \{3\}$$

But these inequalities only involve SDs #2, #3, #7, and #8, and since each of them occurs on the right of one or more of the inequalities, no one of them could be highest in the ordering. Therefore there is no ordering in which A ⇒ B and (A&B) ⇒ C are both probable while A ⇒ C is improbable, hence according to the OMP-counterexample theorem, this inference must satisfy the uncertainty sum condition.

Exercises

1. Construct OMP-counterexamples to each of the inferences in exercise 1 of section 6.3 for which you were able to construct a probability diagram representing its premises as probable and its conclusion as improbable. You may need to use orderings with equivalences, $\approx$, in the tail.
2. Argue that there are no OMP-counterexamples to the following inferences:
   a. $\{A, A \Rightarrow B\} \therefore B$

b. $\{A \vee B, {\sim}A\} \therefore B$
c. $\{A \Rightarrow B\} \therefore A \rightarrow B$
d. $\{A \Rightarrow B, B \rightarrow C\} \therefore A \rightarrow C$
e. $\{A \Rightarrow B, B \Rightarrow A, A \Rightarrow C\} \therefore B \Rightarrow C$

[Note that you only need to consider OMP-orderings of four SDs in the case of inferences a, b, and c, since they only involve the atomic formulas A and B; in carrying out these arguments it may be helpful to recall the point made in footnote 30, that one class of SDs is higher than another in an OMP-ordering if there is at least one member of the first class that is higher in the ordering than all members of the second class.]

⋆3. Argue that if an inference is classically invalid in the sense that it is invalid when $\Rightarrow$ is replaced by $\rightarrow$, then there is an ordering in which all but one SD is equivalent to $\emptyset$, in which all of its premises have probability 1 and its conclusion has probability 0.

⋆4. Argue there are no orderings in which $(A \vee B) \Rightarrow C$ is probable while both of $A \Rightarrow C$ and $(A \vee B) \Rightarrow B$ are improbable.

6.6⋆ Enthymemes and Other Defaults

We now return to a question that was postponed earlier, namely: why do inferences like $A \vee B \therefore {\sim}A \Rightarrow B$, $A \Rightarrow B \therefore {\sim}B \Rightarrow {\sim}A$, and $\{A \Rightarrow B, B \Rightarrow C\} \therefore A \Rightarrow C$, which we have seen have counterexamples even in reasoning in ordinary language, nevertheless have every appearance of being valid, in contrast to 'paradoxes' like ${\sim}A \therefore A \Rightarrow B$ which we recognize immediately as fallacious? This is a serious practical problem, and recently there has been serious inquiry into it because people very often reason in accord with inference-patterns like the above. However, here we can only comment briefly on probabilistic aspects of the problem.

Concerning inference 2, since it does not satisfy the Uncertainty Sum condition it follows that $A \vee B$ can be arbitrarily highly probable while ${\sim}A \Rightarrow B$ has a probability arbitrarily close to zero. The following inequality tells us something about these situations, i.e., ones in which $A \vee B$ is probable but ${\sim}A \Rightarrow B$ is improbable:

$$u(A) \leq \frac{u(A \vee B)}{u({\sim}A \Rightarrow B)}$$

If $A \vee B$ is probable but ${\sim}A \Rightarrow B$ is improbable then $u(A \vee B)$ is small and $u({\sim}A \Rightarrow B)$ is large, and therefore the ratio $u(A \vee B)/u({\sim}A \Rightarrow B)$ must be small. Therefore the inequality tells us that in this situation $u(A)$ must be small and $p(A)$ must be large. Is that likely? Conversational probability logic implies that it is not, because persons don't normally assert disjunctions when their first disjuncts are independently probable. Thus, when a person is told, for instance, that either A or B will attend a meeting, she will conclude without hesitation that if A doesn't attend then B will, because she knows that "either A or B will attend a meeting" wouldn't be asserted if A's attendance was independently probable. This consideration also explains the oddity of the premise "Either the sun will rise tomorrow or it will warm up during the day" in the counterexample to inference 2 that was given in section 6.3, which consists in the fact that a person wouldn't assert this if she could assert "The sun will rise tomorrow" by itself. But, of course, the fact that a disjunction isn't normally asserted when one of its disjuncts can be asserted independently doesn't mean that it is improbable, or even that there may not be special circumstances in which the assertion would be admissible. For instance, someone could ask "Is it the case that either the sun will rise tomorrow or it will warm up during the day?", in which case it would be permissible to answer "Yes, that is true, either the sun will rise tomorrow or it will warm up during the day."

The foregoing considerations should remind the reader of the discussion in section 5.6⋆ of the circumstances in which it is rational to infer L *a posteriori*, when $E \vee L$ has been accepted *a priori* and then $\sim E$ is learned. There it was pointed out that this reasoning is only rational when not only $E \vee L$ but L was also probable given $\sim E$, *a priori*; i.e., when $p(E \vee L)$ and $p(L|\sim E)$ were both high *a priori*. But on the probability conditional assumption, $p(L|\sim E) = p(\sim E \Rightarrow L)$, so we are saying that L can be inferred after $\sim E$ is learned if not only $E \vee L$ but also $\sim E \Rightarrow L$ was probable *a priori*. In effect, we are saying that the dynamical inference is rational in exactly the same circumstances as the static inference, $E \vee L \therefore \sim E \Rightarrow L$, is. And this is essentially inference 2.

Thus, we have reduced the question of the circumstances in which a dynamical inference involving purely factual propositions is rational to the question of when a static inference involving a conditional is rational. A similar reduction can be carried out in other cases as well, e.g., as concerns the circumstances when it is rational to infer B when A is probable *a priori* and then $\sim A$ is learned. And, that will be the case only if $\sim A \Rightarrow B$ is probable *a priori*, which is practically never guaranteed by the fact that A is probable *a priori*.[31] But we will now see that a similar reduction is not possible in the case of inference 3.

The following inequality plays a role in the analysis of inference 3 that is analogous to that of the inequality $u(A) \leq u(A \vee B)/u(\sim A \Rightarrow B)$ in the

[31]This follows from corollaries S2 and S3 in section 5.5, which imply that $\sim A$ cannot belong to the scope of a prior premise, A, that is not perfectly certain, and that the only conclusions that can be inferred from new information in this case are ones that follow from $\sim A$ alone.

analysis of inference 2:

$$u(B) \leq \frac{u(A \Rightarrow B)}{u({\sim}B \Rightarrow {\sim}A)}$$

Again, suppose that A ⇒ B is probable but ~B ⇒ ~A is not probable. In that case u(A ⇒ B) must be small and u(~B ⇒ ~A) must not be small, and therefore the ratio u(A ⇒ B)/u(~B ⇒ ~A) must be small. Therefore u(B) must be small according to the inequality, and so p(B) must be close to 1.[32] Now, the principle of conversational logic that was referred to in section 5.6⋆, that a speaker should not make a weaker, less informative, and less useful statement in circumstances in which it would be possible to make a stronger, more informative, and more useful one, seems to dictate that one shouldn't assert a conditional, A ⇒ B, in circumstances in which its consequent, B, is independently probable. Assuming this, a person hearing an assertion of the form "If A then B" should expect B not to be probable by itself, and therefore she should be justified in contraposing and concluding "If not B then not A."

However, this 'conversational maxim' has less force than the one that holds that A ∨ B shouldn't be asserted when A is probable by itself, since there are many circumstances in which it is legitimate to violate it. For instance, a person who said "Even if it rains the game will be played" wouldn't imply that "The game will be played" couldn't be asserted independently—and one would certainly not contrapose and deduce "If the game isn't played it won't have rained" in this case.[33]

The final point about inference 3 is that the circumstances in which it is rational are not ones in which a corresponding dynamical inference is rational. Specifically, we cannot say that it is rational to infer ~B ⇒ ~A from A ⇒ B if and only if it would be rational to infer ~A *a posteriori* when A ⇒ B is probable *a priori* and ~B is then learned. In fact this

[32]Appendix 5 sketches a geometrical argument due to D. Bamber, that circumstances in which A ⇒ B is probable but ~B ⇒ ~A is not probable must themselves be improbable, and the more probable A ⇒ B is the more improbable they are.

[33]Bennett (1995) discusses 'even if' statements, and the question of whether they are really conditionals. But, whether or not a statement like "Even if it rains the game will be played," is a conditional a person who asserted it would usually be prepared to affirm "if it rains the game will be played."

The assumption that speakers should assert something of the form "Even if A then B" and not "If A then B" when they are in a position to assert B by itself may be questioned in some circumstances. An example due to Bill Adams is "If a giant meteor doesn't strike the house tonight then it will still be standing tomorrow morning", where "The house will still be standing tomorrow morning" is itself very probable. In this case it would be irrational to contrapose and deduce "If the house is not still standing tomorrow morning then a giant meteor will have struck it during the night"—much more probably the house would have burned down than that it would have been struck by a giant meteor if it was not standing in the morning.

is dynamical *modus tollens* reasoning, which is puzzling because A ⇒ B is almost certain to be given up if ~B is learned. But then ~B seems to be the only premise that remains probable *a posteriori*, and ~A doesn't follow from that. Actually, what this shows is that both inference 3 and dynamical *modus tollens* are anomalous in a certain sense, and Appendix 6 suggests that the proper analysis of the latter takes us into the domain of the counterfactual conditional. But now let us turn to inference 7, which is in some ways even more difficult to analyze, and concerning which we will only be able to make a few hopefully suggestive observations.

Looking at the counterexample to $\{A \Rightarrow B, B \Rightarrow C\}$ ∴ $A \Rightarrow C$ (hypothetical syllogism) in diagram 6.4, the reader will note that in addition to depicting A ⇒ B and B ⇒ C as probable while A ⇒ C was improbable, it depicted both A ⇒ (B&~C) and B ⇒ (C&~A) as probable. This is borne out in the interpretation given in section 6.3 where

A = the sun will rise exactly 5 minutes later tomorrow,
B = the sun will rise no more than 5 minutes later tomorrow,
C = the sun will rise less than 5 minutes later tomorrow.

Given the foregoing, A ⇒ (B&~C) and B ⇒ (C&~A) are interpreted as:

A ⇒ (B&~C) = If the sun rises exactly 5 minutes later tomorrow then it will rise no more than 5 minutes later tomorrow and it will not rise less than 5 minutes later tomorrow.

B ⇒ (C&~A) = If the sun rises no more than 5 minutes later tomorrow then it will rise less than 5 minutes later tomorrow and it will not rise exactly 5 minutes later tomorrow.

A ⇒ (B&~C) is clearly a certainty in this case, and B ⇒ (C&~A) is at least a high probability because, given that the sun will rise no more than 5 minutes later tomorrow (than it did today), the chances are that it won't rise exactly 5 minutes later, hence it will actually rise less than 5 minutes later.

Now, the following inequalities imply that whenever A ⇒ B and B ⇒ C are probable but A ⇒ C is not probable then A ⇒ (B&~C) and B ⇒ (C&~A) must be probable:

$$u(A \Rightarrow B) \times u(A \Rightarrow (B\&{\sim}C)) \leq u(A \Rightarrow B) + u(B \Rightarrow C)$$
$$u(A \Rightarrow B) \times u(B \Rightarrow (C\&{\sim}A)) \leq u(A \Rightarrow B) + u(B \Rightarrow C)$$

For instance, these inequalities imply that if both $u(A \Rightarrow B)$ and $u(B \Rightarrow C)$ are no greater than .01 while $u(A \Rightarrow C) \geq .5$ then both $u(A \Rightarrow (B\&{\sim}C))$ and $u(B \Rightarrow (C\&{\sim}A))$ must be less than or equal to .04, hence A ⇒ (B&~C) and B ⇒ (C&~A) must both be at least 96% probable. Intuitively, this suggests that cases in which A ⇒ B and B ⇒ C are probable but A ⇒ C is not probable must be statistical rarities, since it seems highly unlikely that both A ⇒ (B&∼C) and B ⇒ (C&∼A) will also be probable. Assuming this, it should be 'statistically reasonable' to infer A ⇒ C when A ⇒ B and B ⇒ C are probable.

Certain phenomena of conversational logic should also be noted. A person who affirmed a conjunction like

> If the sun rises exactly 5 minutes later tomorrow then it will rise no more than 5 minutes later tomorrow, and, if it rises no more than 5 minutes later tomorrow then it will rise less than 5 minutes later tomorrow.

would be likely to provoke puzzlement, even though she could assert the conjuncts independently, or, perhaps, reversing the order, say something like

> If the sun rises exactly 5 minutes later tomorrow then it will rise no more than 5 minutes later tomorrow. *But*, if it rises no more than 5 minutes later tomorrow, then, I would say, it will actually rise less than 5 minutes later tomorrow.

What is the difference between these assertions? Plausibly, it is that putting the connective 'and' between clauses in a single sentence binds them together more 'intimately' than simply asserting them in another order or independently. This suggests that systematic study of the differences between 'and', 'but', 'moreover', etc., all of which are symbolized in formal logic as '&', should look beyond truth conditions, and at probabilistic aspects of reasoning involving them.

Exercises

1. Prove the following inequalities.
   a. $u(A) \times u(\sim A \Rightarrow B) \leq u(A \vee B)$.
   b. $u(\sim B \Rightarrow \sim A) \times u(B) \leq u(A \Rightarrow B)$.
   ⋆c. $u(A \Rightarrow C) \times u((A \vee B) \Rightarrow B) \leq u(A \Rightarrow B) + u(B \Rightarrow C)$.
2. a. Discuss the circumstances in which it would be reasonable to deduce the conclusion of the non p-valid inference A ∴ B ⇒ A from its premise.
   b. Are these inferences any more reasonable than ones of the form ∼B ∴ B ⇒ A?

6.7⋆ Problems of Inferential Dynamics

The previous chapter stressed the fact that probability change is an important aspect of all inferential processes, deductive or inductive, and in the previous section we saw that the analysis of some dynamical inferences that involve only factual propositions reduces to that of static inferences involving conditionals. However we will now see that serious obstacles stand in the way of developing a dynamical theory of reasoning about conditionals themselves, which the present comments can do no more that point out, as open research problems. To see what these obstacles are, recall the part played by the dynamical uncertainty sum rule stated in Section 5.2 in the dynamical theory of factual reasoning.

According to the dynamical sum rule, if prior premises $\phi_1,\ldots,\phi_n$ together with a new premise or item of information ι entail a conclusion ϕ, then

$$u_1(\phi) \leq u_0(\phi_1/\iota) + \ldots u_0(\phi_n/\iota)$$

where $u_1(\phi)$ is the uncertainty of ϕ after ι is learned and $u_0(\phi_1/\iota),\ldots,$ $u_0(\phi_n/\iota)$ are the conditional uncertainties of $\phi_1,\ldots,\phi_n$ *given* ι, before ι is learned. This implies that, assuming that $\phi_1,\ldots,\phi_n$ were highly probable before ι was learned, and ι belongs to the scopes of all of them, so that the prior conditional probabilities $p_0(\phi_1/\iota),\ldots,p_0(\phi_n/\iota)$ were also high (and there were not 'too many' prior premises), $u_0(\phi_1/\iota)+\ldots+u_0(\phi_n/\iota)$ should be small and therefore $u_1(\phi)$ should be low and $p_1(\phi)$ should be high. But two major difficulties stand in the way of generalizing this to inferences involving conditionals.

The first of these difficulties is that there is no obvious way of generalizing the dynamical sum rule to cases in which the new information, symbolized above as ι, is of conditional form. That is because Bayes' principle, which is presupposed in deriving the dynamical rule, assumes that $p_1(\phi) = p_0(\iota \Rightarrow \phi)$, and our probability conditional theory, which is also presupposed in the derivation, assumes that $p_0(\iota \Rightarrow \phi) = p_0(\phi|\iota) = p_0(\phi\&\iota)/p_0(\iota)$. But if ι is itself a conditional formula then $\phi\&\iota$ is not a formula of our language, and it doesn't have a probability. For example, suppose that the prior premise is $\phi_1 = A$, the new information is $\iota = A \Rightarrow B$, and the conclusion is $\phi = B$. Then according to the dynamical sum rule it ought to be that $p_1(B) = p_0((A \Rightarrow B) \Rightarrow B) = p_0((A \Rightarrow B)\&B)/p_0(A \Rightarrow B)$. But neither $(A \Rightarrow B) \Rightarrow B$ nor $(A \Rightarrow B)\&B$ are formulas, and their probabilities aren't defined. Not only that, we will see in chapter 8$\star$ that they can't be defined without 'reducing the theory to triviality', which would be a fundamental kind of inconsistency in it. Ways of dealing with the problem of have been proposed, but their implications haven't yet been worked out in detail, and we will leave 'updating on conditional information' as an important open research problem that students interested in applications to real life reasoning may wish to consider.[34]

[34]The practical importance of this phenomenon cannot be overstated, since we receive information couched in conditional terms constantly. In fact, it could seem absurd that we can deal 'lamely' with dynamical *modus ponens* reasoning when $A \Rightarrow B$ is the prior

The second and perhaps more serious problem can be illustrated in attempting to give a dynamical analysis of the inference $\{C, \sim A\} \therefore A \Rightarrow B$, where C is assumed to be given *a priori*, then ~A is learned, and the question becomes whether $A \Rightarrow B$ can be inferred *a posteriori*. The new information is not conditional, so the problem is not that of updating on conditional information. But, according to the assumption of the previous chapter, that new information becomes a certainty *a posteriori*, A should have probability 0 after ~A is learned, and therefore $p(A \Rightarrow B)$ should equal 1 *a posteriori*; i.e., the inference $\{C, \sim A\} \therefore A \Rightarrow B$ should be 'dynamically valid'. But this would be a dynamical version of a fallacy of material implication. C is really an irrelevant premise, and if the foregoing argument were right it would prove that $\sim A \therefore A \Rightarrow B$ is dynamically valid, although it is statically invalid, which would be absurd. But of course this results from the idealization that new information should be certain *a posteriori*, and avoiding it requires giving up the idealization. But that means that we cannot assume that probabilities change by conditionalization. i.e., that $p_1(\phi) = p_0(\phi/\iota)$.

The upshot is this: if we are to avoid dynamical versions of fallacies of material implication we must give up Bayes' principle. More will be said about this in section 9.7⋆, but in the meantime, in chapters 7⋆ and 8⋆, we will side-step the problem by confining attention to the principles of 'inferential statics'.

Exercise

1. Jackson (1987: 13) gives the example of a person who believes "If my partner is cheating me I'll never know it," but who doesn't conclude "I'll never know it" when he learns, to his surprise, that his partner is cheating him. Do you think that:
   a. this is a counterexample to Bayes principle?
   b. this is a counterexample to the conditional probability 'thesis' that the probability of "If my partner is cheating me I'll never

---

premise and A is the new information, since in this case it seems plausible to assume that $p_1(B) = p_0(A \Rightarrow B)$. *However*, this assumption rests on Bayes' principle, and the exercise at the end of this section calls that into question in this application.

Adams (1994) proposes a way of dealing with dynamical *modus ponens* and related inferences when the information is received in reverse order, i.e., in which A is the prior premise and $A \Rightarrow B$ is the new information. This follows the suggestion made in section 5.6⋆, that what is really learned when a reasoner is said to acquire information of conditional form is that what she actually learns is an item of factual information, e.g., that someone has asserted a proposition of conditional form. Working this out in detail involves probability distributions over OMP-orderings, and the final section of Adams (1994) points out serious problems that arise in this approach.

know it" equals the ratio:

$$\frac{p_0(\text{My partner is cheating me and I will never know it})}{p_0(\text{My partner is cheating me})}?$$

6.8 Glossary

Conditional Probability Theory The theory that the probabilities of ordinary language indicative conditionals are conditional probabilities.

Diagrammatic Counterexample A diagram depicting the premises of an inference as probable while its conclusion is improbable.

Holding (in an OMP-Ordering) A factual formula ψ holds in an OMP-ordering if it is consistent with the state description that is highest in the ordering, and a conditional $\phi \Rightarrow \psi$ holds in it if the highest state description that is consistent with $\phi \& \psi$ is higher than the highest one that is consistent with $\phi \& {\sim}\psi$.

OMP-Ordering (Order of Magnitude Ordering) An ordering of state-descriptions by the orders of magnitude of their probabilities.

OMP-Counterexample An OMP-ordering in which all of the premises of an inference hold, but the inference's conclusion doesn't hold.

OMP-Counterexample Theorem That if there is an OMP-ordering in which all of an inference's premises hold but its conclusion does not hold then its premises can have probabilities arbitrarily close to 1 while its conclusion has a probability arbitrarily close to 0.

Paradox of Material Implication Paradoxes that are led to by the assumption that an ordinary language conditional "if P then Q" is true unless P is true and Q is false. The best known of these are that inferences of the form "not P; therefore, if P then Q" and "Q; therefore, if P then Q" are valid.

Probability Conditional In the formalism of this and the following chapters, an expression of the form $\phi \Rightarrow \psi$, in which ϕ and ψ are nonconditional formulas. The probability $p(\phi \Rightarrow \psi)$ is assumed to equal $p(\psi|\phi)$.

7⋆

Formal Theory of Probability Conditionals: Derivations and Related Matters

7.1 Aspects of Derivation: Basic Definitions and Equivalences

The last chapter introduced the probability conditional concept and discussed somewhat informally its application to inferences involving conditionals. Now we want to formalize and extend the theory to *chains* of inferences, or, in other words, to *derivations*. Here is an example, in which the conclusion $(A \vee D) \Rightarrow E$ is derived from four 'premises' $A \Rightarrow B$, $A \Rightarrow C$, $A\&B\&C \Rightarrow E$, and $D \Rightarrow E$:

1. $A \Rightarrow B$ Given
2. $A \Rightarrow C$ Given
3. $(A\&B\&C) \Rightarrow E$ Given
4. $D \Rightarrow E$ Given
5. $(A\&B) \Rightarrow C$ Derived from 1 and 2.
6. $(A\&B) \Rightarrow E$ Derived from 3 and 5.
7. $A \Rightarrow E$ Derived from 1 and 6.
8. $(A \vee D) \Rightarrow E$ Derived from 4 and 7.   QED

The derivation starts by asserting the premises as 'givens' in lines 1–4, then it states four things that follow either immediately from the givens, as in line 5, or from the givens plus things previously derived from them, as in lines 6, 7, and 8, and it ends with the conclusion, $(A \vee D) \Rightarrow E$, in line 8. This is simple enough, but we have to specify exactly what rules of inference are allowed in deducing steps 5–8 and others like them in our derivations. These will be stated precisely in the following section, but there are some formal and informal preliminaries to discuss first.

Theories of formal derivation in probability logic involve the following aspects: (1) specifying the *formal language* of the theory, (2) defining precisely *probability* and *uncertainty functions* that apply to formulas of the language, (3) giving a precise *probabilistic* definition of *validity* that applies to inferences formulated in the language, and (4) stating precise *rules of inference* for deducing *probabilistic consequences* of premises in the language. (1)–(3), which are straightforward formalizations of ideas discussed in the previous chapter, will be covered briefly at the end of this section, rules of inference will be discussed in detail in the following one, and the formal theory this leads to will be developed in section 7.3. A formal test for p-validity that is much simpler than using OMP-orderings will be described in section 7.5, and aspects of the 'metatheory' of formal deduction having to do with the completeness of the theory and other matters will be taken up in four concluding sections, 7.6–7.9.

The formal language of the theory will consist of *factual formulas* of ordinary sentential logic like A, ~A, A&B, A ∨ B, A → B, etc., plus *conditional* formulas, $\phi \Rightarrow \psi$, where ϕ and ψ are factual.[1] We also introduce the sentential constants T and F, intuitively symbolizing a tautology and a contradiction, respectively, and we stress again that the probability conditional symbol, ⇒, can only occur as the main connective in any formula in which it occurs. Thus, our language does not contain expressions like ~(A ⇒ B), (A ⇒ B)&(B ⇒ A), or A ⇒ (B ⇒ C).

A probability function *for* the language is any function p() that applies to formulas ϕ, ψ, and $\phi \Rightarrow \psi$ of the language, where $p(\phi)$ and $p(\psi)$ satisfy the Kolmogorov axioms, and $p(\phi \Rightarrow \psi)$ is a conditional probability.[2] The uncertainty function, u( ), associated with p() satisfies the laws that $u(\phi) = 1 - p(\phi)$ and $u(\phi \Rightarrow \psi) = 1 - p(\phi \Rightarrow \psi)$; i.e., uncertainties are 1 minus probabilities, and conversely, probabilities are 1 minus uncertainties. As noted in section 2.6, the uncertainty of a factual formula is the probability of its negation, but we cannot say that $u(\phi \Rightarrow \psi) = p(\sim(\phi \Rightarrow \psi))$ since $\sim(\phi \Rightarrow \psi)$ is not a formula of our language, and it does not have a probability.

An *inference* will be a system, $\{\mathcal{P}_1, \ldots, \mathcal{P}_n\} \therefore \mathcal{P}$, where $\mathcal{P}_1, \ldots, \mathcal{P}_n$ and $\mathcal{P}$ are all formulas of our language,[3] and, as stipulated in section 6.4, we

[1] Strictly, we should allow for different formal languages depending on the atomic formulas like A, B, etc., that 'generate' them, but we will usually ignore this complication.

[2] See sections 2.3 and 4.2. Unconditional probabilities must satisfy the axioms:

K1. $0 \leq p(\phi) \leq 1$,

K2. if ϕ is logically true then $p(\phi) = 1$,

K3. if ϕ logically implies ψ then $p(\phi) \leq p(\psi)$,

K4. if ϕ and ψ are logically inconsistent then $p(\phi \vee \psi) = p(\phi) + p(\psi)$.

It is also stipulated that p(T) = 1 and p(F) = 0.

Conditional probabilities satisfy the condition that $p(\phi \Rightarrow \psi) = p(\phi \& \psi)/p(\phi)$ if $p(\phi) > 0$ and $p(\phi \Rightarrow \psi) = 1$ if $p(\phi) = 0$.

[3] In contrast to Greek letters like 'ϕ' and 'ψ', which are metalinguistic variables that

will say that the system is *probabilistically valid* (p-valid) if it satisfies the uncertainty sum condition; i.e., if

$$u(\mathcal{P}) \leq u(\mathcal{P}_1) + \ldots + u(\mathcal{P}_n)$$

for all uncertainty functions u( ). We also allow that there may not be any premises, i.e., n = 0, in which case $\therefore \mathcal{P}$ is p-valid if $u(\mathcal{P}) = 0$ for any uncertainty function u( ). This is equivalent to the specification that $p(\mathcal{P}) = 1$ for all probability functions p(); i.e., that $\mathcal{P}$ is a *p-tautology* (section 7.8⋆⋆).

Partly recapitulating the discussion in section 6.4., let us comment briefly on this concept of validity.

First, p-validity is a *deduction relation* in the sense that: (1) if a conclusion is p-entailed by a subset of premises then it is p-entailed by the whole set, and (2) if premises $\mathcal{P}_1, \ldots, \mathcal{P}_n$ p-entail conclusions $\mathcal{Q}_1, \ldots, \mathcal{Q}_m$ and $\mathcal{Q}_1, \ldots, \mathcal{Q}_m$ p-entail $\mathcal{P}$, then $\mathcal{P}_1, \ldots, \mathcal{P}_n$ p-entail $\mathcal{P}$. The first property is monotonicity, previously discussed briefly in section 5.3, and the second is sometimes called the *transitivity* of the deduction relation, and both of these properties are essential to the theory of deduction that will be developed in the following section.

Second, p-validity should be contrasted with *classical* validity of the kind considered in courses in elementary logic, where 'if... then... ' statements are symbolized by material conditionals. This is clarified by defining the material conditional $\phi \rightarrow \psi$ to be the *material counterpart* of the probability conditional $\phi \Rightarrow \psi$, (and we will say that the material counterpart of a factual formula is the formula itself), and defining the factual inference $\{\mathcal{P}'_1, \ldots, \mathcal{P}'_n\} \therefore \mathcal{P}'$ to be the material counterpart of $\{\mathcal{P}_1, \ldots, \mathcal{P}_n\} \therefore \mathcal{P}$, if $\mathcal{P}'_1, \ldots, \mathcal{P}'_n$ and $\mathcal{P}'$ are the material counterparts of $\mathcal{P}_1, \ldots, \mathcal{P}_n$ and $\mathcal{P}$, respectively. For instance, the material counterpart of the fallacious inference $A \therefore {\sim}A \Rightarrow B$ is the classically valid inference $A \therefore {\sim}A \rightarrow B$. This suggests the following, the proof of which is left as an exercise:

First Classical Validity Theorem The classical validity of the material counterpart of an inference is a necessary but not a sufficient condition for the inference to be p-valid, but if the inference involves only factual formulas then it is p-valid if and only if it is classically valid.

This theorem is illustrated not only by the fallacies of material implication, but by inferences 1, 2, 3, 6, and 7, discussed in section 6.3. Inference 4, $A \Rightarrow B \therefore A \rightarrow B$, is both p-valid and classically valid, and the fact that its conclusion is factual illustrates another general theorem about the relation between the two kinds of validity that will be stated in section 7.6⋆⋆.

range over factual formulas of our language, script capitals like '$\mathcal{P}$', with or without subscripts, are being used as metalinguistic variables that range over all formulas of the language, including both factual and conditional formulas.

Third There is another 'kind of validity' that applies to inferences involving probability conditionals, which is more closely related to classical validity than p-validity is. That is *weak validity*, which is the property that an inference has if the fact that all of its premises are certain guarantees that its conclusion is certain.[4] Then we can state the

Second Classical Validity Theorem An inference is weakly valid if and only if its material counterpart is classically valid.

For example, the 'fallacious' inference A $\therefore$ $\sim$A $\Rightarrow$ B *is* weakly valid, since if the probability of A is not just close to but actually equal to 1, then the probability of $\sim$A $\Rightarrow$ B must also equal 1.[5]

Significantly, it follows from the combined classical validity theorems that classical, weak, and probabilistic validity are equivalent in their application to purely factual inferences, but distinctions emerge when conditionals come into the picture.

Fourth Other useful equivalences have already been stated, including the following:

First Equivalence Theorem Given an inference I, $\{\mathcal{P}_1,\ldots,\mathcal{P}_n\}$ $\therefore$ $\mathcal{P}$, the following are equivalent:

(1) I is p-valid;
(2) $\mathcal{P}$ holds in all OMP-orderings in which $\mathcal{P}_1,\ldots,\mathcal{P}_n$ hold;
(3) for all $\delta > 0$ there exists $\epsilon > 0$ such that for all probability functions p(), if $p(\mathcal{P}_i) \geq 1-\epsilon$ for $i = 1,\ldots,n$, then $p(\mathcal{P}) \geq 1-\delta$;
(4) it is not the case that for all $\epsilon > 0$ there is a probability function p() such that $p(\mathcal{P}_i) \geq 1-\epsilon$ for i = 1,. . . ,n but $p(\mathcal{P}) = 0$.[6]

Equivalences (2) and (3) follow from the OMP-counterexample theorem stated in section 6.5⋆, and (3) and (4) give 'limit characterizations' of p-validity and invalidity. Thus, according to equivalence (3) $\{\mathcal{P}_1,\ldots,\mathcal{P}_n\}$ $\therefore$ $\mathcal{P}$ is p-valid if and only if $p(\mathcal{P})$ necessarily approaches 1 in the limit when all of the probabilities $p(\mathcal{P}_i)$ approach 1, and according to equivalence (4) if the inference is not p-valid then all probabilities $p(\mathcal{P}_i)$ can approach 1 in the limit, while $p(\mathcal{P}) = 0$. But the contrast between probabilities *approaching* 1 and actually *equalling* 1 has already been pointed out. For instance, according to (4), the probability of the premise of the p-invalid inference

[4]This is called *strict validity* in previous writings by the author. However, the following theorem makes it clear why this kind of validity is not stricter but weaker than p-validity.

[5]Obviously, this depends on the fact that if p(A) = 1 then p($\sim$A) = 0, hence p($\sim$A $\Rightarrow$ B) is stipulated to equal 1. Some reject this assumption, and by implication the second classical validity theorem that depends on it.

[6]The fact that $p(\mathcal{P})$ can actually equal 0, and not just be 'arbitrarily close to 0', as in the OMP-counterexample theorem of section 6.5⋆ (see footnote 27 of chapter 6), is easily established.

A $\therefore$ ~A $\Rightarrow$ B can approach 1 while the probability of the conclusion has probability 0—although the conclusion must have probability 1 when the premise's probability actually equals 1.

Fifth The uncertainty bounds given by the uncertainty sum condition are consistent with the probability of the conclusion of a p-valid inference being less than the probabilities of its premises. For instance, in *modus ponens* inferences, $\{A, A \Rightarrow B\} \therefore B$, when $p(A) = p(A \Rightarrow B) = .9$, $p(B)$ can be less than .9.[7] In this respect p-validity differs from more traditional concepts of validity, e.g., that a conclusion must be 'at least as true' as its premises.[8] On the other hand, we have already pointed out that allowing that conclusions can be less probable than their premises and their uncertainties can accumulate is a step in the direction of realism, because that is 'the way things are' in real life reasoning from premises that are not perfectly certain. Classical logic's idealizing and ignoring this is what leads to the lottery paradox.

Sixth Other conditions that are equivalent to p-validity will be added to the list given in the First Equivalence Theorem, and two important ones will be added in the following sections. One is the completeness of a set of *rules of inference* for deriving p-entailed consequences of premises, which is established in section 7.6⋆⋆ and the other is that a test to be described in section 7.5 yields a necessary and sufficient condition for p-validity.

The next section introduces the rules of inference referred to above.

7.2 Rules of Conditional Inference

Here are the seven rules of inference that define our theory of derivation in conditional logic.[9]

For any factual formulas ϕ, ψ, and η,

---

[7]But, curiously, p(B) must be at least $.9 \times .9 = .81$ in this case, which is actually greater than the probability that is guaranteed by the uncertainty sum condition.

[8]This is true not just of validity in classical bivalent logic, which assumes that statements can only have the truth-values of truth and falsehood, but also of classical *many-valued logics*, which allow for the possibility of intermediate 'degrees of truth', and modern *fuzzy logics* that permit any degrees of truth between 0 and 1—though these values are not probabilities because they assume that the truth-values of compound sentences like A&B must be functions of the values of their parts. McGee (1981) has shown that there is no many-valued logic with a finite number of truth values in which inferences in our language are valid if and only if they are p-valid, and Adams (1995) generalizes this result to a broader class of logics.

[9]Rules equivalent to these but often with different 'labels' are stated at various places in the recent literature in logic and artificial intelligence, cf. Adams (1986) and Makinson and Gärdenfors (1991), both of which include an independent rule of 'disjunctive reasoning'. Exercises at the end of this section ask the student to argue for the *validity* of these rules; i.e., that anything derivable by them must be p-entailed by the premises from which it is derived.

$\Re$1. *Rule G for entering premises*: if $\mathcal{P}$ is a premise of an inference it may be stated on any line.

$\Re$2. *Rule LC of logical consequence*: if ψ is a logical consequence of ϕ then $\phi \Rightarrow \psi$ can be derived from anything.

$\Re$3. *Rule CF of conditional and factual equivalence*: $\mathrm{T} \Rightarrow \phi$ can be derived from ϕ and ϕ can be derived from $\mathrm{T} \Rightarrow \phi$.

$\Re$4. *Rule EA of equivalent antecedents*: If ϕ and ψ are logically equivalent then $\psi \Rightarrow \eta$ can be derived from $\phi \Rightarrow \eta$.

$\Re$5. *Rule DA of disjunctive antecedents*: $(\phi \vee \psi) \Rightarrow \eta$ can be derived from $\phi \Rightarrow \eta$ and $\psi \Rightarrow \eta$.

$\Re$6. *RT of restricted transitivity*: $\phi \Rightarrow \eta$ can be derived from $\phi \Rightarrow \psi$ and $(\phi \& \psi) \Rightarrow \eta$.

$\Re$7. *Rule AR of antecedent restriction*: $(\phi \& \psi) \Rightarrow \eta$ can be derived from $\phi \Rightarrow \psi$ and $\phi \Rightarrow \eta$.

There are three general observations to make, before commenting on the individual rules stated above. The most important one is that the rules presuppose the laws of classical logic, because LC and EA involve the logical consequence and equivalence relations of classical logic that hold between factual formulas ϕ and ψ. However, because we are concentrating on probability conditionals, and we take the classical laws for granted without restating them, the student will not be asked to give classical derivations of factual formulas ψ from other formulas ϕ whenever she or he applies either rule LC or rule EA.

The second point is that presupposing classical logical relations among factual formulas means that our theory is an *extension* of classical logic, and not an alternative to it, since it applies to formulas that the classical theory doesn't apply to, namely probability conditionals.

The last point is that, as would be expected from the equivalence theorems in the previous section, all of our rules are also valid in application to the material counterparts of the conditionals involved. For instance, Rule AR, that $(\phi \& \psi) \Rightarrow \eta$ can be derived from $\phi \Rightarrow \psi$ and $\phi \Rightarrow \eta$, is also valid when '$\Rightarrow$' is replaced by '$\rightarrow$'; i.e., $(\phi \& \psi) \rightarrow \eta$ can be derived from $\phi \rightarrow \psi$ and $\phi \rightarrow \eta$. However, it is significant that $(\phi \& \psi) \rightarrow \eta$ follows from $\phi \rightarrow \eta$ by itself, whereas $(\phi \& \psi) \rightarrow \eta$ depends on both $\phi \Rightarrow \psi$ and $\phi \Rightarrow \eta$. This illustrates another aspect of the equivalence theorems, according to which the validity of the material counterpart of an inference is a necessary but not generally a sufficient condition for the p-validity of an inference involving probability conditionals.

Now we turn to comments on the individual rules $\Re$1–$\Re$7, all but the last of which are illustrated in a derivation to establish the validity of inference 4 of the previous chapter, i.e., to establish $\mathrm{A} \Rightarrow \mathrm{B} \therefore \mathrm{A} \rightarrow \mathrm{B}$, that the probability conditional entails the material conditional. Note that the justifications of the steps in the derivation use their mnemonic labels G–AR, rather than $\Re 1$–$\Re 7$, since they are easier to remember than the numbered rules. Here is the derivation:

1. $A \Rightarrow B$ G.
2. $(A\&B) \Rightarrow (A \rightarrow B)$ By LC.
3. $A \Rightarrow (A \rightarrow B)$ From 1 and 2 by RT.
4. $\sim A \Rightarrow (A \rightarrow B)$ By LC.
5. $(A \vee \sim A) \Rightarrow (A \rightarrow B)$ From 3 and 4 by DA.
6. $T \Rightarrow (A \rightarrow B)$ From 5 by EA.
7. $A \rightarrow B$ From 6 by CF.   QED

The main thing to notice about rule G relates to its function in the derivation, and to the function of the derivation itself. What the derivation did was to establish the probabilistic validity of the inference $A \Rightarrow B \therefore A \rightarrow B$, and Rule G's role was to introduce the inference's premise. However, this role is more restricted than rules for entering premises that the student may have encountered elsewhere in connection with other natural deduction systems, which often allow assumptions that are not 'given' to be entered as premises. For instance, many natural deduction systems include a rule of *reductio ad absurdum* reasoning, in which the negation of a conclusion is assumed for the purpose of proving that it leads to a contradiction. But our theory does not allow indirect reasoning, or the introduction into derivations of any statements that are not either 'given' premises or else deduced from them.[10] This is partly because we cannot even state the negation of a conditional in our formal language, and partly because our probabilistic approach requires us to reexamine the very idea of *assuming* something that is neither true nor probable.

Turning to rule LC, which was used in line 2 of the above derivation to deduce $(A\&B) \Rightarrow (A \rightarrow B)$, the important thing is to note that it is a '0-premise rule', since what it introduces is a 'truth of pure conditional logic', that because the material conditional $A \rightarrow B$ is a classical consequence of the conjunction A&B, the probability conditional $(A\&B) \Rightarrow (A \rightarrow B)$ 'follows no matter what, from anything at all'—i.e., it is a p-tautology. It will be seen that this rule is extremely useful in giving derivations like the ones the student will be asked to work out in the exercises at the end of this section.

Rule CF, which is used in the last line of the derivation above to deduce the factual formula $A \rightarrow B$ from its 'trivially equivalent conditional formula' $T \Rightarrow (A \rightarrow B)$, is significant in being the only rule that allows conditional formulas to be derived from factual ones, or conversely, allows factual formulas to be derived from conditionals. There are three remarks

[10]But section 7.7 states a metatheorem that has the effect of introducing an indirect rule of *conditional reduction to absurdity*, which is a generalization of the classical rule.

to make about this. (1) Formulas of the form $T \Rightarrow \phi$ and 'conditional tautologies' like $A \Rightarrow A$ are the *only* conditionals that are equivalent to factual formulas. (2) It is never necessary to use the rule to establish the validity of 'purely conditional inferences' like $A \Rightarrow (B\&C) \therefore A \Rightarrow B$, that do not involve factual formulas. (3) Certain inferences with factual premises and conditional conclusions like $\{A, B\} \therefore A \Rightarrow B$ are peculiar because their conclusions seem pointless. Who would make a conditional statement when she or he was in a position to state both parts of it? This will be returned to below, and section 7.7 will comment on a metatheorem that shows that something like this must be the case whenever conditional conclusions are derived from factual premises.

Rule EA was used in step 6 of our derivation to deduce $T \Rightarrow (A \rightarrow B)$ from $(A \vee {\sim}A) \Rightarrow (A \rightarrow B)$, because the tautologies T and $A \vee {\sim}A$ are logically equivalent, and according to rule EA either can be substituted for the other in the antecedent of any conditional formula; i.e., conditionals with logically equivalent antecedents are equiprobable, hence equivalent in probability logic. It is interesting that this rule, which presupposes classical logical relations among factual formulas, can be replaced by a 'purely conditional rule':

Rule EA′ of weakened antecedent equivalence: $\psi \Rightarrow \eta$ can be derived from $\phi \Rightarrow \psi$, $\psi \Rightarrow \phi$, and $\phi \Rightarrow \eta$

However, while using EA′ instead of EA might 'purify' our theory, the fact that it involves three premises makes it very inconvenient to use, and we will simplify by using rule EA.[11]

Step 5 in our derivation, in which $(A \vee {\sim}A) \Rightarrow (A \rightarrow B)$ was deduced using Rule DA from $A \Rightarrow (A \rightarrow B)$ and ${\sim}A \Rightarrow (A \rightarrow B)$, is like what is sometimes called a *proof by cases* argument, in which the consequent $A \rightarrow B$ is deduced from antecedent 'cases' A and ~A, and therefore it follows from their disjunction. However our derivation represents the cases A and ~A as antecedents of conditionals $A \Rightarrow ({\sim}A \rightarrow B)$ and ${\sim}A \Rightarrow (A \rightarrow B)$, whereas they would be represented as independent assumptions in standard proof-by-cases arguments.[12] This suggests a general approach to the analysis of reasoning involving assumptions that are neither true nor probable, which cannot be entered into here.

Next, consider rule RT, which was employed in our derivation to go from $A \Rightarrow B$ and $(A\&B) \Rightarrow (A \rightarrow B)$ on lines 1 and 2 to $A \Rightarrow (A \rightarrow B)$ on line 3. In the previous chapter we saw that unrestricted transitivity

[11]The reader will be asked to prove EA′ as an exercise at the end of the following section.

[12]In the present instance the cases are mutually exclusive, but that isn't necessarily the case either in applications of rule DA or in standard proof-by-cases arguments. However, mutual exclusivity is an important special case that is discussed in Adams (1996), which considers *probability preservation properties* of inferences.

inferences, of which $\{A \Rightarrow B, B \Rightarrow (A \rightarrow B)\} \therefore A \Rightarrow (A \rightarrow B)$ would be a special case, are not valid in general, but that restricted inferences like $\{A \Rightarrow B, (A\&B) \Rightarrow C\} \therefore A \Rightarrow C$, which has the same form as inference 8, are always valid. This is the first rule whose unrestricted form is valid for material conditionals, but which is only valid in restricted form for probability conditionals.

Our last rule, AR, is an even more restricted form of a rule that is valid for material conditionals. It is used in line 5 of a derivation that establishes the validity of a rather strange inference that was noted earlier, namely $\{A, B\} \therefore A \Rightarrow B$:

1. A G.
2. B G.
3. $T \Rightarrow A$ From 1 by CF.
4. $T \Rightarrow B$ From 2 by CF.
5. $(T\&A) \Rightarrow B$ From 3 and 4 by AR.
6. $A \Rightarrow B$ From 5 by EA, because A is equivalent to T&A. QED

There are two things worth noting in this derivation. First, the material counterpart of $\{A, B\} \therefore A \Rightarrow B$, namely $\{A, B\} \therefore A \rightarrow B$, only depends on the second premise, since it is one of the fallacies of material implication that $A \rightarrow B$ follows from B. In fact, this is true of the general rule AR from which it is derived, since the material counterpart of AR is that $(\phi\&\psi) \rightarrow \eta$ can be derived from $\phi \rightarrow \psi$ and $\phi \rightarrow \eta$, although $(\phi\&\psi) \rightarrow \eta$ can actually be derived from $\phi \rightarrow \eta$ alone.

Second, as already noted, it seems odd to deduce an ordinary language statement of the form 'If A then B' from statements A and B. But this doesn't make the inference fallacious. The oddity of asserting 'if A then B' when A and B are both probable arises from the fact that it would blatantly violate Grice's maxim of conversational quality, not to assert a weak proposition when a stronger and simpler one can be stated. Nevertheless, when A and B are both probable the probability of B must be high, given A, and if the probabilities of indicative conditionals are conditional probabilities then 'if A then B' must be probable.

We conclude this section with two more derivations, which will lead into the general theory of the consequences of Rules $\mathfrak{R}1$–$\mathfrak{R}7$, which will be developed in the following section. The first derivation establishes the validity of *modus ponens*, $\{A, A \Rightarrow B\} \therefore B$:

1. A G.
2. $A \Rightarrow B$ G.
3. $T \Rightarrow A$ From 1 by CF.
4. $(T\&A) \Rightarrow B$ From 2 by EA.
5. $T \Rightarrow B$ From 3 and 4 by RT.

6. B From 5 by CF. QED

[Note how rule CF is used in lines 3 and 6 to move from a factual formula to a trivial conditional, and then back from a trivial conditional to a factual formula.]

Finally, we establish A&B ∴ B:

1. A&B G.
2. (A&B) ⇒ B By LC.
3. T ⇒ (A&B) From 1, by CF.
4. (T&(A&B)) ⇒ B From 2 by EA (because T&(A&B) is logically equivalent to A&B)[13].
5. T ⇒ B From 3 and 4 by RT.
6. B From 5 by CF. QED

What is special about these two derivations is that the second is essentially a repetition of the first, which differs from the first only in substituting A&B for A in the earlier derivation. The question is: if the first derivation established *modus ponens* 'once and for all', why did it have to be derived all over again the next time it was 'applied'? The answer is that the first derivation only proved *modus ponens* in the particular case in which its premises were A and A ⇒ B, and not when they are A&B and (A&B) ⇒ B. But to establish *modus ponens* once and for all we need to prove a general rule of the form "ψ can be inferred from ϕ and $\phi \Rightarrow \psi$." The next section will show how general rules like this can be derived from rules $\mathfrak{R}1$–$\mathfrak{R}7$.

Exercises

1. Give formal derivations of the conclusions of the inferences below from their premises. No derivation requires more than 7 lines.
   a. A&B ∴ B&A
   b. ∴ (A&B) ⇒ A
   c. A ⇒ ~A ∴ A ⇒ B
   d. A ⇒ (B&C) ∴ A ⇒ B
   e. A ⇒ B ∴ A ⇒ (C → B)
   f. A ⇒ B ∴ A ⇒ (~B → C)
   g. {A ⇒ B, A ⇒ C} ∴ A ⇒ (B&C)
   h. {A ⇒ (B ∨ C), A ⇒ ~B} ∴ A ⇒ C
   i. {A ⇒ B, B ⇒ A} ∴ (A ∨ B) ⇒ (A&B)
   j. {A ⇒ B, B ⇒ A, A ⇒ C} ∴ B ⇒ C

⋆2. More difficult problems. Give formal derivations of the conclusions of the inferences below from their premises:

[13] It is not necessary to include explanations like "because T&(A&B) is logically equivalent to A&B" in formal derivations.

⋆a. $\{A, B\} \therefore A\&B$
⋆b. $\{A \Rightarrow B, \sim B\} \therefore \sim A$
⋆c. $(A \vee B) \Rightarrow \sim A \therefore B \Rightarrow \sim A$
⋆d. $\{A \Rightarrow (B \vee C), (A\&B) \Rightarrow C)\} \therefore A \Rightarrow C$
⋆e. $\{A \Rightarrow B, B \Rightarrow \sim A\} \therefore \sim A$
⋆f. $\{A \Rightarrow (B \vee C), A \Rightarrow \sim B\} \therefore A \Rightarrow C$

7.3 Derived Rules of Inference: Shortcut Derivations

To prove *modus ponens* as a general rule, namely that ψ can be derived from ϕ and $\phi \Rightarrow \psi$, we can substitute metalinguistic variables ϕ and ψ for A and B in the derivation that was given previously to establish the particular case of *modus ponens*:

$\Re$8. *Rule $\Re$8 of modus ponens*: ψ can be derived from ϕ and $\phi \Rightarrow \psi$.
Proof:

p. ϕ
q. $\phi \Rightarrow \psi$
q(a). $T \Rightarrow \phi$ From p by CF.
q(b). $(T\&\phi) \Rightarrow \psi$ From q by EA.
q(c). $T \Rightarrow \psi$ From q(a) and q(b) by RT.
r. ψ From q(c) by CF. QED

Note, though, that the general proof differs from the derivation establishing the validity of the particular inference $\{A, A \Rightarrow B\} \therefore B$ in two respects: (1) 'variable line numbers' p, q, r, etc. are entered in the general proof rather than 1, 2, 3, etc., and (2) the lines that were 'givens' in the particular derivation, namely lines 1 and 2, don't have 'reasons' in the general proof. The reason for using variable line numbers in the general proof is that what the proof shows is how any use of the $\Re$8 rule in a line of a derivation can be replaced by lines that employ only the basic rules, where the lines to be replaced may not be lines 1 and 2, and they can be justified in various ways, not necessarily as 'givens'. This is illustrated in the following 'shortcut derivation', which establishes the validity of the inference $A\&B \therefore B$, and which uses the $\Re$8 rule:

1. A&B G
2. (A&B) ⇒ B By LC.
3. B From 1 and 2 by $\Re$8. QED

Except for line numbering this turns into the 'basic derivation' that was given in the last section to establish $A\&B \therefore B$, just by substituting A&B for ϕ and B for ψ, and 1, 2, and 3 for p, q, and r, respectively, in the proof of rule $\Re$8:

1. A G.

2. A $\Rightarrow$ B   G.
2(a). T $\Rightarrow$ A   From 1 by CF.
2(b). (T&A) $\Rightarrow$ B   From 2 by EA.
2(c). T $\Rightarrow$ B   From 2(a) and 2(b) by RT.
3. B   From 2(c) by CF.   QED

Note that this derivation interpolates lines 2(a), 2(b), and 2(c) between lines 2 and 3 in the shortcut derivation, in order to expand it into a derivation that doesn't use rule $\Re 8$, and just uses the basic rules.

Here is another rule, which has no mnemonic label and which we will simply call rule $\Re 9$:

$\Re 9$. If η is a logical consequence of ψ then $\phi \Rightarrow \eta$ can be derived from $\phi \Rightarrow \psi$.

Proof: suppose that η is a logical consequence of ψ.

p. $\phi \Rightarrow \psi$.
p(a). $(\phi\&\psi) \Rightarrow \eta$.   By LC, since η is a logical consequence of $\phi\&\psi$.
q. $\phi \Rightarrow \eta$.   From p and p(a) by RT.   QED

For example, $\Re 9$ can be used to establish the validity of A $\Rightarrow$ (B&C) $\therefore$ A $\Rightarrow$ B:

1. A $\Rightarrow$ (B&C)   G.
2. A $\Rightarrow$ B   From 1 by $\Re 9$ (since B is a logical consequence of B&C).   QED

The proof of $\Re 9$ shows how to expand this into a derivation in accord with the basic rules, simply by interpolating line 1(a) between lines 1 and 2 of the shortcut derivation:

1. A $\Rightarrow$ (B&C)   G.
1(a). A&(B&C) $\Rightarrow$ B   LC
2. A $\Rightarrow$ B   From 1 and 1(a) by RT.   QED

Both shortcut rules, $\Re 8$ and $\Re 9$, are used in the following derivation, to establish the validity of {A, A $\Rightarrow$ (B&C)} $\therefore$ A $\Rightarrow$ C:

1. A   G.
2. A $\Rightarrow$ (B&C)   G.
3. A $\Rightarrow$ B   From 2 by $\Re 9$, since B is a logical consequence of B&C.
4. B   From 1 and 3 by $\Re 8$.   QED

This is turned into a basic derivation by interpolating line 2(a) between lines 2 and 3, to replace the use of $\Re 9$ in line 3, and interpolating lines 3(a), 3(b), and 3(c) between lines 3 and 4, to replace the use of $\Re 8$:

1. A   G.
2. A $\Rightarrow$ (B&C)   G.

2(a). $(A\&(B\&C)) \Rightarrow B$ By LC] Replacing use of $\Re 9$
3. $A \Rightarrow B$ From 2 and 2(a) by RT.
3(a). $T \Rightarrow A$ From 1 by CF.
3(b). $(TA) \Rightarrow B$ From 3(a) by EA.] Replacing use of $\Re 8$
3(c). $T \Rightarrow B$ From 3(a) and 3(b) by RT.
4. B From 3(c) by CF.

QED

We conclude this section by giving some shortcut rules that are not proved using variable line numbers like p, q, and q(a)–q(c) as in the proofs of $\Re 8$ and $\Re 9$, but using 'ordinary line numbers' instead. This makes it possible to turn ordinary derivations directly into proofs of shortcut rules, although it becomes somewhat more difficult to turn derivations using the shortcut rules into basic derivations. Here is an example in which the derivation given in the last section to establish A&B $\therefore$ B is turned directly into a proof of a shortcut rule, simply by replacing A and B by ϕ and ψ throughout:

$\Re 10$. ψ can be derived from $\phi\&\psi$.

Proof:

1. $\phi\&\psi$
2. $(\phi\&\psi) \Rightarrow \psi$ By LC.
3. $T \Rightarrow (\phi\&\psi)$ From 1, by CF.
4. $(T\&(\phi\&\psi)) \Rightarrow \psi$ From 2 by EA.
5. $T \Rightarrow \psi$ From 3 and 4 by RT.
6. ψ From 5 by CF

QED[14]

The next three rules are direct generalizations of derivations given in the last section, and their proofs will be left as exercises.

$\Re 11$. If η is a logical consequence of ϕ and ψ then it can be derived from them.

$\Re 12$. $\phi \rightarrow \psi$ can be derived from $\phi \Rightarrow \psi$.

$\Re 13$. $\phi \Rightarrow (\psi\&\eta)$ can be derived from $\phi \Rightarrow \psi$ and $\phi \Rightarrow \eta$.

The next two rules are extremely powerful, because by using them it becomes possible to derive the conclusions of p-valid inferences almost directly from their premises.

$\Re 14$. If $\eta\&\mu$ is a logical consequence of $\phi\&\psi$ and $\eta \rightarrow \mu$ is a logical consequence of $\phi \rightarrow \psi$ then $\eta \Rightarrow \mu$ can be derived from $\phi \Rightarrow \psi$.

Proof. Assume that $\eta\&\mu$ is a logical consequence of $\phi\&\psi$ and $\eta \rightarrow \mu$ is a logical consequence of $\phi \rightarrow \psi$

[14]Note that aside from substituting ϕ and ψ for A and B in the derivation of A&B $\therefore$ B, the only way in which the above proof differs from the derivation is that line 1 was not stated as a 'given'. That is because the shortcut rule can be applied to formulas that are not givens.

1. $\phi \Rightarrow \psi$
2. $(\phi\&\psi) \Rightarrow (\eta\&\mu)$ By LC.
3. $\phi \Rightarrow (\eta\&\mu)$ From 1 and 2 by RT.
4. $\phi \Rightarrow \eta$ From 3 by $\Re$9 (because η is a logical consequence of $\eta\&\mu$).
5. $\phi \Rightarrow \mu$ From 3 by $\Re$9 (because μ is a logical consequence of $\eta\&\mu$).
6. $(\phi\&\eta) \Rightarrow \mu$ From 4 and 5 by AR.
7. $((\phi \rightarrow \psi)\&\eta) \Rightarrow \mu$ By LC (because μ must be a logical consequence of $(\phi \rightarrow \psi)\&\eta$).
8. $[(\phi\&\eta) \vee ((\phi \rightarrow \psi)\&\eta)] \Rightarrow \mu$ From 6 and 7 by DA.
9. $\eta \Rightarrow \mu$. From 8 by EA (because $(\phi\&\eta) \vee ((\phi \rightarrow \psi)\&\eta)$ is logically equivalent to η). QED[15]

$\Re$15. $(\phi \vee \eta) \Rightarrow [(\phi \rightarrow \psi)\&(\eta \rightarrow \mu)]$ can be derived from $\phi \Rightarrow \psi$ and $\eta \Rightarrow \mu$. Proof.

1. $\phi \Rightarrow \psi$
2. $\eta \Rightarrow \mu$
3. $\phi \Rightarrow (\phi \rightarrow \psi)$ From 1 by $\Re$9.
4. $(\eta\&{\sim}\phi) \Rightarrow (\phi \rightarrow \psi)$ By LC.
5. $[\phi \vee (\eta\&{\sim}\phi)] \Rightarrow (\phi \rightarrow \psi)$ From 3 and 4 by DA.
6. $(\phi \vee \eta) \Rightarrow (\phi \rightarrow \psi)$ From 5 by EA.
7. $\eta \Rightarrow (\eta \rightarrow \mu)$ From 2 by $\Re$9.
8. $(\phi\&{\sim}\eta) \Rightarrow (\eta \rightarrow \mu)$ By LC.
9. $[\eta \vee (\phi\&{\sim}\eta)] \Rightarrow (\eta \rightarrow \mu)$ From 7 and 8 by DA.
10. $(\phi \vee \eta) \Rightarrow (\eta \rightarrow \mu)$ From 9 by EA.
11. $(\phi \vee \eta) \Rightarrow [(\phi \rightarrow \psi)\&(\eta \rightarrow \mu)]$ From 6 and 10 by $\Re$13. QED

The next section will show how important $\Re$15 is. In the meantime, the following derivation, which establishes the validity of the very important inference $\{(A \vee B) \Rightarrow B, (B \vee C) \Rightarrow C\} \therefore (A \vee C) \Rightarrow C$, shows how powerful rules $\Re$14 and $\Re$15 are:

1. $(A \vee B) \Rightarrow B$ G.
2. $(B \vee C) \Rightarrow C$ G.
3. $[(A \vee B) \vee (B \vee C)] \Rightarrow [((A \vee B) \rightarrow B))\&((B \vee C) \rightarrow C)]$ From 1 and 2 by $\Re$15.
4. $(A \vee C) \Rightarrow C$ From 3 by $\Re$14. QED

There are three things to note here. First, the only rules used in the derivation aside from rule G are $\Re$14 and $\Re$15; i.e., everything is reduced

[15]Note that lines 4 and 5 in this proof are themselves justified by shortcut rule $\Re$9, and therefore expanding a derivation which uses $\Re$14 would require first expanding it to one that uses $\Re$9, and expanding that in turn into a basic derivation.

to stating the premises and applying $\Re14$ and $\Re15$. The second thing is that the main difficulty involved in doing this consists in seeing that

$$[(A \vee B) \vee (B \vee C)]\&[((A \vee B) \rightarrow B))\&((B \vee C) \rightarrow C)]$$

logically entails $(A \vee C)\&C$, and that

$$[(A \vee B) \vee (B \vee C)] \rightarrow [((A \vee B) \rightarrow B))\&((B \vee C) \rightarrow C)]$$

logically entails $(A \vee C) \rightarrow C$, neither of which is obvious from inspection. However, the following section describes a method that makes this easier.

The other thing of interest about the inference $\{(A \vee B) \Rightarrow B, (B \vee C) \Rightarrow C\} \therefore (A \vee C) \Rightarrow C$ is theoretical. Picturing a conditional like $(A \vee B) \Rightarrow B$ in a probability diagram like the one below, you see that the only way it can be pictured as probable is by making A very small in comparison

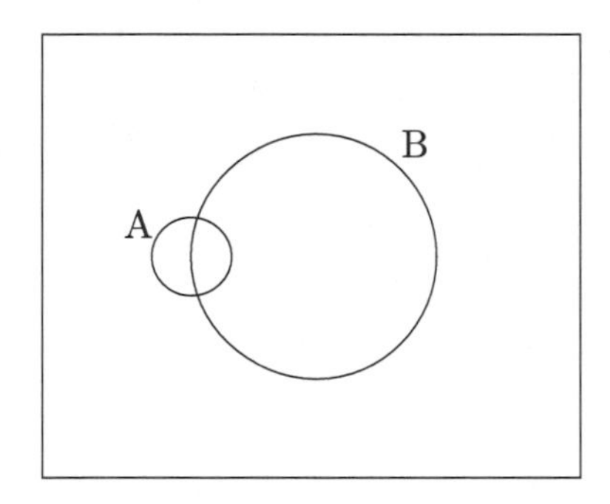

with region B. Therefore, to suppose that both premises of the inference $\{(A \vee B) \Rightarrow B, (B \vee C) \Rightarrow C\} \therefore (A \vee C) \Rightarrow C$ are probable is to suppose that A is small in comparison with B and B is small in comparison with C; hence A ought to be small in comparison with C. In fact, this ties in with the OMP-orderings discussed in section 6.5⋆, since the fact that A is small in comparison with B and B is small in comparison with C suggests that $C \succ B$ and $B \succ A$, which ought to imply that $C \succ A$, i.e., that $(A \vee C) \Rightarrow C$ is probable. The connection between derivability and OMP-orderings leads to a proof that the theory of derivation is *complete*, in the sense that if a conclusion that holds in all OMP-orderings in which the premises of an inference hold, it must be derivable from them in accord with our rules.

Exercises

1. Derive the conclusions of the inferences below from their premises, using shortcut rules $\Re8$–$\Re13$ if desired.
   a. $A \therefore B \rightarrow B$
   b. $(A\&B) \Rightarrow C \therefore A \Rightarrow (B \rightarrow C)$
   c. $\{A, A \Rightarrow B, B \Rightarrow C\} \therefore C$
   d. $\{A \Rightarrow \sim B, B\} \therefore \sim A$

2. Give proofs of rules $\Re 12$ and $\Re 13$.

7.4 Quasi-conjunction

An interesting set of shortcut rules also introduces a 'shortcut operation symbol', $\bigwedge$, for what will be called the *quasi-conjunction* of two conditionals $\phi \Rightarrow \psi$ and $\eta \Rightarrow \mu$.[16] This is defined as follows:

$$(\phi \Rightarrow \psi) \bigwedge (\eta \Rightarrow \mu) =_{df} (\phi \vee \eta) \Rightarrow [(\phi \rightarrow \psi)\&(\eta \rightarrow \mu)].$$

Rules $\Re 16$–$\Re 20$ below bring out some of the properties of quasi-conjunction. These are formulated in terms of 'variable formulas', $\mathcal{P}$, $\mathcal{P}_1$, etc., which can be any formulas of our language, either factual or conditional, and we will write, $\mathcal{P}_1 \bigwedge \mathcal{P}_2$ as the quasi-conjunction of $\mathcal{P}_1$ and $\mathcal{P}_2$. Note, incidentally, that we extend the quasi-conjunction operation to a factual formula ϕ by replacing it by its trivially equivalent conditional, $\mathrm{T} \Rightarrow \phi$. Thus, for instance, $\phi \bigwedge (\psi \Rightarrow \eta)$ is the same as $(\mathrm{T} \vee \psi) \Rightarrow [(\mathrm{T} \rightarrow \psi)\&(\psi \rightarrow \eta)]$, which is equivalent to $\mathrm{T} \Rightarrow (\phi\&(\psi \rightarrow \eta))$. Now we can state rules $\Re 16$–$\Re 20$, of which the proofs of $\Re 17$–$\Re 20$ will be left as exercises:

$\Re 16$. $\mathcal{P}_1 \bigwedge \mathcal{P}_2$ can be derived from $\mathcal{P}_1$ and $\mathcal{P}_2$.

Proof.

Suppose that $\mathcal{P}_1 = \phi \Rightarrow \psi$ and $\mathcal{P}_2 = \eta \Rightarrow \mu$, hence $\mathcal{P}_1 \bigwedge \mathcal{P}_2 = (\phi \vee \psi) \Rightarrow [(\phi \rightarrow \psi)\&(\eta \rightarrow \mu)]$.

1. $\phi \Rightarrow \psi$
2. $\eta \Rightarrow \mu$
3. $(\phi \vee \psi) \Rightarrow [(\phi \rightarrow \psi)\&(\eta \rightarrow \mu)]$ From 1 and 2 by $\Re 15$. QED

$\Re 17$. If $\mathcal{P}_1$ and $\mathcal{P}_2$ are both factual then they can be derived from $\mathcal{P}_1 \bigwedge \mathcal{P}_2$.

$\Re 18$. $\mathcal{P} \bigwedge \mathcal{P}$ can be derived from $\mathcal{P}$ and *vice versa*.

$\Re 19$. $\mathcal{P}_1 \bigwedge \mathcal{P}_2$ can be derived from $\mathcal{P}_2 \bigwedge \mathcal{P}_1$.

$\Re 20$. $\mathcal{P}_1 \bigwedge (\mathcal{P}_2 \bigwedge \mathcal{P}_3)$ can be derived from $(\mathcal{P}_1 \bigwedge \mathcal{P}_2) \bigwedge \mathcal{P}_3$.

Combined, rules $\Re 16$–$\Re 20$ imply that quasi-conjunctions are like ordinary conjunctions in most, though not quite all respects. Most importantly, a quasi-conjunction, $\mathcal{P}_1 \bigwedge \mathcal{P}_2$, is p-entailed by its 'quasi-conjuncts' $\mathcal{P}_1$ and $\mathcal{P}_2$ (by $\Re 16$), and when $\mathcal{P}_1$ and $\mathcal{P}_2$ are factual it p-entails them (by $\Re 17$). And, according to $\Re 18$ the quasi-conjunction of any formula with itself

[16]These are important in so called *conditional event algebras*, which permit 'conditional events' like $\mathrm{A} \Rightarrow \mathrm{B}$ and $\mathrm{C} \Rightarrow \mathrm{D}$ to be compounded to form negations like $\sim(\mathrm{A} \Rightarrow \mathrm{B})$, disjunctions like $(\mathrm{A} \Rightarrow \mathrm{B}) \vee (\mathrm{C} \Rightarrow \mathrm{D})$, and conjunctions of various kinds, including quasi-conjunctions. Cf. Goodman, Nguyen, and Walker (1991), and P. Calabrese (1994).

However, while expressions like $\sim(\mathrm{A} \Rightarrow \mathrm{B})$ and $(\mathrm{A} \Rightarrow \mathrm{B}) \vee (\mathrm{C} \Rightarrow \mathrm{D})$ aren't formulas of our language, the quasi-conjunction $(\mathrm{A} \Rightarrow \mathrm{B}) \bigwedge (\mathrm{C} \Rightarrow \mathrm{D})$ is an abbreviation for $(\mathrm{A} \vee \mathrm{C}) \Rightarrow [(\mathrm{A} \rightarrow \mathrm{B})\&(\mathrm{C} \rightarrow \mathrm{D})]$, which is formula of our language.

is equivalent to the formula, which is also like ordinary conjunction. According to $\Re 19$ and $\Re 20$, the order and grouping of the parts of a quasi-conjunction are immaterial, since any one ordering and grouping of the same quasi-conjuncts p-entails any other.[17] That means that, like ordinary conjunction, we can write a quasi-conjunction of any number of formulas without putting in parentheses, e.g., as $\mathcal{P}_1 \bigwedge \mathcal{P}_2 \bigwedge \mathcal{P}_3 \bigwedge \mathcal{P}_4$.

There is one important respect, however, in which quasi-conjunction differs from ordinary conjunction: When its parts are not factual the quasi-conjunction may not p-entail them.[18] For instance, the quasi-conjunction of $A \Rightarrow B$ and $\sim A \Rightarrow B$ is $(A \vee \sim A) \Rightarrow [(A \rightarrow B)\&(\sim A \rightarrow B)]$, which is equivalent simply to B. But B doesn't p-entail either $A \Rightarrow B$ or $\sim A \Rightarrow B$, since those would be paradoxes of material implication. However, we will see in section 7.6⋆⋆ that although quasi-conjunction isn't quite 'real conjunction', nonetheless it is close enough for a very important purpose.

The last things to note in this section are rules concerning the forms and *ersatz* truth-values of quasi-conjunctions of any number of quasi-conjuncts, depending on the values of their parts. As to form, if $\mathcal{P}_1, \ldots, \mathcal{P}_n$ are the conditionals $\phi_1 \Rightarrow \psi_1, \ldots, \phi_n \Rightarrow \psi_n$,[19] it can be shown that $\mathcal{P}_1 \bigwedge \ldots \bigwedge \mathcal{P}_n$ is equivalent to

$$(\phi_1 \vee \ldots \vee \phi_n) \Rightarrow [(\phi_1 \rightarrow \psi_1)\& \ldots \&(\phi_n \rightarrow \psi_n)].$$

As to truth values, it can easily be shown that the *ersatz* truth value of the quasi-conjunction $\mathcal{P}_1 \bigwedge \ldots \bigwedge \mathcal{P}_n$ is given by the following simple rule: $\mathcal{P}_1 \bigwedge \ldots \bigwedge \mathcal{P}_n$ *has the ersatz value F if any of* $\mathcal{P}_1, \ldots, \mathcal{P}_n$ *has the value F, it has the value T if none of* $\mathcal{P}_1, \ldots, \mathcal{P}_n$ *has the value F and at least one of them has the value T, and it has neither value if none of* $\mathcal{P}_1, \ldots, \mathcal{P}_n$ *has the value T or F.* These rules, which will be applied in section 7.6⋆⋆, are tedious to prove in general, though an exercise at the end of this section asks the student to verify them in the special case of the quasi-conjunction $(A \Rightarrow B)\bigwedge(C \Rightarrow D)$, by actually constructing a 16-line truth-table showing all possible combinations of truth-values of A, B, C, and D.

Exercises

1. Give proofs of rules $\Re 17$, $\Re 18$, $\Re 19$, and $\Re 20$.
2. Construct a 16-line truth-table giving all possible combinations of truth-values of A, B, C, and D, and the *ersatz* truth-values of $A \Rightarrow B$,

---

[17] Rules $\Re 18$–$\Re 20$ imply that the quasi-conjunction operation essentially has the properties of being idempotent ($\mathcal{P} \bigwedge \mathcal{P}$ is equivalent to $\mathcal{P}$), commutative, and associative, and therefore it is what modern abstract algebra calls a Boolean ring.

[18] Actually, we will see in section 7.7 that there doesn't exist a 'real' conjunction operation in our formal language, which is not only entailed by but also p-entails its conjuncts.

[19] Where, as before, if some $i = 1, \ldots, n, \mathcal{P}_i$ is a factual formula, say $\mathcal{P}_i = \psi_i$, we replace it by the trivial conditional equivalent $T \Rightarrow \psi_i$.

C $\Rightarrow$ D, and (A $\Rightarrow$ B) $\bigwedge$(C $\Rightarrow$ D) that correspond to them, and verify that this has the value *F* if either A $\Rightarrow$ B or C $\Rightarrow$ D has the value *F*, and it has the value *T* if neither A $\Rightarrow$ B nor C $\Rightarrow$ D has the value *F* and at least one has the value *T*.

7.5 An Ersatz Truth-table Test for Probabilistic Validity

The test that we will be concerned with shows how to 'survey' the *ersatz* truth values entered in a truth-table containing the premises and conclusion of an inference, to determine whether its conclusion is p-entailed by its premises. Inferences 1–5 below will illustrate the test. These are written with their premises above and their conclusions below a horizontal line, and with their 'validity values' written underneath:

Inference 1	Inference 2	Inference 3	Inference 4	Inference 5
B	A⇒B	B	A, B	(A∨B)⇒∼A, ∼A⇒∼B
A→B	A→B	A⇒B	A⇒B	(A∨B)⇒B
valid	**valid**	**invalid**	**valid**	**valid**

Table 7.1 below lists the *ersatz* truth-values that correspond to each state-description, or combination of truth-values of the atomic parts of inferences 1–5, for all of the formulas involved in them:

	atomic formulas		molecular formulas				
SD	A	B	A→B	A⇒B	(A∨B)⇒∼A	∼A⇒∼B	(A∨B)⇒B
1	T	T	T	*T*	*F*	.....	*T*
2	T	F	F	*F*	*F*	.....	*F*
3	F	T	T	...	*T*	*F*	*T*
4	F	F	T	...	....	*T*	....

Table 7.1

Before describing the truth-table test in detail, let us recall a few facts. (1) Inference 1 is purely factual, the truth-values of its premise and conclusion are filled in for all lines, and it is both classically and probabilistically valid. (2) Inference 2 has a conditional premise and factual conclusion, and while its premise doesn't have a truth value in all lines we have seen that it is valid in both senses. (3) Inference 3 has a conditional conclusion and although there are no lines in which its premise is true and its conclusion is false, the conclusion doesn't have truth values in all lines. Moreover, we saw in section 6.3 that it is classically valid but not p-valid. (4) Inference 4 has factual premises and a conditional conclusion, but we saw in section 7.2 that it is p-valid. (5) Inference 5 should also be p-valid because its conclusion can very easily be derived from its premises:

1. (A ∨ B) ⇒ ∼A G.
2. ∼A ⇒ ∼B G.
3. ((A ∨ B)&∼A) ⇒ B By LC.

4. $(A \vee B) \Rightarrow B$ From 1 and 3 by RT. QED

Now we want to see what the p-valid inferences above have in common with each other, but not with inference 3, which is p-invalid. Consider the following: We will say that a truth-valuation or SD *confirms* a premise or set of premises if it doesn't falsify any premise of the set and it actually verifies at least one, and that it *falsifies* the set if it makes at least one member of it false. For example, consider the set consisting of the two formulas $A \Rightarrow B$ and $(A \vee B) \Rightarrow {\sim}A$. Lines 1 and 2 in table 7.1 both falsify this set, since $(A \vee B) \Rightarrow {\sim}A$ has *ersatz* value *F* in line 1 and both formulas have that value in line 2. On the other hand, line 3 confirms the set since neither formula has the value *F* in that line, and $(A \vee B) \Rightarrow {\sim}A$ actually has the value *T*. Finally, line 4 neither confirms nor falsifies the set, since neither of its members has a value in that case.

The significant thing about confirmation is this: *If an inference is p-valid then any line in which its 'essential' premises are confirmed not only can't falsify its conclusion, it must actually verify it; moreover, if an inference is classically valid and it satisfies this condition then it is p-valid.* This will be stated more precisely below, but first let us check intuitively to see that it holds in the cases of inferences 1–5.

Inference 1, B $\therefore$ $A \rightarrow B$, is classically valid, and the lines in which its premise is confirmed—lines 1 and 3—verify its conclusion; i.e., they make it true. But there is an important thing to notice in this case. That is that this must be true because the inference is classically valid and its conclusion is factual. If B were confirmed but $A \rightarrow B$ were *not* verified in a line, it would have to be falsified in the line because, being factual, it must be one or the other. And, if $A \rightarrow B$ were falsified B couldn't be confirmed since the inference is classically valid. This generalizes a part of the first validity theorem that was stated in section 7.1:

Third Validity Theorem: If the conclusion of an inference is factual then it is p-valid if and only if it is classically valid.[20]

This result is borne out by inference 2, which is also classically valid and which must therefore also be p-valid according to the third validity theorem, since its conclusion is factual.

Turning to inference 3, B $\therefore$ $A \Rightarrow B$, which only replaces the material conclusion of inference 1, $A \rightarrow B$, by the probability conditional $A \Rightarrow B$, we see that although it is classically valid, it doesn't satisfy the requirement that confirming its premise necessarily verifies its conclusion. Thus, B is confirmed in line 3 but $A \Rightarrow B$ is neither verified nor falsified in that

[20]But it should be kept in mind that this depends critically on the fact that our formal language doesn't include compounds like negations of conditionals. For example, the result doesn't imply that "It is not the case that if today is Monday then tomorrow is Wednesday; therefore today is Monday" is valid.

line. This shows that inferences with conditional conclusions are radically different from ones whose conclusions are factual, which arises from the possibility of their conclusions being neither verified nor falsified.

Inference 4 has a conditional conclusion and it is classically valid, but this time confirming the premises guarantees verifying the conclusion, since the set of premises $\{A, B\}$ is only confirmed if A and B are both true, which verifies A $\Rightarrow$ B. Hence, it should be p-valid according to the truth-table test described above.

But inference 5 seems to show that the test doesn't always work, because the second premise of the inference is true and the first premise isn't false in line 4, hence the premise set is confirmed in that line but the conclusion isn't verified. I.e., seemingly contrary to our rule, while the inference is p-valid it is possible to confirm its premise set without verifying its conclusion. What is the trouble?

The answer is that now we must take into account the qualification that test applies to the essential premises of an inference, that if it is p-valid then whatever confirms these premises must verify its conclusion. Now, it happens that the second premise of inference 5 isn't really essential, since, as the reader can see from the derivation given earlier, (A $\vee$ B) $\Rightarrow$ B is actually p-entailed by (A$\vee$B) $\Rightarrow$ ~A alone. Moreover, restricting attention just to the essential premise (A $\vee$ B) $\Rightarrow$ ~A, it is true that confirming it does guarantee verifying (A $\vee$ B) $\Rightarrow$ B. These ideas can be stated more precisely in terms of the concept of *yielding*, as follows.

We will say that a set of premises $\{\mathcal{P}_1, \ldots, \mathcal{P}_n\}$ (where n can equal 0 and the set be empty) yields a conclusion $\mathcal{P}$ if (1) $\mathcal{P}$ is a classical consequence of $\mathcal{P}_1, \ldots, \mathcal{P}_n$, and (2) $\mathcal{P}$ is verified in any case in which $\{\mathcal{P}_1, \ldots, \mathcal{P}_n\}$ is confirmed. Given this, we can restate the truth-table test as follows: *An inference is p-valid if and only if some subset of its premises, possibly the empty set and possibly the whole set, yields its conclusion.* This is another equivalence to add to the First Equivalence Theorem, stated in section 7.1, and which will be incorporated in the second equivalence theorem to be formulated in the next section. But let us end this section with another example, which illustrates both the use and limitations of the truth-table test.

Consider the inference {A $\Rightarrow$ C, B $\Rightarrow$ ~C} $\therefore$ A $\Rightarrow$ ~B, a truth-table for which is outlined in heavy black in table 7.2. The table shows that the test for p-validity is negative. The conclusion, A $\Rightarrow$ ~B, isn't yielded by the total set of premises, {A $\Rightarrow$ C, B $\Rightarrow$ ~C}, since this set is confirmed in line 6 but A $\Rightarrow$ ~B isn't verified in that line. A $\Rightarrow$ ~B isn't yielded by the first premise alone, since A $\Rightarrow$ C is verified in line 1 but A $\Rightarrow$ ~B is false in that line. It isn't yielded by the second premise alone since B $\Rightarrow$ ~C is true but A $\Rightarrow$ ~B is false in line 2. And, finally, it isn't yielded by the empty set of

		SDs			formulas		
lines	distribution	A	B	C	A⇒C	B⇒∼C	A⇒∼B
1	.10	T	T	T	*T*	*F*	*F*
2	0	T	T	F	*F*	*T*	*F*
3	0	T	F	T	*T*	....	*T*
4	0	T	F	F	*F*	....	*T*
5	0	F	T	T	...	*F*	....
6	.90	F	T	F	...	*T*	....
7	0	F	F	T	...	....	....
8	0	F	F	F	...	....	....
		.10	1	.10	.10	.90	0

Table 7.2

premises, since it is false and no member of the empty set is false in lines 1 and 2. Therefore, no subset of the set $\{A \Rightarrow C, B \Rightarrow \sim C\}$ yields $A \Rightarrow \sim B$, so the inference must not be p-valid.

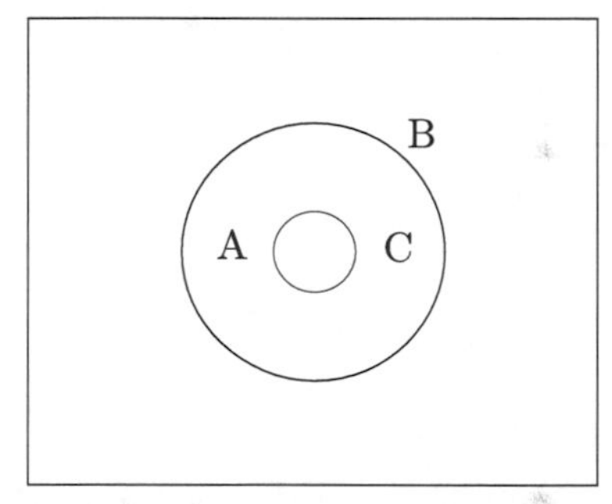

This result is fairly easily established,[21] and it is confirmed by the distribution that borders the table. This in turn determines the long-tailed OMP-ordering $6 \succ 1 \succ 2 \approx 3 \approx 4 \approx 5 \approx 7 \approx 8 \approx \emptyset$, in which $A \Rightarrow C$ and $B \Rightarrow \sim C$ hold but $A \Rightarrow \sim B$ doesn't. The result is further confirmed by the diagrammatic counterexample below the table, which in its turn leads to a real life counterexample:

A = C = there will be a torrential rain tomorrow,

and

B = it will rain tomorrow.[22]

One might well affirm A ⇒ B = "If there is a torrential rain tomorrow it

[21] Dr. Donald Bamber has simplified the method still further, in a way that makes it suitable to for programming on a computer.

[22] Note that when A and C are interpreted identically then the inference reduces to $\{A \Rightarrow A, B \Rightarrow \sim A\} \therefore A \Rightarrow \sim B$, which is essentially contraposition.

will rain" and $B \Rightarrow {\sim}C$ = "If it rains tomorrow it will not be torrential," but not $A \Rightarrow {\sim}B$ = "If there is a torrential rain tomorrow it will not rain."

The example also illustrates the point that although the truth-table test is a fairly quick method for determining p-validity, it is less informative than the diagrammatic and OMP-ordering methods. Merely knowing that it is possible to confirm the set $\{A \Rightarrow C, B \Rightarrow {\sim}C\}$ without verifying $A \Rightarrow {\sim}B$ only tells us that there must exist diagrammatic and real life counterexamples to the inference, but actually finding them is another matter. This will be illustrated in some of the following exercises, in which you are asked not only to apply the truth-table test to an inference but to verify the result, either by drawing a diagrammatic counterexample to it if the test is negative, or by formally deriving its conclusion from its premises if the test is positive.

Exercises

1. Construct a 4-line truth table, fill in the truth-values of the formulas in inferences a–f below, and determine for each of them whether it is classically valid, and which if any of the subsets of its premises yields its conclusion.
   a. $A \rightarrow B \therefore A \Rightarrow B$
   b. $A \vee B \therefore {\sim}A \Rightarrow B$
   c. $A \Rightarrow {\sim}B \therefore B \Rightarrow {\sim}A$
   d. $\{(A \rightarrow B) \Rightarrow (B \rightarrow A), {\sim}(A \leftrightarrow B)\} \therefore A \vee {\sim}B$
   e. $\{(A \vee B) \Rightarrow {\sim}A, (B \vee {\sim}A) \Rightarrow A\} \therefore A \Rightarrow {\sim}B$
   f. $\{A \Rightarrow B, B \Rightarrow {\sim}A\} \therefore {\sim}A$
2. Construct one or more 8-line truth tables and fill in the truth-values of the formulas in inferences a–e below. Then determine for each inference whether any subset of its premises yields its conclusion.
   a. $\{A \Rightarrow C, B \Rightarrow {\sim}C\} \therefore A \Rightarrow {\sim}B$
   b. $\{A \Rightarrow B, A \rightarrow {\sim}B\} \therefore A \Rightarrow C$
   c. $\{A, B \Rightarrow C\} \therefore (A\&B) \Rightarrow C$
   d. $(A \vee B) \Rightarrow {\sim}A \therefore (A \vee B) \Rightarrow B$
   e. $\{A \Rightarrow {\sim}B, B \Rightarrow {\sim}C, C \Rightarrow {\sim}A, {\sim}A \Rightarrow {\sim}A\} \therefore (A \vee B \vee C) \Rightarrow [A \leftrightarrow (B \leftrightarrow C)]$
3. Verify the results of your applications of the truth-table test to the inferences in exercise 2, by constructing diagrammatic counterexamples to the inferences for which the test is negative, and by formally deriving the conclusions of the inferences for which the test is positive from their premises.

7.6★★ Validity and Completeness, with Sketches of Their Proofs

Let us begin by stating our second equivalence theorem, which adds two more equivalences to the first equivalence theorem:

Second Equivalence Theorem: Given an inference I, $\{\mathcal{P}_1, \ldots, \mathcal{P}_n\} \therefore \mathcal{P}$, the following are equivalent:

(1) I is p-valid;
(2) $\mathcal{P}$ holds in all OMP-orderings in which $\mathcal{P}_1, \ldots, \mathcal{P}_n$ hold;
(3) for all $\delta > 0$ there exists $\epsilon > 0$ such that for all probability functions p(), if $p(\mathcal{P}_i) \geq 1-\epsilon$ for $i = 1, \ldots, n$, then $p(\mathcal{P}) \geq 1-\delta$;
(4) it is not the case that for all $\epsilon > 0$ there is a probability function p() such that $p(\mathcal{P}_i) \geq 1-\epsilon$ for $i = 1, \ldots, n$ but $p(\mathcal{P}) = 0$;
(5) there is a subset of $\{\mathcal{P}_1, \ldots, \mathcal{P}_n\}$, which may be empty and may equal $\{\mathcal{P}_1, \ldots, \mathcal{P}_n\}$, that yields $\mathcal{P}$;
(6) $\mathcal{P}$ can be derived from $\mathcal{P}_1, \ldots, \mathcal{P}_n$ by rules $\Re1$–$\Re7$.

From the practical point of view equivalence (5) is the most important result here, since we have seen that it provides a method for determining probabilistic validity that is considerably simpler than the OMP-ordering method described in section 6.4. On the other hand, the remaining sections of this chapter will be concerned with its theoretical implications, beginning with a proof of the claim in equivalence (6), that the theory of derivation based on rules $\Re1$–$\Re7$ is complete and valid in the sense that a conclusion is p-entailed by premises if and only if it is derivable from them by these rules. As with the other aspects of metatheory to be discussed in the later sections, the facts of this theory are of primary concern to us in this work, and the reader may skim their proofs (in any case they are only sketched).

To sketch the argument for completeness and validity we will assume the first four parts of the second equivalence theorem, especially that the p-validity of an inference is equivalent to its conclusion holding in all OMP-orderings in which its premises hold (equivalence 2 of the theorem). Then we will 'argue in a circle', as follows: (i) that being yielded by a subset of premises entails being derivable from them, (ii) that being derivable from premises entails being p-entailed by them, and (iii) that being p-entailed by premises entails being yielded by a subset of them. The three together entail the equivalence of derivability according to $\Re1$–$\Re7$, p-entailment, and being yielded by a subset of premises. Moreover, (2) essentially establishes the validity of the rules of inference, and (1) and (3) together imply their completeness, hence they establish parts 5 and 6 of the second equivalence theorem.

Beginning with a sketch of the proof of (i), we need to argue that if a conclusion, say $\mathcal{P}$, is yielded by premises, say $\mathcal{P}_1, \ldots, \mathcal{P}_n$, it can be derived from them by rules $\Re1$–$\Re7$. This is straightforward, given what we already know about yielding and quasi-conjunction. The quasi-conjunction $\mathcal{P}_1 \bigwedge \ldots \bigwedge \mathcal{P}_n$ is verified if and only if none of its parts are falsified and at least one is verified, and these are precisely the conditions under which the set $\{\mathcal{P}_1, \ldots, \mathcal{P}_n\}$ is confirmed. Moreover, $\mathcal{P}_1 \bigwedge \ldots \bigwedge \mathcal{P}_n$ is falsified if and only if one of its parts is falsified, which are exactly the conditions under which the set $\{\mathcal{P}_1, \ldots, \mathcal{P}_n\}$ is falsified. Therefore $\{\mathcal{P}_1, \ldots, \mathcal{P}_n\}$ yields $\mathcal{P}$ if and only if the fact that $\mathcal{P}_1 \bigwedge \ldots \bigwedge \mathcal{P}_n$ is verified entails that $\mathcal{P}$ is verified, and the fact that $\mathcal{P}$ is falsified entails that $\mathcal{P}_1 \bigwedge \ldots \bigwedge \mathcal{P}_n$ is falsified. To complete the argument all we need to show is that $\mathcal{P}_1 \bigwedge \ldots \bigwedge \mathcal{P}_n$ can be derived from $\{\mathcal{P}_1, \ldots, \mathcal{P}_n\}$, and if the latter yields $\mathcal{P}$ then $\mathcal{P}$ can be derived from it. Rule $\Re16$ shows that quasi-conjunctions can be derived from their parts, and rule $\Re14$ shows that if verifying a formula $\mathcal{P}$ entails verifying $\mathcal{P}$ and falsifying a formula $\mathcal{P}$ entails falsifying $\mathcal{P}$ then $\mathcal{P}$ can be derived from $\mathcal{P}$. This completes the argument for (i).

To argue for (ii), that being derivable from premises entails that what is derived is p-entailed by them, suppose that the conclusion of an inference $\{\mathcal{P}_1, \ldots, \mathcal{P}_n\} \therefore \mathcal{P}$ is derivable from its premises by rules $\Re1$–$\Re7$. It can be argued that $\mathcal{P}$ is p-entailed by showing that its probability can be made arbitrarily close to 1 by requiring the probabilities of $\mathcal{P}_1, \ldots, \mathcal{P}_n$ to be 'close enough' to 1—this is essentially what the 'limit condition' in equivalence 3 says. Now, direct calculation shows that the probability of any formula that follows immediately from $\mathcal{P}_1, \ldots, \mathcal{P}_n$ by one of rules $\Re1$–$\Re7$ can be guaranteed to be arbitrarily close to 1, simply by requiring the probabilities of the formulas from which it is derived to be close enough to 1. For instance, if $\phi \Rightarrow \psi$ is derived from $\phi \Rightarrow \eta$ and $(\phi \& \eta) \Rightarrow \psi$ by RT (rule $\Re6$) then it can be shown by direct though tedious calculation that $\mathrm{u}(\phi \Rightarrow \psi) \leq \mathrm{u}(\phi \Rightarrow \eta) + \mathrm{u}((\phi \& \eta) \Rightarrow \psi)$ (the uncertainty sum condition), hence $\mathrm{u}(\phi \Rightarrow \psi)$ can be made arbitrarily small and therefore $\mathrm{p}(\phi \Rightarrow \psi)$ can be made arbitrarily close to 1 by making $\mathrm{u}(\phi \Rightarrow \eta)$ and $\mathrm{u}((\phi \& \eta) \Rightarrow \psi)$ close enough to 0 and $\mathrm{p}(\phi \Rightarrow \eta)$ and $\mathrm{p}((\phi \& \eta) \Rightarrow \psi)$ close enough to 1. Similar calculations show that the same thing is true of any formula that follows immediately according to any of the other rules.

Beyond showing that an immediate consequence in which one of rules $\Re1$–$\Re7$ is applied just once is p-entailed by the things that it follows from, it can be shown by induction that this is true of inferences in which the rules are applied more than once. For instance, consider a derivation of the conclusion of the inference $(A \Rightarrow B) \Rightarrow {\sim}A \therefore (A \vee B) \Rightarrow B$ (this is the 'essential' part of inference 5, $\{(A \vee B) \Rightarrow {\sim}A, {\sim}A \Rightarrow {\sim}B\} \therefore (A \vee B) \Rightarrow B$, which was commented on in the last section):

1. $(A \vee B) \Rightarrow {\sim}A$ G.
2. $((A \vee B) \& {\sim}A) \Rightarrow B$ By LC

3. $(A \vee B) \Rightarrow B$ From 1 and 2 by RT.

That $(A \vee B) \Rightarrow B$ can be made arbitrarily probable by making $(A \vee B) \Rightarrow {\sim}A$ sufficiently probable can be argued for in two steps. If $(A \vee B) \Rightarrow {\sim}A$ is arbitrarily highly probable then *both* it and $((A \vee B)\&{\sim}A) \Rightarrow B$ must be arbitrarily highly probable, because $((A \vee B)\&{\sim}A) \Rightarrow B$ is itself a p-tautology, hence it has probability 1 'no matter what'. And, if both $(A \vee B) \Rightarrow {\sim}A$ and $((A \vee B)\&{\sim}A) \Rightarrow B$ are arbitrarily highly probable, and $(A \vee B) \Rightarrow B$ follows immediately from them by rule RT, then it must be arbitrarily highly probable,[23] as follows from the uncertainty sum rule.

The argument for (iii), that if a conclusion is p-entailed by premises then it must be yielded by a subset of those premises, is the most difficult. In fact, we will only argue for a particular case of the contrapositive, namely that if the conclusion isn't yielded by a subset of premises then it cannot be p-entailed by the whole set.

Consider the three-premise inference, written here with premises above and conclusion below a line:

$$\frac{A \Rightarrow B, {\sim}(A \leftrightarrow B) \Rightarrow A, {\sim}A \Rightarrow B}{{\sim}(A \vee B) \Rightarrow F}$$

(recall that the atomic sentence F symbolizes a logical contradiction). Table 7.3 below gives the *ersatz* truth-values of the premises and conclusion of the inference in all SDs, and it is easily verified that no subset of the premises yields the conclusion:

		atomic formulas		premises			conclusion
probabilities	SDs	A	B	$A \Rightarrow B$	${\sim}(A \leftrightarrow B) \Rightarrow A$	${\sim}A \Rightarrow B$	${\sim}(A \vee B) \Rightarrow F$
.9	1	T	T	*T*	....	....	....
.09	2	T	F	*F*	*T*	....	....
.009	3	F	T	...	*F*	*T*	....
.001	4	F	F	...	....	*F*	*F*

Table 7.3

What we need to show is that the conclusion cannot be p-entailed, but before focussing this special case, let us introduce two more bits of terminology. We will say that a line or SD is a *weak counterexample* to an inference

[23]This is really a 'limit argument' to the effect that if the value of a quantity z approaches 1 as the values of x and y both approach 1, and the value of y approaches 1 as the value of x approaches 1, then the value of z must approach 1 as the value of x alone approaches 1. In this case $x = p((A \vee B) \Rightarrow {\sim}A)$, $y = p(((A \vee B)\&{\sim}A) \Rightarrow B)$, and $z = p((A \vee B) \Rightarrow B)$.

with premises $\mathcal{P}_1, \ldots, \mathcal{P}_n$ and conclusion $\mathcal{P}$ if it confirms $\mathcal{P}_1, \ldots, \mathcal{P}_n$ without verifying $\mathcal{P}$, and it is a *strong counterexample* if it falsifies $\mathcal{P}$ without falsifying $\mathcal{P}_1, \ldots, \mathcal{P}_n$. The existence of a strong counterexample proves immediately that the inference is classically invalid, *ipso facto* probabilistically invalid, and because of this we look primarily for lines that are strong counterexamples to an inference. However, not finding a strong counterexample doesn't prove that an inference is p-valid, so if we don't find one, we look for weak counterexamples. To see how we proceed in this case, return to table 7.3 and the inference $\{A \Rightarrow B, {\sim}(A \leftrightarrow B) \Rightarrow A, {\sim}A \Rightarrow B\} \therefore {\sim}(A \vee B) \Rightarrow F$.

According to the table there are no strong counterexamples to our inference, but line 1 is a weak counterexample since it confirms the premises by verifying $A \Rightarrow B$ without falsifying either ${\sim}(A \leftrightarrow B) \Rightarrow A$ or ${\sim}A \Rightarrow B$, but it doesn't verify ${\sim}(A \vee B) \Rightarrow F$. This by itself only shows that the total set of premises doesn't yield ${\sim}(A \vee B) \Rightarrow F$, and not that the inference is invalid, since there might be a subset of the premises that yields it. To consider this possibility, we look at the premises that were not verified in line 1, namely ${\sim}(A \leftrightarrow B) \Rightarrow A$ and ${\sim}A \Rightarrow B$. There are no strong counterexamples to the inference $\{{\sim}(A \leftrightarrow B) \Rightarrow A, {\sim}A \Rightarrow B\} \therefore {\sim}(A \vee B) \Rightarrow F$ either, but line 2 is a weak one to it. This still only shows that ${\sim}(A \leftrightarrow B) \Rightarrow A$ and ${\sim}A \Rightarrow B$ don't yield ${\sim}(A \vee B) \Rightarrow F$, and not that the inference is invalid, since the premise that is verified in line 2, ${\sim}(A \leftrightarrow B) \Rightarrow A$, might not be essential. So we look next at the inference from which ${\sim}(A \leftrightarrow B) \Rightarrow A$ is omitted, namely ${\sim}A \Rightarrow B \therefore {\sim}(A \vee B) \Rightarrow F$. There is still no strong counterexample, but now line 3 is a weak one. And, this still only shows that ${\sim}A \Rightarrow B$ doesn't yield ${\sim}(A \vee B) \Rightarrow F$ and not that ${\sim}A \Rightarrow B \therefore {\sim}(A \vee B) \Rightarrow F$ is invalid, since ${\sim}A \Rightarrow B$ could itself be inessential. Therefore we consider the result of leaving this out, which is the 0-premise inference $\therefore {\sim}(A \vee B) \Rightarrow F$, and, at last, we have a strong counterexample, namely line 4. *And*, this combined with the fact that the preceding counterexamples were constructed in the way just described proves that the original inference was invalid, because the probabilities that table 7.3 assigns to these lines make all of the premises probable, while the conclusion has zero probability.[24]

Of course, this argument only applies to a special case, but the reader should be able to apply it to other cases. The general idea is that to construct a probability distribution that is a counterexample to an inference like the one in table 7.3, you begin by looking for a line that is a strong counterexample, and if you find one you stop right there and give that

[24]The similarity between the probabilities in table 7.3 and the probabilities that corresponded to OMP-orderings in section 6.4 is suggestive. In fact, the method described in the present section, of constructing a series of SDs, each of which is a weak or strong counterexample to the inference of a conclusion from a subset of the premises of an inference, is really a method of constructing an OMP-ordering in which all of the premises hold but the conclusion does not.

line probability 1. If you don't find a strong counterexample you look for a line that is a weak one, and if you find it you give it a high probability, say .9, but you then consider the premises that are left after the ones that were verified in the first line are omitted. If you find a line that is a strong counterexample to this inference you give it all the remaining probability, say .1, but if you only find a weak counterexample you give it most of the remaining probability, say .09. And so on, until either you find a subset which has a strong counterexample, to which you give all the remaining probability, or you don't find one, which shows that the inference is p-valid after all.

This concludes the argument for completeness and validity, which emphasizes the role of truth-tables with *ersatz* truth-values in the argument. The following section briefly notes some other metatheoretical applications of the test.

7.7 Other Aspects of Metatheory

Using the fact that the truth-table test yields necessary and sufficient conditions for the p-validity of inferences in our formal language, a number of interesting properties of its logic are fairly easily established. The first and simplest has to do with the role that factual premises can play in inferences with conditional conclusions. We already noted that the inference $\{A, B\}$ $\therefore$ A $\Rightarrow$ B is rather strange, because a person wouldn't ordinarily say "If Anne goes to the party then Ben will too", given the knowledge that both Anne and Ben will go to the party. That would violate Grice's maxim of conversational quality—not to make a less informative assertion in a situation in which a person is in a position to assert a more informative one that is equally simple. Now, the next 'metatheoretical fact' shows that something like this must be the case with any inference that has an essential factual premise and a conditional conclusion. As before, this and other facts are what are most important here, and the reader can skim the arguments that establish them without missing anything essential—though perusing the argument for the following 'factuality-conditionality' property may suggest how other properties like it can be found.

Factuality-conditionality Property: If an inference with a factual premise η and conditional conclusion $\phi \Rightarrow \psi$ is p-valid then either: (1) η is not an essential premise, or (2) the inference with the same premises and conclusion $\phi \& \psi$ is also p-valid.

The argument for this goes as follows. Supposing that η is essential to an inference with conclusion $\phi \Rightarrow \psi$, it must belong to a set of premises, say $\{\eta, \mathcal{P}_1, \ldots, \mathcal{P}_n\}$, that *yields* $\phi \Rightarrow \psi$. Now, assuming that $\{\eta, \mathcal{P}_1, \ldots, \mathcal{P}_n\}$ yields $\phi \Rightarrow \psi$, we can argue that it must also yield the conjunction $\phi \& \psi$. To see this, suppose that $\{\eta, \mathcal{P}_1, \ldots, \mathcal{P}_n\}$ is confirmed in a line of a truth-table. Then $\phi \Rightarrow \psi$ must be verified in this line, hence ϕ and ψ must both be true in it, and therefore $\phi\&\psi$ must be true in it; i.e., $\phi\&\psi$ must be verified in any line in which $\{\eta, \mathcal{P}_1, \ldots, \mathcal{P}_n\}$ is confirmed. Suppose on the other hand that $\phi\&\psi$ is falsified in a line. If $\{\eta, \mathcal{P}_1, \ldots, \mathcal{P}_n\}$ were not falsified in the line it would actually have to be confirmed, because none of its members would be false and η would actually have to be true, since it is factual. But we have seen that confirming $\{\eta, \mathcal{P}_1, \ldots, \mathcal{P}_n\}$ entails verifying $\phi\&\psi$ as well, contradicting the supposition that it was falsified. *Ipso facto*, confirming $\{\eta, \mathcal{P}_1, \ldots, \mathcal{P}_n\}$ entails verifying $\phi\&\psi$ and falsifying $\phi\&\psi$ entails falsifying $\{\eta, \mathcal{P}_1, \ldots, \mathcal{P}_n\}$, and therefore $\{\eta, \mathcal{P}_1, \ldots, \mathcal{P}_n\}$ yields $\phi\&\psi$, i.e., $\{\eta, \mathcal{P}_1, \ldots, \mathcal{P}_n\}$ also p-entails $\phi\&\psi$. QED

Of course this result raises philosophically difficult questions about the 'factual basis' of conditional assertions, which cannot be entered into here, though they will be alluded to in the final section of this chapter.[25]

The next point is also related to general philosophical issues. Sometimes it is felt that the antecedents of conditionals should be relevant to their consequents, and that what is wrong with 'fallacies' like "Anne will go to the party; therefore, if Boston is a large city then Anne will go to the party" is that Boston's being a large city is irrelevant to Anne's going to the party.[26] Now, we already know that the inference A $\therefore$ B $\Rightarrow$ A is not p-valid, but we can make a more general claim:

Relevance result: A conditional $\alpha \Rightarrow \phi$ can only be p-entailed by premises $\mathcal{P}_1, \ldots, \mathcal{P}_n$ that don't contain an atomic sentence α if $\alpha \Rightarrow \phi$ is a *conditional tautology*; i.e., it follows from the empty set of premises.

The argument for this is also very simple. Suppose that $\mathcal{P}_1, \ldots, \mathcal{P}_n$ are all essential, hence confirming this set implies verifying $\alpha \Rightarrow \phi$, and therefore making both α and ϕ true. But if α is an atomic sentence that doesn't occur in any premise in $\mathcal{P}_1, \ldots, \mathcal{P}_n$, then no matter what the truth values of the atomic sentences in $\mathcal{P}_1, \ldots, \mathcal{P}_n$ are, α can have any truth-value at all, including F. QED

Again, the general philosophical import of this cannot be entered into, but at least we see that we can't infer probability conditionals from premises, factual or conditional, that are wholly irrelevant to their antecedents.

Now consider the fact that we have defined a quasi-conjunction operation in our formal language, which is the same as true conjunction when it applies to factual formulas but which differs from it in application to conditionals. It is natural to ask whether we can improve on quasi-conjunction, and define a true conjunction operation that is not only p-entailed by, but

[25] Cf. Edgington 1991.

[26] Cf. Weingartner and Schurz 1986.

also p-entails its parts. The following result shows that the answer is No; in particular, there is no formula of our language, $\phi \Rightarrow \psi$,[27] that is both p-entailed by the conjunction of A $\Rightarrow$ B and $\sim$A $\Rightarrow$ B, and which also p-entails these formulas:

Undefinability of true conjunction: There is no formula $\phi \Rightarrow \psi$ of our language that has the properties: (1) that $\{A \Rightarrow B, \sim A \Rightarrow B\} \therefore \phi \Rightarrow \psi$ is p-valid, and (2) that both $\phi \Rightarrow \psi \therefore A \Rightarrow B$ and $\phi \Rightarrow \psi \therefore \sim A \Rightarrow B$ are p-valid.

Again this follows very simply from facts about the truth-table test for p-validity. Suppose that $\phi \Rightarrow \psi$ had properties (1) and (2) above. Obviously that could only happen if $\{A \Rightarrow B, \sim A \Rightarrow B\}$ yielded $\phi \Rightarrow \psi$, and $\phi \Rightarrow \psi$ yielded both A $\Rightarrow$ B and $\sim$A $\Rightarrow$ B. But if that were the case then $\{A \Rightarrow B, \sim A \Rightarrow B\}$ would have to yield both A $\Rightarrow$ B and $\sim$A $\Rightarrow$ B. But in fact it yields neither A $\Rightarrow$ B nor $\sim$A $\Rightarrow$ B. (Though it p-entails them.) QED

Of course, this result only proves that 'true conjunctions' can't be expressed in our language, and not that the language couldn't be 'enriched' by adding formulas like $(A \Rightarrow B)\&(\sim A \Rightarrow B)$ to it. That is a deeper question that will be returned to in the next chapter.

The last point to be noted in this section is related to the fact that we have not included any *indirect* rules among our rules of derivation, $\mathfrak{R}1$–$\mathfrak{R}7$, and in particular we have no rule of *reductio ad absurdum*, in which, in order to show that a conclusion follows from given premises, you assume the negation of the conclusion and deduce a contradiction from it. As said, the reason for not having such a rule is that our language doesn't contain the negations of all of the formulas in it, and therefore a *reductio ad absurdum* rule cannot even be stated in a form that assumes that all formulas have negations. However, we can state the following:

Conditional reductio ad absurdum (conditional RAA) rule: An inference $\{\mathcal{P}_1, \ldots, \mathcal{P}_n\} \therefore \phi \Rightarrow \psi$ is p-valid if and only if the inference $\{\mathcal{P}_1, \ldots, \mathcal{P}_n, \phi \Rightarrow \sim\psi\} \therefore \phi \Rightarrow F$ is p-valid.

This follows straightforwardly though somewhat tediously from things we already know about the relation between probabilistic validity and the yielding relation, and we will only sketch the argument for the claim that if $\{\mathcal{P}_1, \ldots, \mathcal{P}_n\} \therefore \phi \Rightarrow \psi$ is p-valid then $\{\mathcal{P}_1, \ldots, \mathcal{P}_n, \phi \Rightarrow \sim\psi\} \therefore \phi \Rightarrow F$ must be p-valid. Suppose, then, that $\{\mathcal{P}_1, \ldots, \mathcal{P}_n\} \therefore \phi \Rightarrow \psi$ is p-valid, and in fact that $\{\mathcal{P}_1, \ldots, \mathcal{P}_n\}$ yields $\phi \Rightarrow \psi$. This implies by a somewhat complicated argument that $\{\mathcal{P}_1, \ldots, \mathcal{P}_n, \phi \Rightarrow \sim\psi\}$ must yield $\phi \Rightarrow F$; i.e., any SD confirming $\{\mathcal{P}_1, \ldots, \mathcal{P}_n, \phi \Rightarrow \sim\psi\}$ must verify $\phi \Rightarrow F$, and any SD falsifying $\phi \Rightarrow F$ must falsify $\{\mathcal{P}_1, \ldots, \mathcal{P}_n, \phi \Rightarrow \sim\psi\}$. In fact, as-

[27] We can always restrict our considerations to conditional formulas $\phi \Rightarrow \psi$, since any factual formula ψ is p-equivalent to $T \Rightarrow \psi$.

suming that $\{\mathcal{P}_1,\ldots,\mathcal{P}_n\}$ yields $\phi \Rightarrow \psi$, this follows from the fact that the set $\{\mathcal{P}_1,\ldots,\mathcal{P}_n,\phi \Rightarrow {\sim}\psi\}$ cannot be confirmed. To see this, suppose that $\{\mathcal{P}_1,\ldots,\mathcal{P}_n,\phi \Rightarrow {\sim}\psi\}$ were confirmed by an SD, or in a line of a truth table—it is easy to see that in that case either $\{\mathcal{P}_1,\ldots,\mathcal{P}_n\}$ would have to be confirmed or $\phi \Rightarrow {\sim}\psi$ would have to be verified. Now, if $\{\mathcal{P}_1,\ldots,\mathcal{P}_n\}$ were confirmed $\phi \Rightarrow \psi$ would have to be verified, since $\{\mathcal{P}_1,\ldots,\mathcal{P}_n\}$ yields it; but that would falsify $\phi \Rightarrow {\sim}\psi$ and therefore falsify $\{\mathcal{P}_1,\ldots,\mathcal{P}_n,\phi \Rightarrow {\sim}\psi\}$. On the other hand, if $\phi \Rightarrow {\sim}\psi$ were verified then $\phi \Rightarrow \psi$ would be falsified, which would entail that $\{\mathcal{P}_1,\ldots,\mathcal{P}_n\}$ would be falsified, since it yields $\phi \Rightarrow \psi$; and again that would falsify $\{\mathcal{P}_1,\ldots,\mathcal{P}_n,\phi \Rightarrow {\sim}\psi\}$.

Having shown that $\{\mathcal{P}_1,\ldots,\mathcal{P}_n,\phi \Rightarrow {\sim}\psi\}$ cannot be confirmed, all that has to be shown to complete the argument that $\{\mathcal{P}_1,\ldots,\mathcal{P}_n,\phi \Rightarrow {\sim}\psi\}$ $\therefore$ $\phi \Rightarrow$ F is p-valid is that any SD falsifying $\phi \Rightarrow$ F must falsify $\{\mathcal{P}_1,\ldots,\mathcal{P}_n,\phi \Rightarrow {\sim}\psi\}$. So, suppose that $\phi \Rightarrow$ F is falsified in some SD or line in a truth table. That would mean that ϕ was verified in that line, and therefore either $\phi \Rightarrow \psi$ or $\phi \Rightarrow {\sim}\psi$ would have to be falsified. If $\phi \Rightarrow {\sim}\psi$ were falsified then $\{\mathcal{P}_1,\ldots,\mathcal{P}_n,\phi \Rightarrow {\sim}\psi\}$ would also be falsified. On the other hand, If $\phi \Rightarrow \psi$ were falsified then some member of $\{\mathcal{P}_1,\ldots,\mathcal{P}_n\}$ would have to be falsified, since it yields $\phi \Rightarrow \psi$, and therefore $\{\mathcal{P}_1,\ldots,\mathcal{P}_n,\phi \Rightarrow {\sim}\psi\}$ would also have to be falsified in that case.

The other parts of the argument, to show that the result holds even when $\{\mathcal{P}_1,\ldots,\mathcal{P}_n\}$ doesn't yield $\phi \Rightarrow \psi$, and also that the p-validity of $\{\mathcal{P}_1,\ldots,\mathcal{P}_n,\phi \Rightarrow {\sim}\psi\}$ $\therefore$ $\phi \Rightarrow$ F entails the p-validity of $\{\mathcal{P}_1,\ldots,\mathcal{P}_n\}$ $\therefore$ $\phi \Rightarrow \psi$, will be left as exercises at the end of this section. But there are four points to note before turning to that.

One point is that the conditional RAA rule reduces to standard *reductio ad absurdum* in the case of inferences with factual conclusions. In that case the ϕ in the conclusion $\phi \Rightarrow \psi$ may be assumed to be T, hence $\phi \Rightarrow \psi$ is equivalent to ψ and ${\sim}\psi$ is equivalent to T $\Rightarrow {\sim}\psi$. Moreover, $\phi \Rightarrow$ F is T $\Rightarrow$ F, which is equivalent to a factual contradiction, F.

Second, what we might call the 'conditional denial' of $\phi \Rightarrow \psi$, namely $\phi \Rightarrow {\sim}\psi$, does in fact act like ordinary truth-functional negation or denial in many respects, e.g., in the context of *reductio ad absurdum* arguments. But, like quasi-conjunction, the conditional denial $\phi \Rightarrow {\sim}\psi$ isn't a 'true' negation of $\phi \Rightarrow \psi$, in part because $\phi \Rightarrow \psi$ and $\phi \Rightarrow {\sim}\psi$ don't p-entail anything at all; e.g., unless $\phi \Rightarrow \psi$ is essentially factual $\{\phi \Rightarrow \psi, \phi \Rightarrow {\sim}\psi\}$ doesn't p-entail the factual contradiction, F, hence it p-entails every sentence. The question of whether there exists a formula in our language that acts in all respects like truth-functional negation will be left open, though we will return in the following chapter to the matter of whether the language could be enriched to include one.

The third point is that the conclusion of the conditional RAA argument, $\phi \Rightarrow$ F, expresses a practical or probabilistic certainty; namely that ϕ is practically certain to be false. The logic of this kind of certainty as well as probabilistic versions of logical truth, logical consistency, and logical equivalence will be discussed briefly in the following section.

Finally an application of the conditional RAA rule shows that derivations employing it have to be complicated by distinguishing between givens that are premises of inferences, and assumptions in which conditional denials are introduced 'for the purposes of the argument'.[28] An example is involved in a derivation to establish the p-validity of the inference $A \Rightarrow (A \rightarrow B) \therefore A \Rightarrow B$:

1. $A \Rightarrow (A \rightarrow B)$ G.
2. $A \Rightarrow \sim B$ Assumption.
3. $(A\&\sim B) \Rightarrow (A \rightarrow B)$ From 1 and 2 by AR.
4. $((A\&\sim B)\&(A \rightarrow B)) \Rightarrow F$ By LC.
5. $(A\&\sim B) \Rightarrow F$ From 3 and 4 by RT.
6. $A \Rightarrow F$ From 2 and 5 by RT.
7. $A \Rightarrow B$ By conditional RAA, since $A \Rightarrow F$ follows from $A \Rightarrow \sim B$ plus the premise. QED

Exercises

1. Derive the conclusions of the following inferences from their premises, using conditional RAA where convenient:
   a. $A \Rightarrow B \therefore A \Rightarrow (A \rightarrow B)$
   b. $(A \leftrightarrow B) \Rightarrow A \therefore (A \leftrightarrow B) \Rightarrow B$
   c. $(A \vee B) \Rightarrow \sim A \therefore (A \vee B) \Rightarrow B$

⋆2. Show that the conditional RAA rule holds even when $\{\mathcal{P}_1, \ldots, \mathcal{P}_n\}$ doesn't yield $\phi \Rightarrow \psi$.

⋆3. Argue that if both $A \Rightarrow B$ and $A\&\sim B$ p-entail $\phi \Rightarrow \psi$ then $\phi \Rightarrow \psi$ must be a conditional tautology; i.e., $p(\phi \Rightarrow \psi) = 1$ for all probability functions p().

7.8⋆⋆ P-tautology, Equivalence, and Consistency

The concepts to be discussed in this section, namely those of *p-tautology*, *p-equivalence*, and *p-consistency*, are the 'probabilistic surrogates' of the classical concepts of tautologousness, equivalence, and consistency. For our purposes they are less important than p-validity, but they are intrinsically interesting, and the difficulty of analyzing them probabilistically is similar to that of analyzing p-validity. That has to do with the fact that the conditional formulas to which the concepts are expected to apply don't have

[28] In spite of the fact that *reductio ad absurdum* reasoning is commonly used in mathematics, its 'logic' is often not made clear even in the systems of natural deduction that formalize it.

truth-values, even *ersatz* ones, in all SDs, and therefore we are forced to replace the classical definitions, which are stated in terms of truth-values, by something else. Not only that, we cannot assume that generalized concepts of consistency, tautology, and equivalence are interdependent and mutually definable in the way they are classically. For instance, since our language doesn't contain the negations of all of the formulas in it, we cannot define an inference in it to be p-valid if and only if the negation of its conclusion is p-inconsistent with its premises. And, given this, while the concepts of p-consistency and p-validity may be related, they cannot be as simply related as they are in classical logic. What follows will sketch hopefully plausible probabilistic characterizations of these and related p-concepts, starting with p-tautologousness, but we shall not study the consequences of these definitions with the same thoroughness with which we have studied those of our definition of p-validity.

Instead of speaking of a formula like $A \Rightarrow A$ as *logically true*, which suggests that it has truth values in all SDs, let us say that it is a p-tautology if it is certain, 'no matter what'. For instance, $(A\&B) \Rightarrow A$ is a p-tautology in this sense, since $p((A\&B) \Rightarrow A) = p(A\&(A\&B))/p(A\&B) = 1$ (and $p((A\&B) \Rightarrow A) = 1$ if $p(A\&B) = 0$), which is what justifies deriving this formula from anything at all, by rule LC, since A is a logical consequence of A&B. This generalizes as follows:

p-tautologousness condition: $\phi \Rightarrow \psi$ is a p-*tautology*, in the sense that $p(\phi \Rightarrow \psi) = 1$ for all probability functions p(), if and only if ψ is a logical consequence of ϕ.

The argument for this is trivial. If ψ is a logical consequence of ϕ then $\phi\&\psi$ is logically equivalent to ψ, and by Theorem 4 of chapter 2, $p(\phi\&\psi) = p(\psi)$, hence $p(\phi \Rightarrow \psi) = p(\phi\&\psi)/p(\phi) = 1$ (unless $p(\phi) = 0$, in which case $p(\phi \Rightarrow \psi) = 1$ by definition). On the other hand, if ψ is not a logical consequence of ϕ then according to theorem 8 of chapter 2, there must exist a probability function p() such that $p(\phi) = 1$ and $p(\psi) = 0$, and since $\phi\&\psi$ logically implies ψ, $p(\phi\&\psi) \leq p(\psi) = 0$ (by Kolmogorov axiom K3). Therefore $p(\phi \Rightarrow \psi) = p(\phi\&\psi)/p(\phi) = 0/1 = 0$. QED

P-equivalence is equally simple to characterize: *$\phi \Rightarrow \psi$ and $\eta \Rightarrow \mu$ are p-equivalent if and only if $p(\phi \Rightarrow \psi) = p(\eta \Rightarrow \mu)$ for all probability functions* p(). Then we can state the following:

p-equivalence condition: $\phi \Rightarrow \psi$ and $\eta \Rightarrow \mu$ are p-equivalent if and only if each p-entails the other. It follows that factual formulas are p-equivalent if and only if they are classically equivalent, and if they are not p-equivalent then one of them can be arbitrarily probable while the other has probability 0.

Examples of p-equivalent formulas are ϕ and $T \Rightarrow \phi$, and $A \Rightarrow B$ and

$A \Rightarrow (A\&B)$. P-equivalent nonfactual formulas are not intersubstitutable, e.g., $T \Rightarrow B$ cannot be substituted for B in $A \Rightarrow B$ because that would yield $A \Rightarrow (T \Rightarrow B)$, which isn't a formula in our language. However, they can be substituted for each other as premises or conclusions of inferences. Thus, if the inference $\{\mathcal{P}_1, \ldots, \mathcal{P}_n\} \therefore \mathcal{P}$ is p-valid and $\mathcal{P}'_1, \ldots, \mathcal{P}'_n$ and $\mathcal{P}'$ are p-equivalent to $\mathcal{P}_1, \ldots, \mathcal{P}_n$ and $\mathcal{P}$, respectively, then $\{\mathcal{P}'_1, \ldots, \mathcal{P}'_n\} \therefore \mathcal{P}'$ is p-valid. The obvious practical use of the p-equivalence condition is to simplify conditional formulas $\phi \Rightarrow \psi$ by replacing their antecedents and consequents by simpler formulas that are classically equivalent to them. For instance $(A\&(A \vee B)) \Rightarrow {\sim\sim}B$ can be simplified just to $A \Rightarrow B$, since $A\&(A \vee B)$ is logically equivalent to A and ${\sim\sim}B$ is logically equivalent to B.

But probabilistic consistency is considerably more difficult to characterize, especially because it has more than one sense. In one sense $A \Rightarrow B$ and $A \Rightarrow {\sim}B$ are consistent, since they are both certain when $p(A) = 0$, and therefore $p(A \Rightarrow B) = p(A \Rightarrow {\sim}B) = 1$.[29] But there seems to be another sense, for instance in which "If Anne goes to the party then Ben will go, and if she goes to the party then Ben won't go" would be 'absurd' even though $p(A \Rightarrow B)$ and $p(A \Rightarrow {\sim}B)$ can both equal 1, and it is important to make this sense clear. A way of doing this is as follows.

One significant thing about the *set* of formulas $\{A \Rightarrow B, A \Rightarrow {\sim}B\}$ is that it is unconfirmable because, while it is possible for neither $A \Rightarrow B$ nor $A \Rightarrow {\sim}B$ to be false, any SD that verifies one of them has to falsify the other. This implies that no 'normal' probability function, that attaches a positive probability to A,[30] can make both $A \Rightarrow B$ and $A \Rightarrow {\sim}B$ highly probable. In fact, if p() is normal in this sense then $p(A \Rightarrow B) + p(A \Rightarrow {\sim}B) = 1$, and therefore $p(A \Rightarrow B)$ and $p(A \Rightarrow {\sim}B)$ cannot both be greater than $\frac{1}{2}$. Though this isn't logical inconsistency, since $p(A \Rightarrow B)$ and $p(A \Rightarrow {\sim}B)$ both equal 1 if $p(A) = 0$, it is another kind of probabilistic inconsistency, since we can exclude the possibility $p(A) = 0$ as a practical possibility.[31] Let us give a more precise characterization of this kind of probabilistic consistency.

To begin with we will assume that the formulas we are talking about are all *normal* in the sense either that they are factual or they are conditionals with logically consistent antecedents,[32] hence the sets, $\{\mathcal{P}_1, \ldots, \mathcal{P}_n\}$, whose

[29] Again, this depends on our arbitrary stipulation that $p(A \Rightarrow B) = 1$ when $p(A) = 0$, which may be questioned. However, even theories such as that of McGee (1994), which admit 'infinitesimal probabilities' and the possibility that $p(A \Rightarrow B) < 1$ when $p(\Lambda) = 0$ (cf. appendix 2) imply that $p(\phi \Rightarrow \psi) = 1$ when ϕ is a logical contradiction.

[30] This kind of probability function is sometimes called *regular*, rather than *normal*.

[31] Recall the argument of Hume cited earlier, that it is irrational to be perfectly certain about matters of fact, e.g., as to whether Anne will attend the party.

[32] Assuming that a factual formula ϕ is equivalent to the trivial conditional $T \Rightarrow \phi$, it is trivially normal. Adams (1975) restricts its formal language and probability to normal formulas and probability functions.

consistency we are trying to characterize only contain normal formulas. Then we will say that $\{\mathcal{P}'_1,\ldots,\mathcal{P}'_n\}$ *is p-consistent if every nonempty subset of this set is confirmable.* In the end we will generalize and allow $\mathcal{P}_1,\ldots,\mathcal{P}_n$ to include nonnormal formulas, but first we can state some equivalences that help to justify our definition of p-consistency, whose proofs will be left as exercises:

Equivalents to p-consistency. If $\mathcal{P}_1,\ldots,\mathcal{P}_n$ are normal formulas then the following are equivalent:

1. $\{\mathcal{P}_1,\ldots,\mathcal{P}_n\}$ is p-consistent;
2. $\mathcal{P}_1 \bigwedge \cdots \bigwedge \mathcal{P}_n$ is p-consistent;
3. if $\mathcal{P}_1,\ldots,\mathcal{P}_n$ are all factual the set is classically consistent;
4. $\mathcal{P}_1,\ldots,\mathcal{P}_n$ does not p-entail a normal formula of the form $\phi \Rightarrow$ F;
5. defining a probability function p() to be normal if it only assigns probability 0 to logically false factual formulas, for all $\epsilon > 0$ there exists a normal probability function p() such that $p(\mathcal{P}_i) > 1-\epsilon$ for $i = 1,\ldots,n$;
6. it is not the case that $p(\mathcal{P}_1)+\ldots+p(\mathcal{P}_n) \leq n-1$ for all normal probability functions p();
7. there exists a *normal* OMP-ordering (one in which no SD is equivalent to F) in which $\mathcal{P}_1,\ldots,\mathcal{P}_n$ all hold.

Three examples illustrate the concept of p-consistency defined above, and its equivalents. These are applied to the three-premise sets $\{$C, A $\Rightarrow$ ~B, A $\Rightarrow$ B$\}$, $\{$A $\Rightarrow$ B, B $\Rightarrow$ C, A $\Rightarrow$ ~C$\}$, and $\{$A $\Rightarrow$ B, (A&B) $\Rightarrow$ C, A $\Rightarrow$ ~C$\}$, all of the members of which are entered into table 7.4, below:

probabilities	SDs	A	B	C	A⇒~B	A⇒B	B⇒C	(A&B)⇒C	A⇒~C
δ	1	T	T	T	*F*	*T*	*T*	*T*	*F*
.1-3δ	2	T	T	F	*F*	*T*	*F*	*F*	*T*
δ	3	T	F	T	*T*	*F*	...	....	*F*
δ	4	T	F	F	*T*	*F*	...	....	*T*
.9-3δ	5	F	T	T	....	...	*T*	....	....
δ	6	F	T	F	....	...	*F*	....	....
δ	7	F	F	T	....	...	...	....	....
δ	8	F	F	F	....	...	...	....	....

Table 7.4

Given the table above, the following points are easily established:

(1) The set $\{$A $\Rightarrow$ B, (A&B) $\Rightarrow$ C, A $\Rightarrow$ ~C$\}$ is p-inconsistent, since there is no line in which at least one of its members is true and none of them is false. $\{$C, A $\Rightarrow$ B, A $\Rightarrow$ ~B$\}$ is also p-inconsistent because, though

it is confirmed in line 5, where C is true but A $\Rightarrow$ B and A $\Rightarrow$ ~B are neither verified nor falsified, its nonempty subset $\{A \Rightarrow B, A \Rightarrow {\sim}B\}$ is not confirmed in any line.

(2) The set $\{A \Rightarrow B, B \Rightarrow C, A \Rightarrow {\sim}C\}$ is p-consistent, since it is confirmed in line 5, and it is easily verified that each of its nonempty subsets is confirmed in some line.

(3) The probabilities generated by the distribution in the left-hand column are normal, since all lines have positive probabilities, though the lines with probability δ come arbitrarily close to 0 as δ approaches 0. Moreover, the distribution gives the formulas in the p-consistent set $\{A \Rightarrow B, B \Rightarrow C, A \Rightarrow {\sim}C\}$ the probabilities $(.1 - 2\delta)/.1$, $(.9-2\delta)/(1-4\delta)$ and $(.1-2\delta)/.1$, respectively, which approach arbitrarily closely to 1, .9, and 1 as δ approaches 0. Exercise 2 at the end of this section asks you to construct distributions in which the probabilities of A $\Rightarrow$ B, B $\Rightarrow$ C, and A $\Rightarrow$ ~C all approach 1 arbitrarily closely.

(4) The distribution gives the formulas A $\Rightarrow$ B, (A&B) $\Rightarrow$ C, and A $\Rightarrow$ ~C the probabilities $(.1-2\delta)/.1$, $\delta/(.1-2\delta)$ and $(.1-2\delta)/.1$ respectively, which, when $\delta = .01$, sum to $.8+.125+.8 = 1.725 \leq n-1 = 2$. This illustrates equivalence (6).

(5) That A $\Rightarrow$ B, (A&B) $\Rightarrow$ C, and A $\Rightarrow$ ~C are p-inconsistent is obviously related to the fact that the restricted transitivity inference $\{A \Rightarrow B, (A\&B) \Rightarrow C\} \therefore A \Rightarrow C$ is p-valid. This generalizes in a way that more than anything else justifies defining p-consistency in this way.

p-consistency and p-validity: An inference whose premises and conclusion are normal formulas is p-valid if and only if its premises are p-inconsistent with the conditional denial of its conclusion.

Concluding the discussion of p-consistency, let us note how the definition of p-consistency can be generalized to apply to nonnormal formulas. The idea is to define an arbitrary set of formulas, $\{\mathcal{P}_1, \ldots, \mathcal{P}_n\}$, to be p-consistent if *all nonempty subsets of its normal members can be confirmed.* Given this, it follows that the set $\{F \Rightarrow A, A \Rightarrow B\}$, including the nonnormal formula F$\Rightarrow$ A, is p-consistent, since its only normal member A $\Rightarrow$ B can be confirmed, but $\{F \Rightarrow A, A \Rightarrow F\}$ is p-inconsistent since its normal member A $\Rightarrow$ F cannot be confirmed. And, generalizing the p-consistency and validity condition above, we can add that a conclusion $\mathcal{P}$ is p-entailed by any set of formulas if and only if it is p-entailed by the set of members of them that are normal, and $\mathcal{P}$ must be p-entailed if it is nonnormal.

Exercises

1. Determine which sets of formulas below are p-consistent, and for the sets that are p-consistent give probability distributions (not necessarily normal) in which all of their members have probability at least .9.
   a. $\{{\sim}A \rightarrow A\}$
   b. $\{{\sim}A \Rightarrow A\}$
   c. $\{(A \vee B) \Rightarrow {\sim}A\}$
   d. $\{A \Rightarrow B, B \Rightarrow {\sim}A\}$
   e. $\{A \Rightarrow B, (A \vee B) \Rightarrow {\sim}B\}$
   f. $\{(A \vee B) \Rightarrow {\sim}A, (A \vee B) \Rightarrow {\sim}B\}$
   g. $\{A \Rightarrow B, B \Rightarrow C, C \Rightarrow {\sim}A\}$
   h. $\{A \Rightarrow C, {\sim}A \Rightarrow C, A \Rightarrow (B\&{\sim}C)\}$
   i. $\{A \Rightarrow B, B \Rightarrow A, {\sim}A \Rightarrow {\sim}B, {\sim}B \Rightarrow A\}$
2. a. Describe distributions over the eight SDs in table 7.3 that generate probabilities for the formulas $A \Rightarrow B$, $B \Rightarrow C$, and $A \Rightarrow {\sim}C$ that approach 1 arbitrarily closely.
   b. Argue that no distribution can generate probabilities for formulas $A \Rightarrow B$, $B \Rightarrow C$, and $A \Rightarrow {\sim}C$ that are all *equal* to 1.

⋆3. In the seven equivalents to p-consistency listed below (assuming that $\mathcal{P}_1, \ldots, \mathcal{P}_n$ are normal):
   (1) $\mathcal{P}_1, \ldots, \mathcal{P}_n$ are p-consistent;
   (2) $\mathcal{P}_1 \bigwedge \cdots \bigwedge \mathcal{P}_n$ is p-consistent;
   (3) if $\mathcal{P}_1, \ldots, \mathcal{P}_n$ are all factual the set is classically consistent;
   (4) $\mathcal{P}_1, \ldots, \mathcal{P}_n$ does not p-entail a normal formula of the form $\phi \Rightarrow$ F;
   (5) for all $\epsilon > 0$ there exists a normal probability function p() such that $p(\mathcal{P}_i) > 1 - \epsilon$ for $i = 1, \ldots, n$;
   (6) it is not the case that $p(\mathcal{P}_1) + \ldots + p(\mathcal{P}_n) \leq n-1$ for all normal probability functions p();
   (7) there exists a normal OMP-ordering (one in which no atom is equivalent to F) in which $\mathcal{P}_1, \ldots, \mathcal{P}_n$ all hold.

   a. Argue that (1) implies (2).
   b. Argue that (2) implies (3).
   c. Argue that (3) implies (4).
   d. Argue that (4) implies (5).
   ⋆e. Argue that (5) implies (6).
   f. Argue that (6) implies (7).
   ⋆g. Argue that (7) implies (1).

⋆4. Comment further on Lewis Carroll's Barbershop Paradox argument given in exercise ⋆8 in section 6.3, and in particular whether the propositions ${\sim}B \Rightarrow {\sim}C$ and ${\sim}B \rightarrow C$ are inconsistent, and whether the *reductio ad absurdum* reasoning involved would be valid if they were inconsistent.

7.9⋆⋆ On Probabilistic Certainty Formulas

The formula A ⇒ F 'says' that A is certainly false, since the only way it can have positive probability is for A to have probability 0. By the same token ∼A ⇒ F expresses the fact that A is practically certainly true, i.e., it is a *probabilistic* or *p-certainty formula*, since the only way it can have any positive probability is for A to have probability 1. Let us abbreviate ∼A ⇒ F as **C**(A), and, generalizing, define **C**(A ⇒ B) $=_{df}$ (A&∼B) ⇒ F. **C**(A ⇒ B) expresses the fact A ⇒ B is probabilistically certain, and the only way it can have positive probability is for A ⇒ B to have probability 1. Note that if $\mathcal{P}$ is normal and p() is a normal probability function then p(**C**($\mathcal{P}$)) = 0; i.e., normal formulas are not normally certainties. This section will make some brief comments about the logic of these kinds of formulas.

The following facts are easy to prove:

(1) **C**($\mathcal{P}$) p-entails $\mathcal{P}$, but $\mathcal{P}$ p-entails **C**($\mathcal{P}$) only if $\mathcal{P}$ is itself a p-certainty formula.

(2) **C**($\mathcal{P}_1 \bigwedge \mathcal{P}_2$) is p-equivalent to **C**($\mathcal{P}_1$) $\bigwedge$ **C**($\mathcal{P}_2$).

(3) If $\mathcal{P}_1, \ldots, \mathcal{P}_n$ are normal and p-consistent they do not p-entail **C**($\mathcal{P}$) for any $\mathcal{P}$ that is not a p-tautology.

(4) {**C**($\mathcal{P}_1$),...,**C**($\mathcal{P}_n$)} p-entails **C**($\mathcal{P}$) if and only if the inference {$\mathcal{P}_1$, ..., $\mathcal{P}_n$} ∴ $\mathcal{P}$ is classically valid; i.e., it is valid when all of the formulas involved are replaced by their material counterparts. It follows that:
   a. if {$\mathcal{P}_1, \ldots, \mathcal{P}_n$} p-entails $\mathcal{P}$ then {**C**($\mathcal{P}_1$),...,**C**($\mathcal{P}_n$)} p-entails **C**($\mathcal{P}$), but the converse is not always true, e.g., **C**(A → B) p-entails **C**(A ⇒ B) but A → B does not p-entail A ⇒ B,
   b. but when all of the formulas involved are factual then {$\mathcal{P}_1, \ldots, \mathcal{P}_n$} p-entails $\mathcal{P}$ if and only if {**C**($\mathcal{P}_1$),...,**C**($\mathcal{P}_n$)} p-entails **C**($\mathcal{P}$).

Conditions (1) and (2) say that the p-certainty operator acts like a necessity operator,[33] except that it cannot be embedded in larger formulas like ∼**C**(A ⇒ B). Condition (3) says that nontautologous certainty formulas are not p-entailed by consistent noncertainty formulas. Roughly, normal premises do not entail the certainty of anything but tautologies. Condition (4) says that, restricted to unembedded certainties, the logic of certainty is the same as classical logic. This formalizes a point that we have made repeatedly, namely that classical logic is the logic of certainty, and in application to factual formulas it is also valid for high probability. However, it differs in application to conditionals, since while **C**(A → B) p-entails **C**(A ⇒ B), A → B does not p-entail A ⇒ B.[34]

[33] Probabilistic aspects of modality are discussed at greater length in Adams (1997).

[34] This follows from Condition (4). Clearly it depends on the stipulation that p(A ⇒ B) = 1 when p(A) = 0. Thus, given that p(A → B) = 1 − p(A) + p(A&B), hence

Another way to look at condition (4) is to say that while classical logic preserves certainty, probability logic preserves certainty and high probability, though classical logic preserves both in application to purely factual formulas.[35]

7.10 Glossary

Conditional Denial The conditional denial of $\phi \Rightarrow \psi$ is $\phi \Rightarrow \sim\psi$. This is not a 'true negation' because A $\Rightarrow$ B and A $\Rightarrow$ ~B do not entail a contradiction.

Conditional *Reductio Ad Absurdum* The law that a conclusion $\phi \Rightarrow \psi$ is p-entailed by premises $\mathcal{P}_1, \ldots, \mathcal{P}_n$ if and only if $\phi \Rightarrow$ F is p-entailed by $\mathcal{P}_1, \ldots, \mathcal{P}_n$ together with $\phi \Rightarrow \sim\psi$. This reduces to standard *reductio ad absurdum* when all formulas involved are factual.

Confirmation An assignment of truth values confirms a set of premises $\mathcal{P}_1, \ldots, \mathcal{P}_n$ if it does not falsify any of its members, and it verifies at least one of them.

p-Consistency A set of formulas is p-consistent if every nonempty subset of its normal formulas is confirmable. This is equivalent to classical consistency in the case of sets of factual formulas, and in the conditional case it is equivalent to the property that all formulas of the set can have arbitrarily high normal probabilities.

p-Equivalence Two formulas are p-equivalent if they have equal probability 'no matter what', and that is equivalent to each of them p-entailing the other.

p-Tautology A formula is a p-tautology if it has high probability 'no matter what'. That is equivalent to having probability 1 no matter what, and to being derivable from anything by Rule LC.

Probabilistic Certainty (p-Certainty) Formula The formulas $\mathbf{C}(\phi) = \sim\phi \Rightarrow$ F and $\mathbf{C}(\phi \Rightarrow \psi) = (\phi \& \sim\psi) \Rightarrow$ F 'say' that ϕ and $\phi \Rightarrow \psi$ are probabilistically certain, since the only way they can have any probability is for them to have probability 1, and that is the case only when $p(\phi) = 1$ and $p(\phi \Rightarrow \psi) = 1$. All logical certainties (logical truths) are p-certainties, but not all p-certainties are logical certainties.

Normal Formula All factual formulas are normal and conditionals are normal if their antecedents are consistent.

Normal Probability Function A probability function is normal if it attaches positive probability to all logically consistent formulas.

$1-p(A)+p(A\&B) = 1$ it follows that $p(A\&B) = p(A)$ and $p(A \Rightarrow B) = p(A\&B)/p(A) = 1$.

[35] Still other properties that are 'preserved' by different inferences involving conditionals, beyond certainty and high probability, are discussed in Adams (1996).

Probabilistic Validity (p-Validity) An inference, $\{\mathcal{P}_1, \ldots, \mathcal{P}_n\} \therefore \mathcal{P}$, with premises $\mathcal{P}_1, \ldots, \mathcal{P}_n$ and conclusion $\mathcal{P}$, is defined to be p-valid if it satisfies the **Uncertainty Sum Condition**.

Quasi-Conjunction The quasi-conjunction of $(\phi \Rightarrow \psi)$ and $(\eta \Rightarrow \mu)$ is the formula $(\phi \Rightarrow \psi) \bigwedge (\eta \Rightarrow \mu) = (\phi \vee \eta) \Rightarrow [(\phi \rightarrow \psi) \& (\eta \rightarrow \mu)]$. This reduces to ordinary conjunction when all formulas are factual (assuming that formulas of the form $\mathrm{T} \Rightarrow \phi$ are 'essentially factual'), but it is not a 'true conjunction' in other cases, since, while $\phi \Rightarrow \psi$ and $\eta \Rightarrow \mu$ p-entail $(\phi \Rightarrow \psi) \bigwedge (\eta \Rightarrow \mu)$, $(\phi \Rightarrow \psi) \bigwedge (\eta \Rightarrow \mu)$ does not always p-entail $\phi \Rightarrow \psi$ and $\eta \Rightarrow \mu$.

Uncertainty Sum Condition The inference $\{\mathcal{P}_1, \ldots, \mathcal{P}_n\} \therefore \mathcal{P}$ satisfies the uncertainty sum condition if

$$\mathrm{u}(\mathcal{P}) \leq \mathrm{u}(\mathcal{P}_1) + \ldots + \mathrm{u}(\mathcal{P}_n)$$

holds for all uncertainty functions. This condition defines **Probabilistic Validity**.

Yielding Premises $\mathcal{P}_1, \ldots, \mathcal{P}_n$ yield a conclusion $\mathcal{P}$ if any system of truth values that *confirms* $\mathcal{P}_1, \ldots, \mathcal{P}_n$ verifies $\mathcal{P}$ and any system that falsifies $\mathcal{P}$ falsifies $\mathcal{P}_1, \ldots, \mathcal{P}_n$. This relation is the basis of the truth-table test for p-validity: $\mathcal{P}_1, \ldots, \mathcal{P}_n$ p-entail $\mathcal{P}$ if and only if a subset of $\mathcal{P}_1, \ldots, \mathcal{P}_n$ (possibly the whole set, possibly the empty set) yield $\mathcal{P}$.

8★

Truth, Triviality, and Controversy

8.1 Problems, and Some History

We have already noted a number of 'anomalies' in the theory of conditionals developed in the previous two chapters, including the following: (1) Unlike unconditional statements like "Jane will not take logic" or "Jane will take ethics", statements like "If Jane doesn't take logic she will take ethics" are not held to be true or false—they are merely probable or improbable. (2) Unconditional statements, e.g., "Jane will take logic" can be 'compounded', first into "Jane will not take logic" and then into "If Jane doesn't take logic she will take ethics," but the conditional cannot enter into still larger compounds like "It is not the case that if Jane doesn't take logic she will take ethics"—or, more exactly, this cannot be symbolized in our formal theory. (3) Inferences involving these conditionals, such as "Jane will take either logic or ethics; therefore, if she doesn't take logic she will take ethics", are not held to be *valid* or *invalid*, depending on whether it is possible for the premises to be true while their conclusions are false; instead we have focussed on the vaguer question of whether it is possible for their premises to be probable while their conclusions are improbable.

We also pointed out that these anomalies are related. If a conditional like "If Jane doesn't take logic she will take ethics" does not have an ordinary truth value then the 'meaning' of a statement like "It is not the case that if Jane doesn't take logic she will take ethics" cannot be defined in the usual way, as being true if and only if "If Jane doesn't take logic she will take ethics" is false. And, if it does not have a truth-value then the ordinary definition of validity in terms of possible truth and falsehood cannot be applied to inferences involving it. We have repaired the deficiency in defining validity by characterizing it in another way when conditionals are involved, essentially by substituting *probable* and *improbable* for *true* and *false* in the 'orthodox definition'—i.e., by requiring that it shouldn't be possible for an inference's premises to be probable while its conclusion is improbable. We still have not repaired the deficiency in failing to give a precise meaning to compounds like "It is not the case that if Jane doesn't take logic she will take ethics," and we have side-stepped the problem by focussing exclusively on reasoning involving simple, uncompounded conditionals. In fact this problem will only be commented on cursorily in appendix 4.

But here we want to consider the question: do these problems arise only because so far logicians have not arrived at a generally accepted and satisfactory theory of the truth of statements like "If Jane doesn't take logic she will take ethics"? It has also been noted that the debate over this subject has had a very long history. As already noted in chapter 1, even in ancient Greece the controversy stirred up over it by Stoic logicians was widely discussed. Diodorus Cronus and his pupil Philo held two of the leading views under dispute, and, as the satirist Callimachus put it, "Even the crows on the rooftops are cawing over the question as to which conditionals are true."

Even in the 1930s there was substantial discussion of a view that was in some ways closer to the one held by Diodorus, namely C. I. Lewis' idea (Lewis and Langford, 1932) that a conditional expresses a strict kind of implication, symbolized with a special symbol ◄, according to which ~L ◄ E means that it is impossible for ~L to be true and E to be false. More recently still, the prominent logician W. V. Quine wrote "The mode of composition described in the table constitutes the nearest truth-functional approximation to the conditional of ordinary discourse" (Quine, 1958: 15), which seems to suggest that material, truth-table conditionals only *approximate* ordinary language conditionals like "If Jane doesn't take logic she will take ethics." This is supported by some of the formal results stated in the previous chapter, which show that so far as the validity of inferences is concerned, the material conditional can only be a 'bad' approximation when it is the conclusion of an inference, and then only in special cases. The following section will show that Quine was also right in claiming that the material conditional is the best truth-functional approximation of the ordinary language conditional.

8.2 Truth-functionality

A conditional, any conditional, is truth-functional if the truth-value of the whole is uniquely determined by the truth-values of its parts. How the value of the whole is determined can always be shown in a truth-table:

atomic formulas		conditionals: material	probability	undetermined	Stalnaker
L	E	L→E	L⇒E	L⟹E	L>E
T	T	T	*T*	.....	T
T	F	F	*F*	.....	F
F	T	T	.....	.....	T or F
F	F	T	.....	.....	T or F

Table 8.1

L → E is the material conditional, and the column under it is filled in with 'T's and 'F's accordingly. The 'probability conditional', L ⇒ E, has *ersatz* truth values *T* and *F* in the top two rows, and blanks in the rows in which L is false. L ⟹ E is an 'undetermined conditional', whose rows could be filled in with any one of sixteen possible combinations of 'T's and 'F's, which will be returned to immediately below. And, L > E is a nontruth-functional 'Stalnaker conditional', which will be discussed in the following section, and which has 'T or F' entered in the last two rows to indicate that even when the truth-values of L and E are fixed, the truth-value of L > E could still be either T or F. But our question for the moment is this: Assuming that the undetermined conditional L ⟹ E is truth-functional, what truth-values should be filled in in its rows? It will now be argued that the 'best' truth-values are precisely those of the material conditional.

One way to determine the 'best' combination of truth-values is by elimination, that is, by trying them out and eliminating the combinations that are 'obviously' wrong. Now, everyone agrees that if a conditional's antecedent is true and its consequent is false then the whole conditional must be false, and therefore we can eliminate all combinations that have 'T's in the second row. The material conditional fills in all the other rows with 'T's, but perhaps some other way would be better still. It is generally agreed that 'T' is the only plausible entry in the top row, because a statement like "If Jane takes logic she will take logic" cannot be false, so it must be true when both parts are true. Therefore the only cases that are undecided are those in which the antecedent is false, in the two bottom rows. In fact, however, the argument that L ⟹ E must be true when L and E are both true also implies that it must be true when they are both false, since "If Jane takes logic she will take logic" must be true when "Jane will take logic" is false. Therefore, the only thing left to determine is what the truth value of L ⟹ E ought to be when L is false and E is true. Can it be false in that case?

Suppose that after it hadn't rained and therefore it hadn't poured either, someone looked at the ground and said "If it rained it didn't pour". The antecedent of this conditional would be false and its consequent would be true in these circumstances, and we would probably say that the person who uttered the statement would have spoken the truth. If so, there must be at least one true conditional whose antecedent is false and whose consequent is true. And, if the truth is determined by the truth-values of the conditional's antecedent and consequent then *all* conditionals with false antecedents and true consequents must be true, i.e., 'T' must be filled in in the truth-table for L $\Longrightarrow$ E in the row in which L is false and E is true. Thus, we seem to have established the 'rightness' of the material truth values for the ordinary language conditional, for all combinations of truth-values of its parts.

But the foregoing presupposes that the truth of a conditional is determined by the truth-values of its parts, more or less in the way that the truth of negations, disjunctions, and conjunctions is determined. That certainly simplifies logic, by making it possible to use truth-tables to determine the validity of any inference whose premises and conclusion are truth-functional. But it begs the fundamental question: Is the ordinary language conditional truth-functional?

8.3 Truth-conditionality: Stalnaker's Theory

Not all compound sentences are truth-functional. For instance, the truth of "Jane will take logic after she takes ethics" depends not only on Jane's taking logic and ethics but also on the order in which she takes them. Therefore we cannot determine the truth of "Jane will take logic after she takes ethics" just by knowing that she will take both logic and ethics, and we would not know how to fill in the top row in a truth-table for this sentence:

atomic		temporals	
L	E	L after E	E after L
T	T	T or F	T or F
T	F	F	F
F	T	F	F
F	F	F	F

Table 8.2

Now, it is easy to see that neither Diodorus Cronus' nor C. I. Lewis' theories of conditionals are truth-functional. The theory that L strictly implies E, L◀E, is true if it is not possible for L to be true and E to be false implies that "If today is Friday then today is Friday" is true because, although its antecedent and consequent may both be false, it is not possible for one to be true and the other false. On the other hand, although both parts of "If today is Friday then we will have fish" may be false, the whole statement could be false because it is possible for today to be Friday without our having fish. But the fact that the strict conditional is not truth-functional doesn't mean that it is not logical. In fact, assuming that we can be clear about the 'possibilities' in terms of which Lewis defines strict implication, his theory is still truth-conditional in the sense that the conditions under which statements of the form L ◀ E are true are precisely defined, and therefore it can be a subject of precise logical study.[1]

But being precise and truth-conditional isn't necessarily being right, and, because there is more than one truth-conditional theory of conditionals, the question arises: Which, if any, is right? We will conclude this section by sketching the outlines of one such theory that looks as 'right' as any theory of this kind can possibly be.

What is wrong with Lewis' strict implication as a theory of the meaning of conditionals like "If Jane doesn't take logic she will take ethics" is not that it isn't truth-functional or that it isn't precise, but that it seems to make the truth of these statements matters of pure logic. We feel strongly that the rightness of the statement should depend on whether, as a matter of fact, Jane has to take either logic or ethics. Now, in a paper "A theory of conditionals", R. Stalnaker (1968) developed a truth-conditional theory akin to Lewis' that is able to incorporate the element of factuality that Lewis' theory lacks, and which has other desirable features as well. What follows gives a very brief and rough sketch of the theory.

We begin, as in Lewis' theory, by considering a 'universe of possible worlds' among which our 'actual world', w, is one.[2] For instance, Jane might end up taking logic and not ethics in the actual world, but there could have been a world, say w′, in which she ended up taking ethics and not logic. Moreover—and this is crucial to Stalnaker's theory—certain possible worlds are *closer* to our world than others. Thus, world w′ in which Jane ends up taking ethics but not logic might be closer than one in which she ends up taking neither ethics nor logic. Furthermore, what is closer may depend on 'the facts', e.g., that Jane had to take at least one of logic or ethics. This is pictured geometrically in diagram 8.1.

Here the actual world is pictured as point labelled w, in the 'space' of worlds or points that are located at different places in the 'universe' inside the diagram, in which the circles L and E contain the worlds in which Jane

---

[1]In fact, strict conditionals have been studied intensively, first by C. I. Lewis himself (cf. Lewis and Langford 1932), then by many others. The most important work along this line has been that of S. Kripke (1963; 1965) which establishes the foundations of the 'possible worlds' approach to possibility that is exploited by Stalnaker (1968) and D. Lewis (1973) in the nontruth-functional but truth-conditional theories of conditionals that will be discussed briefly later in this section.

[2]The 'Leibnizian' construal of Lewis' theory in terms of possible worlds is most prominently developed in the work of S. Kripke cited in the previous footnote.

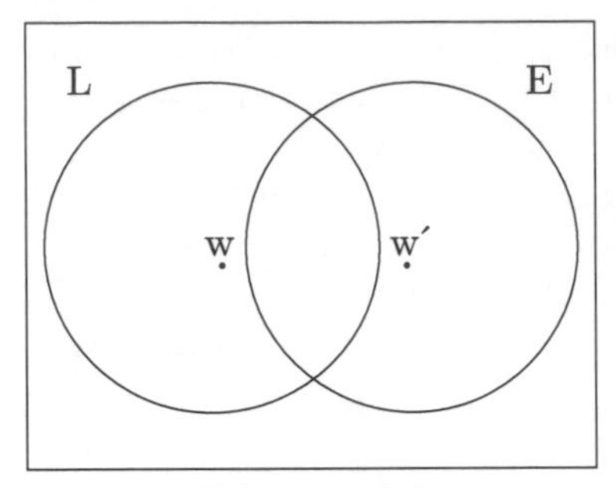

Diagram 8.1

takes logic and in which she takes ethics, respectively. The significant thing is that while Jane is pictured as taking logic and not ethics in the actual world, she takes ethics and not logic in the world w′, which is closer to the actual world than any world outside of both circles, in which she takes neither logic nor ethics. Now we are in a position to account for conditionals like "If Jane doesn't take logic she will take ethics."

"If Jane doesn't take logic she will take ethics" turns out to be true according to Stalnaker's theory, not because there is no world in which she takes neither logic nor ethics, but because *she takes ethics in the world nearest to the actual world in which she doesn't take logic.* That is just what Diagram 8.1 shows, since the world nearest to world w in which Jane doesn't take logic is on the edge of region L, next to w′, and that is in the class of worlds in which she takes ethics. This also accounts for many of the facts that we have already noted, concerning the validity of inferences involving conditionals.

We have argued that "If Jane doesn't take logic she will take ethics" ought not to follow from "Either Jane will take logic or she will take ethics". Diagram 8.2 represents a situation in which "Either Jane will take logic or she will take ethics" is true in the actual world, w, because she takes logic in w, but "If she doesn't take logic she will take ethics" is false in w because she doesn't take ethics in the world nearest to w in which she doesn't take logic. Hence the inference "Either Jane will take logic or ethics; therefore, if she doesn't take logic she will take ethics" is invalid in 'Stalnaker logic' because there is a possible world in which its premise can be true while its conclusion is false.

The well-known fallacies of material implication also turn out to be invalid in Stalnaker's theory, since diagram 8.2 pictures "Jane will take logic" as true while "If Jane doesn't take logic she will take ethics" is false. Similarly, "Jane won't take ethics; therefore, if she doesn't take logic she won't take ethics" is invalid in diagram 8.1 because its premise is true in world w but its conclusion is false in that world, since Jane does take ethics

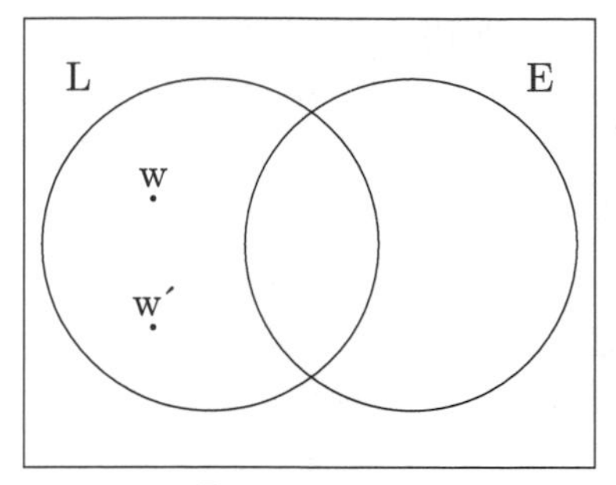

Diagram 8.2

in the world nearest to w in which she doesn't take logic.

Thus, Stalnaker's theory agrees with the probabilistic theory in its evaluation of certain inferences, but it does not take the radical step of redefining validity in terms of probability and improbability rather than in terms of truth and falsity. While not truth-functional, this theory is truth-conditional, which also means that it does not run into problems defining logical compounds like negations of conditionals. Even more, it can be proved that in its application to inferences involving noncompounded conditionals, the theory yields the same results as the probabilistic theory.[3] We seem to get the same gain at far less cost! The next section points out that there is still a problem, however, but first the reader is urged to attempt the following exercises in applying the Stalnaker theory.

Exercises

1. For each of the inferences below, give a diagram like diagrams 8.1 and 8.2 in which the classes of worlds in which propositions A and B are true are represented by regions and possible worlds are represented by points, and in which the premises of the inferences are pictured as true according to Stalnaker's theory, while their conclusions are false.
   a. $B \therefore A \Rightarrow B$
   b. $\sim(A \Rightarrow B) \therefore A$
   c. $A \leftrightarrow B \therefore A \Rightarrow B$
   d. $A \Rightarrow \sim B \therefore B \Rightarrow \sim A$
   e. $\{A \Rightarrow B, \sim A \Rightarrow \sim B\} \therefore B \Rightarrow A$

⋆2. For each of the following inferences, either argue that it is valid according to Stalnaker's theory or give a diagram like diagrams 8.1 and 8.2 in which the classes of worlds in which propositions A and B (or A, B, and C in the case of inferences involving all three propositions) are true are represented by regions and possible worlds are represented

[3]This is proved in Adams (1977) and in Gibbard (1981).

by points, and in which the premises of the inferences are pictured as true according to the theory, while their conclusions are false.

a. $A \Rightarrow \sim B \therefore \sim(A \Rightarrow B)$

⋆b. $\therefore (A \Rightarrow B) \vee (A \Rightarrow \sim B)$[4]

c. $(A \Rightarrow B) \Rightarrow B \therefore \sim A \Rightarrow B$

d. $A \Rightarrow C \therefore (A\&B) \Rightarrow C$

e. $(A \vee B) \Rightarrow C \therefore A \Rightarrow C$

f. $A \Rightarrow (B \Rightarrow C) \therefore (A\&B) \Rightarrow C$

g. $(A\&B) \Rightarrow C \therefore A \Rightarrow (B \Rightarrow C)$

8.4 The Problem of Probability

While the Stalnaker theory has the same consequences as the probabilistic theory so far as the validity of inferences is concerned, at least when the inferences don't involve compounds of conditionals, the probabilities assigned to conditionals by it are not conditional probabilities, and therefore it disagrees with the probability conditional theory at the level of probability. The difference between the probabilities of Stalnaker conditionals and conditional probabilities is made obvious in comparing the probability of the Stalnaker conditional $L > E$ with the probability of E given L in diagram 8.3.

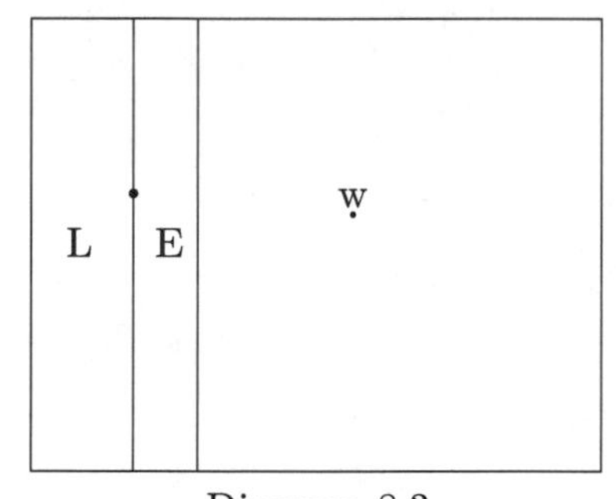

Diagram 8.3

Here L and E correspond to vertical strips whose only common points are on the boundary between them, and the proportion of L that lies inside E, which corresponds to the conditional probability of E given L, is infinitesimally small. On the other hand, the worlds in which $L > E$ is true are ones like w, which are such that the nearest world to w in the L-strip,

[4]The formula $(A \Rightarrow B) \vee (A \Rightarrow \sim B)$ expresses what Stalnaker calls the *law of the conditional excluded middle*, whose validity in application to *counterfactual conditionals* is a point in dispute between himself and David Lewis. Cf. the papers of Lewis and Stalnaker in Harper, Stalnaker, and Pearce (1981); also see appendix 6 for comments on counterfactual conditionals.

corresponding to the dot on the border of L, are in E, and these include all worlds except those in the interior of the L-strip. But these are almost all worlds, and therefore the probability of L > E being true, i.e., the probability of being in a world in which it is true, is very close to 1.

It looks as though we have to choose: Do we accept probability conditionals and give up truth-conditionality, with all that entails, or do we accept Stalnaker conditionals and reject the thesis that the probability of a conditional is a conditional probability? Of course, we may say that there are two kinds of conditionals, probability conditionals symbolized, for instance, as ~L ⇒ E, and Stalnaker conditionals symbolized as ~L > E, which are subject to the same 'logic' as far as the validity of inferences is concerned, but which have different probabilities. The real question is: What kind of conditional is an ordinary language statement like "If Jane doesn't take logic she will take ethics?" This question will be taken up in the following chapter, when we consider how the probability of such a statement ought to be measured. For now, however, we will concentrate on a prior question.

Are probability conditionals and Stalnaker conditionals the only options? Have we ruled out the possibility of 'having our cake and eating it' by defining a truth-conditional conditional, like Stalnaker's, but agreeing with the probability conditional on probabilities? Now we will see that the answer is "no," and this follows from certain of the most important formal 'results' of modern probability logic.

8.5 Triviality Results

The original and most famous *triviality results* were derived by David Lewis in a celebrated paper "Probabilities of conditionals and conditional probabilities", Lewis (1976). However since this paper appeared alternative proofs have been given, often of slightly varying results, and here we will give a version due to A. Hájek (1994) that is somewhat weaker but which is considerably easier to follow (Lewis' own argument is set forth in detail in appendix 3).

There is one fundamental respect in which the probability conditional and the Stalnaker theories differ. That is that, as we have supposed, p(L > E), the probability of a Stalnaker conditional, is assumed to be the probability of being in a world in which L > E is true, whereas the probability of L ⇒ E is assumed to be the ratio of the probability of being in a world in which L and E are both true to the probability of being in one in which L is true. Pictured as in diagram 8.3, the probability of L > E is the area of the region of worlds like w whose nearest point in L is in E, while the probability of L ⇒ E is the proportion of region L that lies inside region E. The probability of a truth-conditional conditional is pictured as an *area* while that of a probability conditional is pictured as a *ratio of areas.* But assuming that these are pictures of the same thing leads to triviality results, as follows.

Suppose now that $L \Longrightarrow E$ is any truth-conditional conditional, not necessarily the Stalnaker conditional, whose probability necessarily equals the probability of being in a world in which it is true, and which also equals p(L&E)/p(L). The class of worlds in which $L \Longrightarrow E$ is true could then be represented by another region such as the circle labeled $L \Longrightarrow E$ in diagram 8.4.[5]

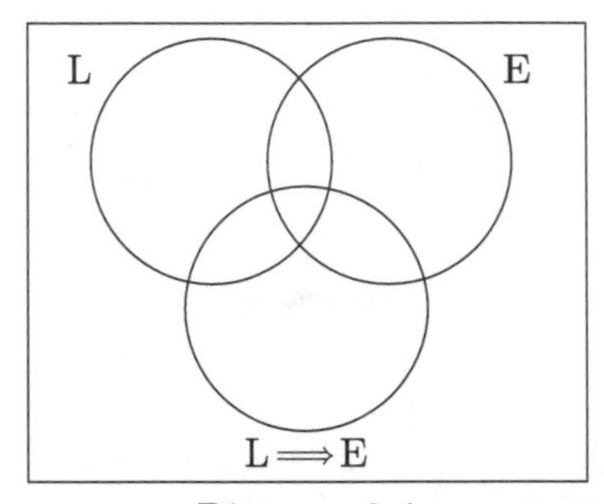

Diagram 8.4

But if $p(L \Longrightarrow E)$ necessarily equalled p(L&E)/p(L) then the area of the $L \Longrightarrow E$ region should equal the proportion of the area of L that lies in E. But the following consideration suggests that this isn't very plausible.

It is not implausible to assume there are only a finite number of possible worlds in our universe,[6] thickly distributed like pixels on a computer screen,

[5]If $L \Longrightarrow E$ were the Stalnaker conditional, hence true in worlds w such that the nearest L world to w was an E world, and nearness was measured geometrically, then the region of worlds in which $L \Longrightarrow E$ was true would be the round-nosed, cone-shaped region shaded in the diagram below, consisting of points such that the nearest point to them in the L circle was in the E circle. But diagram 8.4 neither assumes that $L \Longrightarrow E$ is a Stalnaker conditional nor that distances between worlds are measured geometrically.

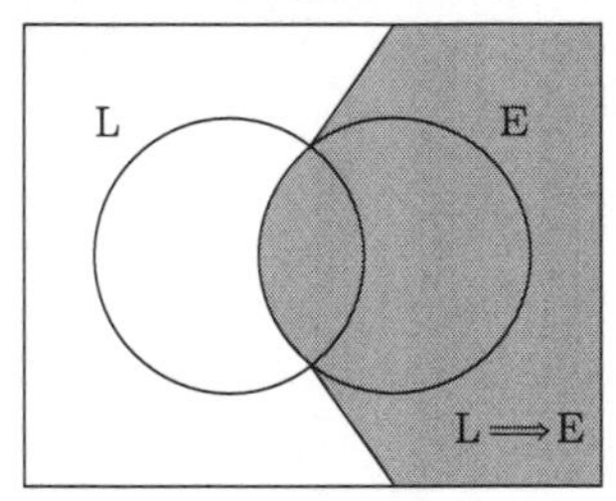

[6]Actually, 'possible worlds theorists' never do assume this, but this is not essential to the present argument.

and that the area of a region is determined by the number of dots or pixels it contains. Now, suppose there are a million pixel-worlds in all, represented by a million tiny dots inside the diagram, and regions L and E each contain 300,000 dots, while their intersection contains 100,000 dots. In that case $p(E/L) = p(L\&E)/p(L)$ should equal $100,000/300,000 = 1/3$, and if that equalled the probability of $L \Longrightarrow E$ then the region corresponding to $L \Longrightarrow E$ should contain exactly 1/3 of the million dots or pixel-worlds. But no region contains exactly 1,000,000/3 pixel-worlds, so $p(L \Longrightarrow E)$ *couldn't* correspond to an area and equal $p(L \Longrightarrow E)$.

What the foregoing suggests is that calling ratios of probabilities, like $p(L\&E)/p(L)$, 'probabilities' may be a mistake in terms, like calling a ratio of areas an area. Or, as David Lewis puts it, conditional probabilities, which are ratios of probabilities are 'probabilities in name only'. The next chapter will consider how 'right' probabilities *should* be determined, and return to the question of whether conditional probabilities are rightly called 'probabilities'. But we will conclude this section by making a preliminary comparison between the foregoing argument and David Lewis' triviality argument, which will be set forth in detail in appendix 3.

Lewis' argument does not assume that there are only a finite number of possible worlds, but it ends up with the absurdity that *if the probability of a truth-conditional conditional,* $p(L \Longrightarrow E)$, *always equalled* $p(L\&E)/p(E)$, *then there could be at most four possible probability values!* That is what triviality means: While trivial probabilities might satisfy the formal laws of probability, and the Kolmogorov axioms in particular, they would exclude 'real' probabilities like the chances of drawing 0, 1, 2, 3, or 4 aces in a row from a shuffled deck of cards, which are generally assumed to be 12/13, 1/13, 1/221, 1/5525, and 1/270725, respectively—five different values.

The foregoing seems to suggest that we cannot have our cake and eat it, and we have to choose. If we accept the conditional probability thesis we have to give up truth-conditionality, and if we accept truth-conditionality we have to give up the conditional probability thesis.[7] The choice is not easy because there are strong reasons for accepting each. To suppose that there is a 1 in 17 chance that a second ace will be dealt if one ace has already been dealt implicitly assumes the Conditional Probability Thesis. But if we accept that and give up truth-conditionality we can't say that "If the first card is an ace then the second card will also be an ace" is true or false in the standard 'logical' sense of truth and falsity. Nor can we claim

[7] Actually, though persons who are concerned with this subject tend to assume that David Lewis's triviality arguments prove the logical incompatibility of the conditional probability thesis and truth-conditionality, other possibilities exist. Appendix 3 outlines one such possibility, and appendix 4 describes another one due to McGee (1989) which has been widely debated recently. The final pages of section II.2 of Adams (1975) compare and contrast four attitudes to the triviality arguments.

that "It is not the case that if the first card is an ace then the second card will also be an ace" must be true or even clearly probable.

The final section of chapter 9 will return to the 'triviality theme', bringing pragmatic considerations to bear on it, but we will end this one with comments on a deeper problem, concerning the 'very idea of truth'.

8.6⋆ The Controversy about Truth: Theories of Truth

What *is* truth? In some ways the answer is right in front of us, and in other ways it is as distant as the far side of the moon. Who does not know how to determine whether someone who says "The first card will be an ace" has spoken the truth? Who does know how to determine whether someone who says "If the first card is an ace then the second card will also be an ace" has spoken the truth? These matters are as much debated among philosophers today as they were when Francis Bacon wrote his celebrated essay "Of Truth" 370 years ago.[8] However, conditionals like "If the first card is an ace then the second card will also be an ace" put the problem of truth in a special light.

One theory of truth holds that the way to determine the truth of a statement is to 'look at the facts'. This is a *correspondence theory of truth*,[9] and it means, for example, that to determine whether "The first card will be an ace" is true, one looks to see what the first card is. It is 'factually true' if it *is* an ace and factually false if it is something else, like a king. This is the sort of truth that science is often said to be concerned with, and it stresses the idea that for a statement to be true there must be something in the observable world that makes it true.

But that doesn't seem to be the case with the statement "If the first card is an ace then the second card will also be an ace." What 'facts' can we look at to determine whether this statement is true, if the first card is, say, a king? It might be said that we can examine the deck and see what the second card would have been, if the first card had been an ace. But this is very doubtful, since if the first card had been an ace and not a king, the second card might have been the king.[10] And, it is significant that

[8]The first sentence of this essay is often quoted and worth repeating. It goes: "What is truth? said jesting Pilate, and would not stay for an answer." Pilate is, of course, Pontius Pilate, the Roman Procurator of Judea, who 'washed his hands' of the case against Jesus, thus signifying his acceptance of the demand by the ecclesiastical authorities of Jerusalem for Jesus' crucifixion (Gospel according to Matthew, 27:24).

[9]Cf. chapter 6 of Schmitt (1995).

[10]The thesis that the truth of "If the first card is an ace then the second card will also be an ace" can be determined 'retrospectively' by considering what would have been the case if the next card had been an ace really assumes that the subjunctive "If the first card had been an ace then the second card would also have been an ace" is a kind of past tense. This is debatable, but in any case, what determines the truth of the subjunctive is no more obvious than what determines the truth of the indicative conditional it is

experimental science does not generally concern itself with hypothetical propositions.

Now, other theories of truth have been advocated forcefully over the past century, including a so called *coherence theory*,[11] and a *pragmatic theory* that will be commented on briefly in the following chapter, but there is one that has been very widely held in recent years, and which merits special comment here because of the difficulties it has in application to conditional statements. This is the so called *semantic conception of truth* that was brought into prominence by the famous logician Alfred Tarski, in a celebrated paper "The concept of truth in a formalized language" (1956).[12] The most fundamental theme of Tarski's approach is that however truth is defined, the definition should entail *statements of Form T*, i.e., statements of the form:

(T) S is true, if, and only if p.

where S 'names' an assertive sentence, and p is the sentence it names. An example might be:

(T1) "The first card will be an ace" is true, if, and only if the first card will be an ace.[13]

---

allegedly a past tense of.

[11]Very roughly, this is the theory that true statements 'cohere' with one another, and false ones do not. Cf. chapter 4 of Schmitt (1995).

[12]Originally published (in Polish) as "O pojęciu prawdy w odniesieniu do sformalizowanych nauk dedukcyjnych" ("On the notion of truth in reference to formalized deductive sciences") in *Ruch Filozoficzny* (1930-1); German translation in Tarski (1936). See also the informal paper, Tarski 1944. Among recent works applying Tarski's fundamental ideas to natural languages, see McGee (1991).

[13]This would normally be stated as

(T1′) "The first card will be an ace" is true, if, and only if the first card is an ace.

in which there is a slight difference between the 'quotation name' on the left, " 'The first card will be an ace' " and the sentence it names, "The first card is an ace," on the right. Tarski's theory was meant to apply to sentences in languages like those of formal logic that don't involve differences of tense, like the difference between 'will be' and 'is'. Other logicians, especially Donald Davidson and his students, have attempted to apply Tarski's 'program' to natural languages, but the application involves difficulties, one of which is its application to conditionals, to be noted below.

Whether formalized or not, the difference between the expression with the quotation marks

"The first card will be an ace,"

and the expression without them

The first card will be an ace

is subtle but extremely important to Tarski's theory (or rather, to its application to natural languages). Placing quotation marks around an expression turns it into a name (more exactly, a description) of the expression that is quoted, which implies that what is named is a 'thing' with such and such properties, including truth and falsehood.

This truism hardly seems worth stating, and yet convention T, which requires that a definition of truth should entail it, has been violated surprisingly often.[14]

But now consider how convention T applies to the conditional:

(T2) "If the first card is an ace then the second card will also be an ace" is true, if, and only if, if the first card is an ace then the second card will also be an ace.

Assuming that (T2) is entailed by a definition it ought to be logically true, and therefore it should follow from the Kolmogorov axioms that:

p("If the first card is an ace then the second card will also be an ace" is true)
= p(If the first card is an ace then the second card will also be an ace).[15]

The quantity on the left side of this equation, namely p("If the first card is an ace then the second card will also be an ace" is true), is the probability that whatever the expression " 'If the first card is an ace then the second card will also be an ace' " names has the 'property' of being true. And, that probability must be the sum of the probabilities of the possible worlds in which the thing named has that property. But according to the triviality results that cannot be a conditional probability, and if it necessarily equals p(If the next card is an ace then the following card will be too) that cannot be a conditional probability either. Conversely, *if* the conditional probability thesis is true then it follows from the triviality results that the semantic conception of truth, and in particular convention T, cannot be right, at least in its application to conditional statements.[16]

This is not the place to elaborate on the consequences for philosophy of accepting the conditional probability thesis, and by implication giving up the semantic conception of truth, but it should be clear that very fun-

[14]For example, Bertrand Russell criticized William James' pragmatic theory because it seemed to suggest that a statement like "The first card will be an ace" could be true even though the first card wasn't an ace—because it might be better to believe it than to believe its contrary (Russell, 1945: 817–8).

[15]The assumption that statements that are equivalent in virtue of the meanings of the terms in which they are stated are equiprobable is a generalization of theorem 4 of section 2.3, which, as originally stated, only applies to statements of formal logical languages

[16]But this doesn't mean that another theory which has often been confused with it cannot be applied to conditionals. That is the so called *redundancy theory* formulated by F. P. Ramsey (1927), which says, for instance, that the "It is true that" in "It is true that if the first card is an ace then the second card will also be an ace" is, logically, a redundant addition to "If the first card is an ace then the second card will also be an ace." The essential difference is that Tarski's theory holds that truth is a property of the things it applies to, whereas Ramsey's theory holds that adding "It is true that" to an affirmation is only a way of 'endorsing' it, and that is not ascribing a property to it.

damental issues are at stake.[17] For this reason, the status of the thesis is controversial as of this writing. David Lewis may have described the present situation best, when he wrote in his paper "Probabilities of conditionals and conditional probabilities":

> I have no conclusive objection to the hypothesis that indicative conditionals are nontruth-valued sentences, governed by a special rule of assertability that does not involve their nonexistent probabilities of truth. I have an inconclusive objection, however: The hypothesis requires too much of a fresh start. It burdens us with too much work to be done, and wastes too much that has been done already.

We have not, here, tried to settle the issue between truth-conditionalism and conditional probabilism; we have only tried to point out some of the things that are at stake in it. The following chapter introduces another factor in the controversy.

8.7 Glossary

Correspondence Theory of Truth The theory that the truth of propositions consists in their corresponding with 'facts' about the world.

Coherence Theory of Truth The theory that the truth of propositions consists in their agreeing with or 'cohering' with accepted propositions.

Diodorus Cronus' Theory of Conditionals The theory of the Stoic logician Diodorus Cronus (d. 307 B.C.) that "it neither is nor ever was capable of having a true antecedent and a false consequent."

Hayek's Triviality Result Due to Alan Hayek; roughly, that if there are only finitely many possible worlds then the thesis that the probability of a conditional is a conditional probability is inconsistent with the assumption that it is the probability of being in a world in which it is true.

Pragmatic Theory of Truth Roughly, that the truth of propositions consists in their 'usefulness'; associated with philosophies of Charles Sanders Peirce, William James, and John Dewey.

Semantic Conception of Truth Associated with the philosopher-logician Alfred Tarski. The key 'thesis' of Tarski's theory is that, however truth is defined, the definition should entail all 'sentences of form T':

S is true, if, and only if p,

[17]For instance, issues related to theories of knowledge that hold some version of the view that knowledge is justified true belief. If a conditional like "If the first card is an ace then it is an ace" can't be true in an 'ordinary' sense, then according to such theories it cannot be known, nor can belief in it be justified if justification consists in having good reason for thinking it is true.

where S 'names' an assertive sentence, and p is the sentence it names.

Stalnaker's Theory of Conditionals Due to R. Stalnaker; roughly, that a conditional "If P then Q" is true if Q is true in the nearest possible world in which P is true.

Strict Conditionals Conditionals "If P then Q" that are true if and only if it is not possible for P to be true and Q to be false. Associated with philosopher-logician C. I. Lewis.

Triviality Results Mathematical results tending to show that the thesis that the probabilities of conditionals are conditional probabilities is inconsistent with generally accepted 'hypotheses' about probabilities.

Truth-Conditionality A sentence is truth conditional if the conditions under which it is true are precisely defined.

Truth-Functionality A sentence that is formed from parts that are sentences is truth-functional if its truth-value is determined by the truth-values of its parts.

9

Practical Reason

9.1 Introduction

This chapter brings us back to the practical side of reasoning, which was originally commented on in chapter 1, but which is not generally considered in courses in formal logic or in writings on probability logic. This has to do with the effect that arriving at conclusions can have on the decisions and actions of the persons who arrive at them. This was illustrated in chapter 1 in the case of a student—a 'decision maker' or 'DM'—who reasons "Either Jane will take logic or she will take ethics, and she won't take ethics; therefore, she will take logic", and who then, wanting to take the same class as Jane, decides to take logic. The example showed the practical importance of arriving at the truth, since if the DM's conclusion that Jane would take logic was right then acting on it by taking logic should result in his attaining what he wants, namely to be in the same class as Jane, while if it was wrong he would not attain it because he would take logic while Jane took something else. This is something that is not considered in 'pure reason', which generally considers that whatever truth is, persons should aim at it for its own sake, apart from the practical advantages that may yield.

But the last chapter raised the question of what truth is in the case of statements like "If the first card is an ace then the second card will also be an ace", and this chapter will suggest that the fundamental difference between propositions like that and 'factual' propositions like "Jane will take logic" has to do with the practical advantage of being right about them. It is practically advantageous that the factual proposition should 'correspond to the facts', but a more subtle 'probabilistic correspondence' determines the advantage that may accrue from arriving at a conclusion of conditional form. Moreover, when we examine this we come to understand more clearly the nature of the probabilities that define the correspondence. Until now these have simply been 'probability values', whose properties, such as that of satisfying the Kolmogorov axioms, we have taken for granted, much as we have taken for granted the properties of the truth-values that pure formal logic theorizes about.

To arrive at an understanding of the probabilistic correspondences that determine the 'utility values' of conclusions like "If the first card is an ace then the second card will also be an ace" it is most convenient to proceed as follows. (1) We will begin by considering how opinions as to matters of fact, e.g., that Jane will take logic, affect actions, and how the success of these actions may depend on the rightness of the opinions. These matters will be discussed in sections 9.2–9.4. (2) Next we will consider how a DM's *degrees of confidence* in his opinions as to matters of fact affect his actions. This will be discussed in sections 9.5–9.7. (3) Section 9.8 generalizes to consider how the DM's degrees of confidence in opinions expressed *conditionally*, e.g., that if Jane doesn't take logic she will take ethics, affect his actions. (4) Finally, section 9.9⋆⋆ considers what degrees of confidence in either factual or conditional opinions would be 'best', judged in terms of the success of the actions to which they lead, and how these are related to probabilities. Let us make some preliminary comments before entering into details.

One point is that in considering how degrees of confidence, as contrasted with 'all or nothing beliefs', affect actions, we will draw on currently widely discussed theories of decision making under uncertainty that are often applied as adjuncts to the theory of games and in the behavioral sciences, particularly economics and psychology. Extending this to decisions based on degrees of confidence in conditional propositions, we will apply Richard Jeffrey's *Logic of Decision* (Jeffrey, 1983)—though not in a way that Jeffrey himself envisages. It should be said that this discussion only requires us to sketch the most rudimentary aspects of the concepts and formalism of theories of decision making under uncertainty and of Jeffrey's extension, since the detail that is usual in texts on these subject is not necessary for present purposes.

Final comments pertain to the topic discussed in section 9.9⋆⋆, concerning the practical utilities of the degrees of confidence whose influence on actions is considered in sections 9.5–9.8. There it will be suggested that 'real probabilities', like the 50% chance that a coin that is about to be tossed will fall heads, are the 'facts' to which subjective degrees of confidence, which act like probability estimates, should correspond, and something like this also applies to degrees of confidence in propositions like "If the first card is an ace then the second card will also be an ace" (however, section 9.5 points out something simpler that 'ideal' degrees of confidence in factual propositions can correspond to). But these ideas, or better, 'reflections', due entirely to the author, are highly unorthodox, as is the account given here of the 'real probabilities' to which degrees of confidence should correspond. For this reason, section 9.9⋆⋆ is the most 'philosophical' and speculative of this book, and the reader can omit it without missing anything that is currently accepted by workers in the field of probability logic.

Now we turn to the utility of being right in factual opinions, and begin by considering in more detail the case of a DM acting on the conclusion "Jane will take logic."

9.2 Practical Inference: A Qualitative Representation

Diagram 9.1 below pictures relations between various aspects of the DM's practical reasoning, in which his 'premises' are his belief or thought, that Jane would take logic, and his want or desire to take the same course that she does, and whose conclusion is the action of taking logic:

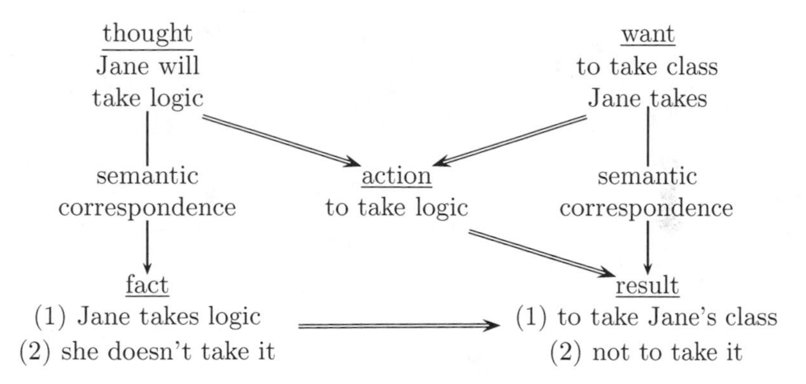

Diagram 9.1 Pragmatic Diagram

The 'decision picture' above involves the following five 'aspects' or 'factors': (1) *The thought* (conclusion, upper left) that Jane will take logic, (2) *The want* (desire, upper right) to take the same class as Jane, (3) *The action* (decision, center) to take logic,[1] (4) *the fact* (lower left) of what Jane actually does, either to take logic or not to, and (5) *the result* (consequences lower right), which is either for the DM to be in Jane's class, or for him not to be in it. These factors are connected in two kinds of ways: *causally*, as shown by double-barred arrows, and *semantically* as shown by single-barred arrows. There are two connections of each kind.

The first causal connection is between three things: (1) *the thought*, (2) *the want*, and (3), *their 'resultant'*, namely the action. This is sometimes called a *practical inference*, or *practical syllogism*, but it is not what is usually regarded as an inference in pure logical theory, since neither the

[1]This assumes that there is no 'gap between decision and action', and that deciding to take logic automatically leads to taking it. But for a variety of reasons this may not be so, and the problem of filling the gap in a realistic way is a serious one, which we will not try to deal with here.

want nor the action is a proposition that can be true or false or more or less probable.[2] Nevertheless it is the result of a mental process, and it can be regarded as more or less 'rational'. Most importantly for our purposes, the connection between the thought and the action is what gives the thought the practical importance that it has.

The second causal influence also involves three things: (3) *the action*, (4) *the facts*, and (5), *the result*, which is jointly caused by the action and the facts.[3] Like the action, the result can be regarded as a resultant of two *forces*, the action itself, which was the resultant of a belief and a desire, and 'the facts'.

In addition to the two causal processes, the diagram also pictures two semantic correspondences, which are represented by vertical single-barred arrows. One is between what the DM thinks and 'the facts', and this determines whether the thought is *true*. If the thought is that Jane will take logic then it is true if she *takes* logic.[4] The second correspondence is between what the DM wants and the result, and this determines whether the action succeeds. If the DM wants to be in the same class as Jane and he ends by *being* in that class, then he gets what he wants.[5] This is an instance of a *pragmatic principle* that is of fundamental importance in motivating the application of 'principles of right reason'.

9.3 A Pragmatic Principle

Although diagram 9.1 represents the DM's beliefs as corresponding to the facts, they don't necessarily correspond to them, and when they don't the actions to which they lead don't have the results desired. For example, signing up for logic will not result in the DM getting what he wants, namely to be in the same class as Jane, if what he believes, namely that Jane will take logic, doesn't correspond to the facts. This conforms to the following general principle:

Qualitative Pragmatic Principle Consequences of actions correspond to what DMs want if the beliefs they act on correspond to the facts, and

[2]Practical syllogisms or inferences are widely discussed in the current theories of action (cf. Davidson (1980) especially Essays 1 and 6). Although they are construed here as purely irrational phenomena, which do not involve reason or calculation, this will be reconsidered when we generalize to allow for degrees of confidence and of utility values.

[3]It might be said that being in the same class as Jane is the result purely of the DM's taking logic, but to be accurate this is produced by the DM's taking logic *and* Jane's taking it.

[4]This is a version of convention T, which is central to the semantic conception of truth commented on in section 8.6⋆, and which will be discussed further in the next section.

[5]We could call this a 'semantic conception of success', but it is important to stress the time relation between the want, which comes earlier, and the outcome that follows later and determines success. It is quite possible that the want should change in the interval, and the DM might arrive at an outcome that she or he no longer wanted.

not if they don't. More briefly: actions succeed if the beliefs they are based on are true, and not if they are false.[6]

Three points are important to note concerning this principle, the first of which has to do with the relation between it and the theories of truth that were commented on in section 8.6⋆. The principle is related to the pragmatic theory of truth, a rough version of which is that true propositions are ones that it is 'beneficial to believe'. This is sometimes attributed to Charles Sanders Peirce, but William James and John Dewey also maintained versions of the doctrine, which differ in their conceptions of what is beneficial to believe, and, as noted in footnote 14 of the previous chapter, James' version violates convention T. However, the qualitative Pragmatic Principle is not a theory or definition of truth; in fact, it actually presupposes convention T. To say, e.g., that the DM's action of taking logic attains his aim of getting into Jane's class if his belief that Jane will take logic is true, presupposes that "Jane will take logic" is true if and only if Jane *takes* logic—convention T.[7] By the same token, the pragmatic principle involves a correspondence conception of truth, because the DM's belief that Jane will take logic is true if it corresponds to the fact that she takes it.

The second point is that, as already stated, the qualitative principle provides a practical motive that is not taken into account in pure logical theory, for inquiry that aims at discovering the truth. A DM wishing to take the same class as Jane is well advised to inquire and try to find out what that class will be. He is also well advised to reason and 'put two and two together', if he learns from one source that Jane will take logic or ethics and from another that she won't take ethics.

The last point is a qualification of the previous one, namely that the practical motive for inquiry is *transitory*. There is an immediate practical need to know at the time of acting (more exactly, at the time of deciding to act), which becomes moot when that time has passed, or the object of inquiry is remote from practical considerations. Of course, knowledge of 'remote' things such as past happenings can have practical utility—for in-

[6] Anticipations of this principle can be found in F.P. Ramsey's paper "Facts and Propositions," (1927) but so far as the author knows this version of it was first stated in chapter III of Adams (1975).

Obviously this principle is closely related to the practical syllogism. It is easy to see that even in simple, limited applications, e.g., to cases like that of choosing what class to take, the principle is at best a rough, first approximation, just as the practical syllogism is. Nevertheless, something like the principle must be presupposed if the pure reason that leads to beliefs is to be connected to the practical reason that leads to actions. We will see, however, that the principle is not valid even as a first approximation in application to conditional propositions, and this is linked to why, arguably, a correspondence conception of truth doesn't apply to them.

[7] Modified as in footnote 13 of chapter 8⋆ to allow for the difference in tense between "Jane will take logic" and "Jane takes logic".

stance, it can be useful to know where an object has been left on a former occasion—but a general rule relating this kind of knowledge to the practical advantage of persons who have it is less easy to state. Other kinds of knowledge, for example of generalities such as that sugar is sweet, have still other practical uses, and still other rules describe these uses (appendix 9 comments briefly on the values of certain kinds of general knowledge). The lesson to be learned is that different kinds of knowledge have different practical uses and different pragmatic principles apply to them, which theories of practical reason ought to explore.[8] And, among these kinds of knowledge are propositions expressed in conditional form, which it is part of our present business to explore.

Finally, clearly the qualitative pragmatic principle is only a first approximation even in its application to actions based on simple factual beliefs such as that Jane will take logic. Among the many complications that might be considered, we will refine the principle to allow for the fact that wants and confidence are matters of degree. As a preliminary to this, we will rearrange the five factors in the decision 'picture' given in diagram 9.1.

9.4 Decision Matrices

Matrix 9.1 below rearranges the factors in diagram 9.1:

	possibilities			
	Jane takes logic		she doesn't takes logic	
	thought (believed)	fact (actual)	thought (not believed)	fact (not actual)
possible acts: **take logic**	want **to be in Jane's class**		don't want not to be in Jane's class	
don't take logic	don't want not to be in Jane's class		don't want not to be in Jane's class	
	Results			

Matrix 9.1 (qualitative)

This involves the same factors as in diagram 9.1, but by scrambling their order it puts them together in a more compact way. The DM believes that

[8]The suggestion is that a general logical theory, including practical and pure logic, should concern itself not only with different kinds of propositions, atomic, molecular, conditional, general, and so on, to which only one kind of 'value' applies, namely truth, but also with the different kinds of value that propositions of these kinds may have.

Jane will take logic and it is a *fact* that she takes it, and because there is a semantic correspondence between the belief and the facts they are put together as 'possibilities' at the top of the picture, although the belief comes before and the facts come after the action, and the facts together with the action lead to the result. Similarly, wanting to be in Jane's class and being in Jane's class are entered in the top left cell of the matrix, though the DM wants to be in it before he *is* in it. Scrambling the factors in this way also scrambles the causal arrows that connect them, so that the belief and the want arrows lead from the top and from the cell in the middle to the action on the left, while the action and the fact arrows lead from the left and the top to the result in the cell, as shown.

There are three other things to note in matrix 9.1.

First, the result, which is the resultant of the causal process that leads to it from the act and the facts, is located at the intersection of the act row and actual fact column, i.e., it is located where the act-arrow pointing to the right and the fact-arrow pointing downwards meet (and which is highlighted in the matrix). It should be noted that, given the DM's act of taking logic and the actual fact of Jane's taking it, we have pictured the result following as a certainty. However, section 9.7⋆ will allow the possibility that there might be some uncertainty about this, and then degrees of confidence in propositions like "If Jane and I both take logic then we will be in the same class" will enter the picture.

Second, possible results are represented just as wanted or not wanted, as though being in Jane's class were all that mattered. But that doesn't take into account the possibility that the desirability of this result could depend on how it was arrived at. For instance, if the DM were not going to be in Jane's class then he might prefer not to sign up for logic, and therefore the result of not being in Jane's class wouldn't be quite as unwanted in the lower row of the matrix as it would be in the upper one. However, if we are to make this explicit we have to make wants a matter of degree, or utility, and this will be taken up in the following section.

Third, we have seen that the practical inference that leads from the DM's beliefs and wants to his action is represented by arrows running from beliefs at the top and wants in the cells of the matrix to the acts on the left. But when we generalize to allow for the fact that wants and beliefs may be matters of degree, we will depict the inference as a process of calculating the *expected utilities* of the acts,[9] and the qualitative matrix 9.1 must then be replaced by the quantitative matrix that we will now consider.

[9]This isn't meant to imply that calculating isn't a causal process, though if it is one it is much more complicated than simply 'adding' beliefs and wants to determine actions.

9.5 First Generalization: Degrees of Confidence and Utility[10]

The previous section noted two factors in the qualitative decision matrix that can be matters of degree: namely *values of results*, or utilities, and *degrees of confidence.* Entering these values into the decision matrix in place of their qualitative counterparts leads to a *quantitative matrix*, as follows:

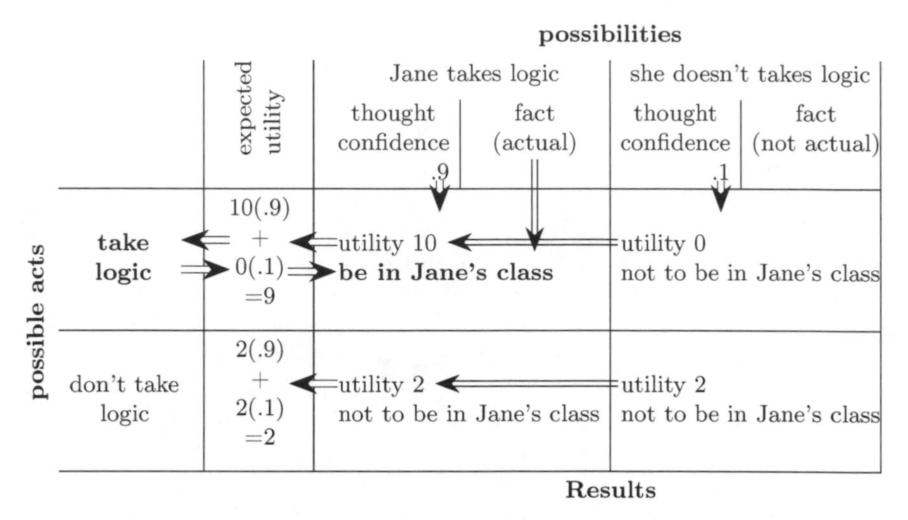

Matrix 9.2 (quantitative)

Matrix 9.2 changes the qualitative picture given in matrix 9.1 in three major respects: (1) *Degrees of confidence* are written at the top of the matrix in place of unqualified 'belief' or 'disbelief', (2) *Numerical utilities* are written into the cells of the matrix in place of unqualified 'wants' and 'don't wants', and (3) Instead of picturing acts as arrived at simply as resultants of the 'forces' of beliefs and wants, they are arrived at by calculations according to a complex formula that is written into a column headed 'expected utility' to the left of the matrix. Section 9.7⋆ will comment on how degrees of confidence and numerical utilities are defined or measured, after we have discussed the expected utility formula, but we should first note that the degrees of confidence and utilities that are entered in matrix 9.2 are intuitively plausible.

The matrix represents the DM as having a degree of confidence of .9 that Jane will take logic and a degree of confidence of .1 that she won't,

[10]The subjects of this and the two succeeding sections are treated in far greater detail in textbooks on mathematical decision theory, cf. Luce and Raiffa (1957), especially chapter 1, and Jeffrey (1983).

and though we haven't yet specified precisely how these values might be determined, at least they should seem plausible given that the DM believes that Jane will take logic. The DM should have some confidence in what he believes, and therefore his degree of confidence in what he believes should be significantly higher than his degree of confidence in its opposite. In fact, is not unreasonable to suppose that he would take 90% odds on a bet that Jane would take logic.[11]

The matrix also represents the DM as attaching a utility of 10 to being in Jane's class, which is significantly higher than the utility he attaches to not being in it, which is 0 if that class is logic and 2 if it isn't. Again, we haven't specified precisely how these values might be determined, but they do correspond roughly to the fact that the DM wants to be in Jane's class. Given that the DM wants to be in this class, he should attach a significantly higher value to being in it than to not being in it—perhaps he would be willing to pay $10 to insure his getting into the class that Jane was going to take. But note that we have represented the value that he attaches to not getting into Jane's class if that class is logic as different from the value he attaches to not getting into it if it isn't logic. Both of these are lower than the value of the 'best case' result of taking logic when Jane takes it, which is 10. But the 'worst case' result of taking logic when Jane doesn't take it has a value of 0, while the intermediate case result of not taking logic, and probably not getting into the class Jane does take, has a value of 2. This is closer to the worst case than it is to the best case because the DM doesn't have as strong a 'feeling' about the desirability of the different unwanted results as he does about the difference between them and the result he does want.

Let us now consider where this leads when it comes to the DM choosing what class to take, which involves calculating the *expected utilities* that are entered into the left hand column of matrix 9.2. As noted above, choosing an action is no longer represented as a simple causal inference of the form **belief + want → action**, but rather of calculating the expected utilities of the possible actions, and choosing the action with the highest expected utility. The general rule for calculating expected utilities goes as follows:

Expected utility rule. The expected utility of a possible action is the sum of the utilities of its possible results, weighted by the DM's degrees of confidence in their occurrence.

The application of this rule is diagrammed in matrix 9.2 by drawing a caret, '▼', pointing downwards from the degrees of confidence at the top of the matrix towards the utilities that they weight in the cells below—for instance, a '▼' points down from 'confidence .9' at the top towards the

[11] Appendix 1 discusses in detail a 'betting odds' approach to measuring degrees of confidence.

'utility 10' in the cell below. Then, after the utilities have been multiplied by the degrees of confidence that weight them, all of the weighted utilities in a row are added together to give the expected value of the action that the row corresponds to, as shown by linking these products with left-pointing arrows, '$\Longleftarrow$'. For instance, the expected utility of the action of taking logic is equal to the utility of being in Jane's class, namely 10, weighted by the degree of confidence that taking logic will have that result, which is .9, added to the utility of not being in her class, which is 0, weighted by the degree of confidence that taking logic will have *that* result, which is .1. Therefore, the expected utility of taking logic is $10 \times .9 + 0 \times .1 = 9$. Similarly, the expected value of not taking logic is $2 \times .9 + 2 \times .1 = 2$, as shown by the left-pointing arrow in the lower row of the matrix. And, since this is lower than the expected utility of taking logic the DM should take logic, which is itself indicated by the right-pointing arrow, '$\Longrightarrow$' from the act with the higher expected utility towards the result of being in Jane's class, in the lower part of the top row of the matrix.

It is significant that the quantitative calculation in matrix 9.2 leads to the same conclusion that the qualitative one did in matrix 9.1, namely to take logic, but the quantitative calculation gets there in a more roundabout way than the qualitative one did. Actually, the qualitative route can be regarded as the 'limit' that the quantitative route approaches, as the degrees of confidence involved approach certainties. For instance, if the DM were certain that Jane would take logic his confidence in that would be 1 and his confidence in her not taking it would be 0, so the expected utility of taking logic would be the certain value of $10 \times 1 + 0 \times 0 = 10$, and the expected utility of not taking logic would be $2 \times 1 + 2 \times 0 = 2$, and he would take the direct route to logic. But now we will see that it is sometimes necessary to take the roundabout route.

Matrix 9.3 has the same result utilities as the ones in matrix 9.2, but it reverses the degrees of confidence involved.

Computing expected utilities on the basis of the utility values given here, the DM should not take logic since its expected utility is 2 while the expected utility of taking logic is only 1. The important point that this illustrates is that this result depends on the fact that the utility of not being in Jane's class is slightly less if that class is logic than it would be if it were something else, since if the two values were the same, say both were equal to 0, the expected utility of taking logic would have been 1 as before, but the expected value of not taking logic would have been 0. This is a case in which a small difference in the utilities of different unwanted results affects a decision, and deciding which action to take requires calculating their expected utilities.

Another important thing that matrix 9.3 illustrates is the DM's making

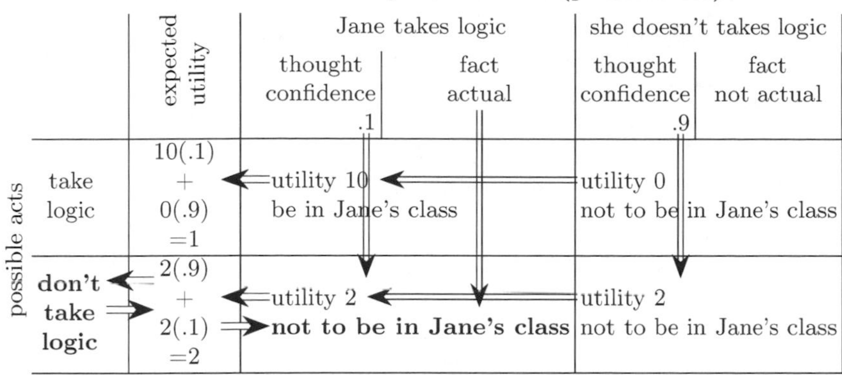

Matrix 9.3

the wrong decision by not taking logic, because Jane actually takes it, and if the DM had taken it too then he would have ended up with what he wanted—to be in Jane's class. Moreover, what led the DM to the wrong decision was having degrees of confidence that didn't correspond to the facts. That is, while the DM had confidence .9 that Jane would *not* take logic, she actually took logic. This illustrates an important *degree of confidence pragmatic principle*, which can be regarded as a generalization of the qualitative pragmatic principle stated in section 9.3, and which reduces to that principle in the limit:

Degree of Confidence Pragmatic Principle (rough). Other things being equal, decisions based on degrees of confidence that correspond to the facts, in the sense that things that DMs have high degrees of confidence in are actually the case, have at least as good or better results than ones that do not.[12]

This gives a motive for DMs to have degrees of confidence that correspond to the facts in the sense they have high degrees of confidence in things that are actually the case and correspondingly low ones in things that are not the case.[13] The rough principle will be returned to in section 9.9⋆⋆, but we will comment first on *critical* degrees of confidence.

[12]This principle implies that 'pragmatically best' degrees of confidence are ones that correspond to the facts in the specified sense, but it doesn't imply that they are necessarily wrong if they don't correspond to the facts in that sense. Thus, though the principle implies that it is pragmatically best to have a high degree of confidence in a proposition of the form "The coin will fall heads" if the coin actually falls heads, it doesn't mean that it is wrong to judge that the coin is as likely to fall tails as heads.

[13]As before, the motive is transitory, since the correspondence is between the degrees of confidence that prevail at the time that DMs act, and the facts that 'come to pass' at later times.

9.6 Critical Degrees of Confidence

Suppose that the degrees of confidence given in matrix 9.2 are altered once more:

		possible facts (possibilities)			
		Jane takes logic		she doesn't takes logic	
possible acts	expected utility	thought confidence .2	fact actual	thought confidence .8	fact not actual
take logic?	10(.2) + 0(.8) =2	utility 10 be in Jane's class		utility 0 not to be in Jane's class	
don't take logic?	2(.2) + 2(.8) =2	utility 2 not to be in Jane's class		utility 2 not to be in Jane's class	

Matrix 9.4

Calculating on the basis of these data, the expected utilities of taking and of not taking logic are both equal to 2. In this case the rule 'choose the action with the highest expected utility' doesn't apply, and the DM can choose at random. We also see that given the utilities in the table, confidence values of .2 and .8 are 'critical', because shifting them either up or down leads to unequal expected utilities and to a clear choice between the possible actions. This is pictured graphically in diagram 9.2.

The diagram plots possible degrees of confidence in Jane's taking logic as points from 0 to 1 on a line of unit length, with confidence values that would lead the DM to choose *not* to take logic corresponding to points in the interval stretching from the left end of the line to the 0.2-point. Confidence values that would lead the DM to *take* logic then correspond to points in the interval to the right of the 0.2-point to the right end of the line, and the critical confidence of 0.2 separates the 'don't take logic' and the 'take logic' intervals:

degrees of confidence that Jane will take logic

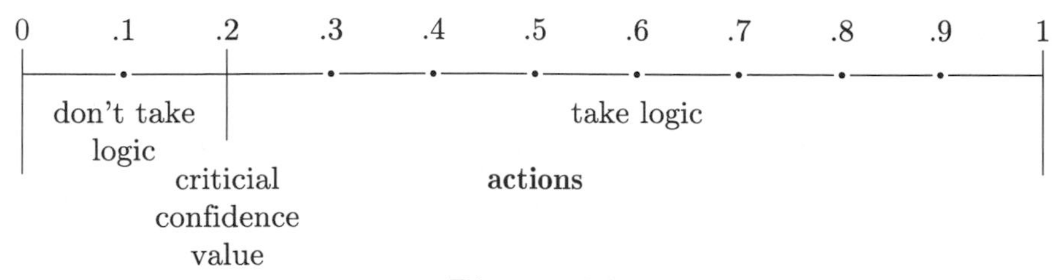

Diagram 9.2

There are two points to make about this diagram. One is that it shows that the DM really doesn't need to determine his exact degree of confidence in Jane's taking logic; all he needs to know is whether it is greater than or less than the critical degree of .2. For instance, just being 'pretty sure' that Jane will take logic is enough to allow him to make his decision, and he only needs to balance probabilities if he thinks the chances of her taking it are pretty small.

The second point is that the sizes of the 'don't take logic' and the 'take logic' intervals and the critical confidence value that separates them depend on the DM's utility values, which in this case are the ones that are given in Matrices 9.2–9.4. If the utility of not taking logic had been 2 rather than 0 in the upper right cell, then the critical confidence value would have shifted all the way to 0 at the left end of the line,[14] and if it had been −2 in that cell the critical value would have been .33.[15]

It is even possible to represent degrees of confidence and critical degrees of them when there are three or more possible actions. For instance, suppose that the choice is between taking logic, ethics, or history, and the utilities of the possible results are given in matrix 9.5[16]:

[14] When the critical degree of confidence lies at either end of the interval then one of the actions *dominates* the other, in the sense that its result is as good as or better than the other, no matter what the facts are. In this case the dominance principle discussed in the following two sections applies, and it is not really necessary to calculate expected utilities in order to decide what action to take.

[15] Because utility is measured on an *interval scale*, like temperature, it can have a negative value, analogous to −10° on the Fahrenheit scale. This will be returned to in section 9.7⋆.

[16] Here the DM is pictured as acting as though his 'primary utilities' are for being in or not being in Jane's class, while the acts that might or might not lead to these results can be thought of as having 'secondary utilities' of 0, 1, and 2, for taking logic, ethics, and history, respectively. Then the 'composite utility' of a result, say of being in the *same* class as Jane if that results from taking history, is the sum of the primary utility of being in Jane's class and the secondary utility of taking history, which is 10+2 = 12. However, not all resultant utilities can be 'factored' in this way

		Jane takes		
possible acts	expected utility	logic conf. .5 actual	ethics conf. .3 not actual	history conf. .2 not actual
take logic	10(.5) + 0(.3) + 0(.2) =5	utility 10 same class	utility 0 different class	utility 0 different class
take ethics	1(.5) + 11(.3) + 1(.2) =4	utility 1 different class	utility 11 same class	utility 1 different class
take history	2(.5) + 2(.3) + 12(.2) =4	utility 2 different class	utility 2 different class	utility 12 same class

Matrix 9.5

Before commenting on the degrees of confidence that would be critical in this case, it is worth noting that while there isn't a high degree of confidence in any of the three possibilities, i.e., none of them is believed, it is still possible to calculate expected utilities and decide what action to take on the basis of the degrees of confidence and the utilities that are given.[17] Moreover, the degree of confidence pragmatic principle still applies in a generalized form: given that Jane does take logic the DM made the right choice by taking it too, and what led him to do that was having the highest degree of confidence in the thing that actually happened, namely that Jane would take logic.

We can also calculate *critical* degrees of confidence. Letting Conf(ϕ) be the DM's degree of confidence in a proposition ϕ, if Conf(L) = 13/30, Conf(E) = 10/30, and Conf(H) = 7/30, all three actions would have expected utilities of 130/30 = 4.33 and the DM would have to choose at random between them. If Conf(L) = Conf(E) + .1, then whatever Conf(H) is, the expected utility of taking logic and taking ethics would be equal, and if that is greater than the value of history the DM would still have

[17]That the expected values of the actions are in inverse order to the intrinsic values of the classes is because the degrees of confidence that Jane will take these classes are in that order, possibly because the DM thinks that Jane prefers logic to ethics and ethics to history.

to make a random choice.[18] Similarly, if Conf(L) = Conf(H) + .2 then the expected utilities of taking logic and taking history will be equal, and if Conf(E) = Conf(H) + .1 the expected utilities of taking ethics and taking history will be equal.

Finally, it is a matter of curiosity to see that degrees of confidence can still be represented diagrammatically, even when there are three of them involved.[19] However, in this case they cannot any longer be represented on a *line*, and instead they have to be represented in a *two*-dimensional 'space', namely an equilateral triangle:

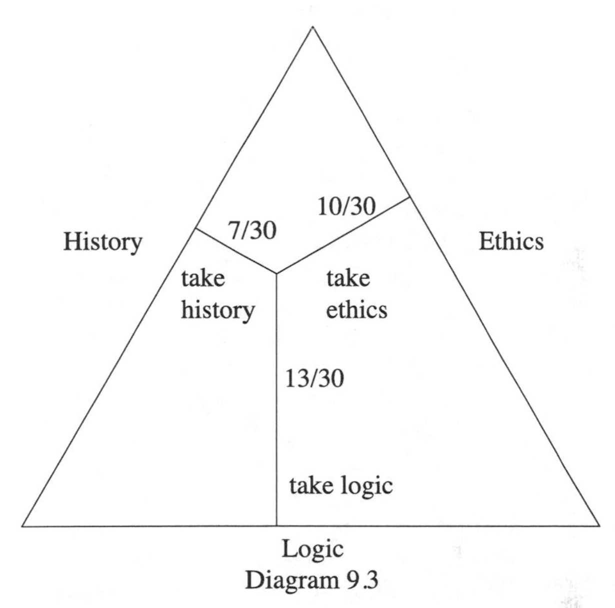

Diagram 9.3

Now the DM's degrees of confidence in Jane's taking logic, ethics, and history are represented by distances along lines from a point inside the triangle to the sides labelled 'logic', 'ethics', and 'history', which, assuming that the triangle has altitude 1, have to total 1.[20] Here the point is chosen so that the distances to the three sides, 13/30, 10/30, and 7/30 correspond

[18]This is determined by replacing the degrees of confidence of .5, .3, and .2 in Jane's taking logic, ethics, and history respectively, by variables x, y, and z in the formulas $(10 \times .5) + (0 \times .3) + (0 \times .2)$ and $(1 \times .5) + (11 \times .3) + (1 \times .2)$, which are the expected utilities of the DM's taking logic and taking ethics, yielding $(x \times .5) + (y \times .3) + (z \times .2)$ and $(x \times .5) + (y \times .3) + (z \times .2)$. Equating these gives

$$(x \times .5) + (y \times .3) + (z \times .2) = (x \times .5) + (y \times .3) + (z \times .2),$$

and assuming that $x + y + z = 1$, it follows that $x = y + .1$.

[19]It is not possible to represent them in a two dimensional diagram if there are four or more possibilities.

[20]The coordinates of the point representing a 'distribution' of degrees of confidence

to the critical degrees of confidence given above, and the lines in which Conf(L) = Conf(E)+.1, Conf(L) = Conf(H)+.2, and Conf(E) = Conf(H)+ .1 radiate out from the point, roughly as shown. The three regions that these lines separate, labelled 'take logic', 'take ethics', and 'take history' then correspond to the degree of confidence distributions in which the DM would take these subjects.

We now return to the question of how numerical measures of degrees of confidence and subjective utilities may be determined.

9.7⋆ Measuring Confidence and Utility

Let us concentrate on how the degrees of confidence and utility in matrix 9.5 are measured. How does the DM determine that the 'value' to him of taking logic if Jane takes logic is 10 units of utility, while the value to him of taking ethics if Jane takes ethics is 11 units? We have seen that in certain 'critical' cases he has to be able to do this in order to decide which class to take himself. But confidence and utility are not like height and weight, which can be determined by objective means. They are subjective, like feelings of heat and cold that differ from person to person, which we seem to have no objective means of measuring. Now, as of this writing, the problem of measuring any kind of subjective 'feeling' is very controversial,[21] and this is true of utility and degrees of confidence in particular, but what follows will discuss aspects of one widely accepted approach to the problem.

To begin with, without necessarily assuming that subjective values are numerically measurable, we are entitled assume that a DM can tell whether he prefers one result to another, or whether they are equally preferable. That is analogous to supposing that a person can tell whether one thing feels hotter than another, or they feel equally hot, without necessarily assuming that subjective hotness can be measured numerically. Similarly, we can suppose that the DM can say which of two actions he would prefer to take, even if he isn't prepared to assign them numerical expected utilities.[22] Let us see where this leads us.

---

x, y, and z in Jane's taking logic, ethics, and history, in rectangular coordinates whose origin is the center of the base of the triangle and whose vertical axis passes through the triangle's apex, are calculated as follows. The height of the point above the horizontal axis is x, and the distance of the point from the vertical axis is $|y - z|/\sqrt{3}$.

[21]The literature on this is enormous, particularly in psychophysics. The work of Krantz and others (Krantz et al., 1971; Luce et al., 1989; Suppes et al., 1990) to be cited in later footnotes gives a comprehensive introduction to the mathematical side of this subject, and many books, e.g., Baird and Noma (1978) give excellent introductions to its more practical aspects.

[22]This does not necessarily lead to circularity in applying the theory, where the DM has to rank possible actions in order to measure the utilities that are involved in calculating the expected utilities that determine the ranking. Certain rules of choice or 'axioms' of the theory can be applied directly, like the dominance principle stated below, and it is sufficient to follow them in order to be able to measure utilities, degrees of confidence,

Return to matrix 9.5, but now suppose that all we know is that the DM wants to be in Jane's class, but he doesn't care which class he is in so long as it is the one that Jane takes, or which one he is in if she doesn't take it. Matrix 9.6 below represents this by replacing the specific degrees of confidence given in matrix 9.5 by unknown degrees of confidence x, y, and z, and specific utility values given in matrix 9.5 by unknown utilities, s and d, for being in the *same* class as Jane, and for being in a *different* class. The new matrix also adds an *unknown act* whose significance will be returned to below:

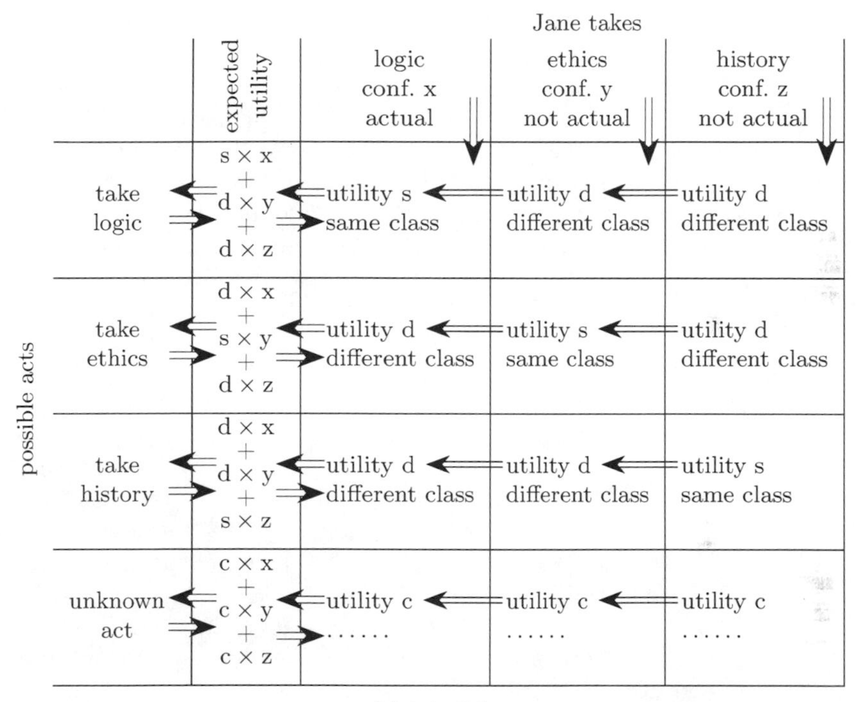

Matrix 9.6

As stated, we are not assuming that utilities and degrees of confidence are known *a priori*, hence they are represented as unknowns in the matrix, written as x, y and z, and s and d, with the sole requirement that $x+y+z = 1$,[23] and that s is greater than d—that is, the DM prefers being in Jane's class to not being in it. And, the utility of the possible results of the new

---

and expected utilities.

[23]It should be stressed that while for measurement purposes, degrees of confidence are assumed to satisfy the Kolmogorov axioms (which justifies the label subjective probabilities that is commonly used for them), they should not be confused with the factual 'real probabilities' to which they should, 'ideally', correspond.

'unknown act' is assumed to have a constant, unknown value c. Given this, the expected utilities of the actions are not given in the expected utility column as specific numbers, but as abstract formulas that show how these numbers depend on the unknowns x, y, and z, and s, d, and c.

But now suppose that the DM is indifferent between all four of the possible actions. That would mean that their expected utilities were all equal, and therefore that

$$sx + dy + dz = dx + sy + dz = dx + dy + sz = cx + cy + cz = c.$$

The first equality implies that $(s - d)x = (s - d)y$, and therefore x must equal y since $s > d$. By the same reasoning, the second equality implies that y must equal z, and together these imply that x, y and z should all be equal. I.e., if all that the DM cares about is getting into the same class as Jane and he is indifferent as to which class he takes, he must think that the chances of Jane's taking these classes are all equal. If these chances add up to 1, then each of them must equal 1/3, so we have 'measured' the degrees of confidence that DM has in Jane's taking logic, ethics, or history.

We can even measure one of the utilities in matrix 9.6. The numerical values s and d can't be determined because they act like an arbitrarily chosen *unit* and *zero point* of utility measurement, but because of this we can arbitrarily stipulate that $s = 1$ and $d = 0$.[24] The value of c can then be calculated. Assuming that x, y and z are all equal to 1/3, $cx+cy+cz$ must equal c, and since c is the expected utility of the unknown act and the DM is indifferent between it and the other three acts, all of these must have the same expected utility. Therefore, since the expected utilities of each of the other acts is $1/3s + 2/3d = 1/3$ and $s = 1$ and $d = 0$, c must also equal 1/3.

We won't try to systematize the method of confidence and utility measurement just illustrated because that involves considerable mathematical difficulty,[25] but there are two important points to make about the procedure. One is that it assumes that the DM calculates expected utilities according to the expected utility rule, which means that the theory that DMs choose on the basis of expected utilities implicitly *defines* these utilities.

The other point is that while degrees of confidence and utility are implicitly defined by the theory, the theory isn't merely a definition. This follows from the fact that there are certain nondefinitional principles that DMs who act in accord with the theory have to conform to. One of these is the following:

---

[24]Again, it is because utility is an interval scale like temperature that a zero and unit of subjective value can be chosen arbitrarily. Cf. footnote 15.

[25]The general theory is developed at length in many works on the theory of fundamental measurement. Cf. especially chapter 8 of Krantz et al. (1971).

Dominance principle. If for every possibility, the result of one action has higher utility than the result of another, then the first action should be preferred to the second.

For example, suppose that all of the 'unwanted' results in matrix 9.2 had had utility 0. In that case, each result in the 'take logic' row would have had a higher utility than the corresponding result in the 'don't take logic' row. Given this, the dominance principle would have implied that the DM should take logic, because that is the only action that has a chance of getting him into Jane's class, and that is all that matters to him. This is not a purely definitional rule, and since it follows from the expected utility theory, the theory it follows from can't be purely definitional either. Other nondefinitional rules also follow from the theory, and a fundamental *metatheoretical* problem is to give a 'complete axiomatization', or list of these principles, which shows exactly how much of the theory is definitional and how much of it is 'substantive' in that it transcends mere definition.[26]

The interesting thing about dominance and the other substantive, nondefinitional principles that are implied by the theory of expected utility is their seeming self-evidence, in contrast to the abstract rule that DMs should choose the action with the highest expected utility. But now we will see that even if the dominance principle seems self-evident, there are situations in which it is doubtful.

9.8 Decision Making in the Nonindependence Case

Now consider the problem of a DM who has to choose whether to take up smoking, and to make this choice he takes into consideration the possibility of getting cancer.[27] We might try representing this decision problem by a matrix like matrix 9.7 below, in which plausible result utilities are written into its cells, but degrees of confidence are written in as unknowns.

[26]See again works on fundamental measurement theory such as chapter 8 of Krantz, et al. (1971), cited in footnote 25. The fact that scientific theories that implicitly define concepts that enter into them aren't necessarily 'true by definition' has caused a great deal of confusion and controversy in the philosophy of science. Some of the most famous of these controversies have arisen in the philosophy of physics, one of them being about the whether the Copernican theory that earth moves and the sun is at rest is merely an implicit definition of motion and rest. Modern theories of measurement make a start on answering these questions, although many unanswered ones remain.

[27]As the reader is probably aware, whether or not smoking is harmful to health was long a controversial matter, and no less a personage as the famous statistician R.A. Fisher strongly argued that it was not (cf. Fisher 1959). The gist of Fisher's argument was that the positive correlation that exists between smoking and cancer could just as well be explained by the assumption, not that smoking causes cancer, but rather by the fact that the same preexisting factors that cause persons to get cancer also cause them to smoke. There has been an enormous amount of writing in recent years on the problem of distinguishing causal connections from statistical correlations, a good introduction to which is given in Suppes (1970).

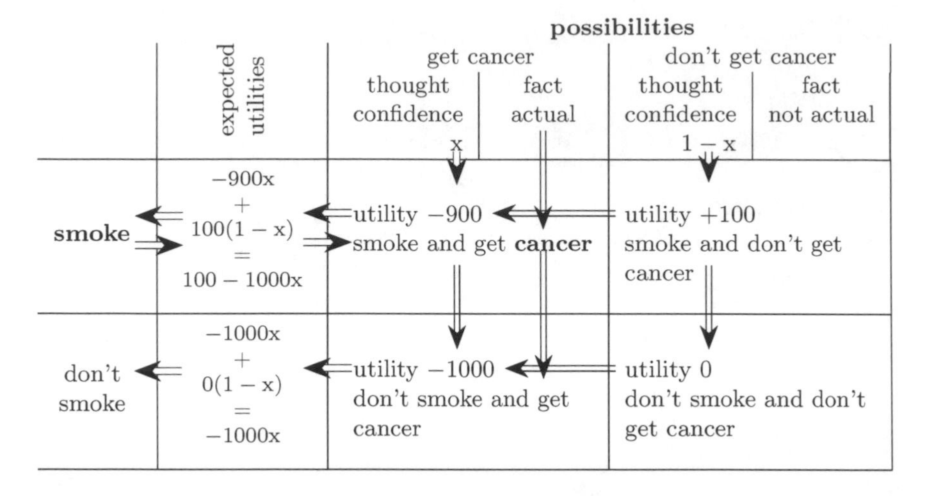

Matrix 9.7

The striking thing about this matrix is that it represents a dominance situation, and according to the dominance principle the DM ought to smoke, because the result of smoking is 100 utiles better than not smoking, whether or not the DM gets cancer. As you can see, no matter what x is, the expected utility of smoking, which is 100 − 1000x, must also be 100 utiles higher than the expected utility of not smoking, which is −1000x.[28]

But this reasoning is specious.

Arguing that people ought to smoke because whether or not they get cancer, they would be happier if they smoked than if they didn't is fatalistic because it assumes that smoking doesn't influence the chances of getting cancer. But present medical evidence is overwhelming that smoking *does* influence these chances, and this needs to be taken into account in the theory of expected utility. However, doing so requires us to go right back to the beginning and add another causal influence that wasn't in our original pragmatic diagram.

Now the action of smoking is causally connected not just to the result but to the fact of whether the DM gets cancer. And, this implies that the action influences the result in two ways: (1) *Directly*, by way of the arrow that goes directly from it to the result, and (2) *Indirectly*, by way of the arrow that goes first to the facts, and then by the arrow that goes from the facts to the result. This also means that further changes have to be

[28]It is always possible to describe the outcomes in the cells of a matrix as 'logical sums', i.e., conjunctions of the acts and facts that combine to produce them, as is done in this case. However, their utilities aren't necessarily the arithmetic sums of the utilities of the acts and facts that produce them, as is the case both here and in matrix 9.5.

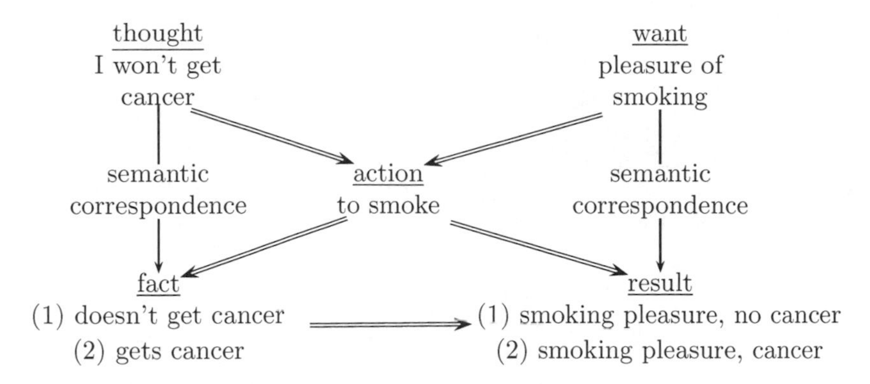

Diagram 9.4 Revised Pragmatic Diagram

made in the picture, besides adding a new causal arrow. However, these are more conveniently pictured in the revised decision matrix 9.8, below, which incorporates the influence of acts on facts by showing the DM's degrees of confidence as being *conditional* on the acts.

Now, instead of supposing that the DM has a single degree of confidence, x, that he will get cancer, which is entered at the top of the 'get cancer' column in matrix 9.7, we will assume that he has one degree of confidence, x, that he will get cancer if he smokes, and another degree of confidence, y, that he will get it if he doesn't. And since x and y now depend on the DM's actions, they are entered in the cells of the matrix, along with the utilities of the results, and not at the top:

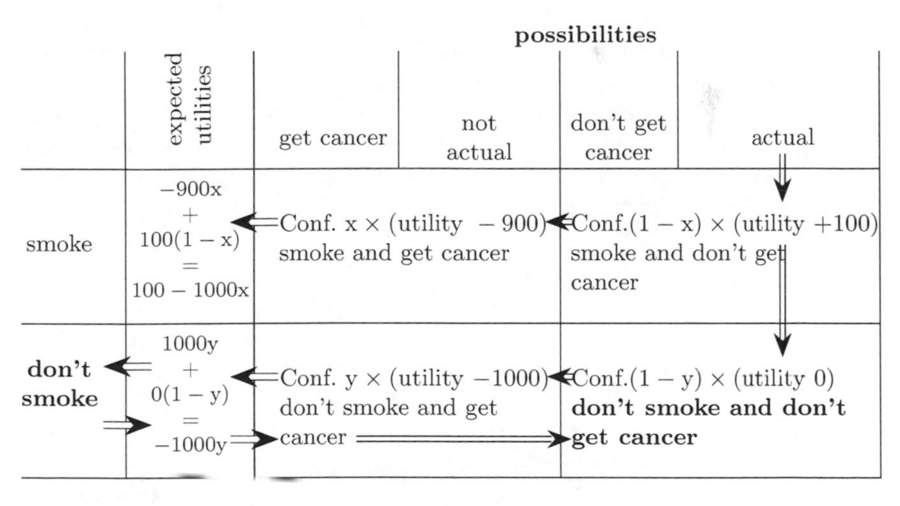

		possibilities			
	expected utilities	get cancer	not actual	don't get cancer	actual
smoke	−900x + 100(1 − x) = 100 − 1000x	Conf. x × (utility − 900) smoke and get cancer		Conf.(1 − x) × (utility +100) smoke and don't get cancer	
don't smoke	1000y + 0(1 − y) = −1000y	Conf. y × (utility −1000) don't smoke and get cancer		Conf.(1 − y) × (utility 0) **don't smoke and don't get cancer**	

Matrix 9.8

Although matrix 9.8 still pictures a dominance situation, smoking doesn't necessarily have the highest expected utility if these values are computed from degrees of confidence and utilities in the same way as before. For instance, if the DM thought he had a 30% chance of getting cancer if he smoked and a 10% chance of getting it if he didn't, then x would equal .3 and y would equal .1, and the expected value of smoking would be −200 while the expected value of not smoking would be −100, which, although it is negative, is greater than −200. Therefore the DM ought not to smoke.

Interestingly, while the changes in the decision matrix that we have just described are formally very simple, they are associated with more profound changes in the assumptions that underlie the formalism, and we will end this section with comments on two of them. One is that because the picture of the relation between degrees of confidence, utilities, and decisions has been modified, we also have to modify the implicit definitions that were given in section 9.7⋆, in order to measure numerical degrees of confidence and utilities. And, it is curious that in spite of the smallness of the change in the formalism, the modification that is required to measure these quantities is very substantial—so much so that no finite number of observations can determine any specific numerical measure with perfect accuracy—although they can yield better and better approximations. However, this subject is too complicated to enter into here.[29]

Even more important is the fact that the degrees of confidence x and y that are entered in the cells of matrix 9.8 no longer simply concern 'the facts', but rather the relation between the DM's actions and the facts. Thus, now x is not the DM's degree confidence that he will get cancer, but rather his degree of confidence that he will get it if he smokes. Therefore the DM's thought should not be represented 'unconditionally' as "I won't get cancer", as in diagram 9.4, but rather conditionally, as "If I smoke I will get cancer, and if I don't smoke I won't". But these thoughts do not stand in simple semantic correspondences with the facts on the lower left of the diagram, and therefore the vertical arrow from the thoughts to the facts has to be eliminated. And if that is so then we cannot any longer maintain the pragmatic principles that were stated in sections 9.4 and 9.5, and we cannot say that decisions based on degrees of confidence that correspond to the facts have at least as good or better consequences than ones that don't. If the DM decides not to smoke and he doesn't get cancer, then his smoking isn't a fact that can be observed, and the belief that the DM acts on, which he expresses to himself as "If I smoke I will have a 30% chance of getting cancer" can't correspond to such a fact.

This brings us back to matters that were discussed in section 8.6⋆, concerning conditional statements not corresponding to facts. The final section

[29]Cf. chapters 8 and 9 of R. Jeffrey (1983).

of this chapter will continue this discussion, and end by commenting on how beliefs such as are expressed by "If I smoke I will have a 30% chance of getting cancer" might correspond to a different kind of 'fact'. This will throw light on the meaning and justification of conditional statements, but, as stated, the views to be outlined will be 'philosophical' and controversial, and they will only be sketched briefly.

9.9★★ Unorthodox Reflections on Practical Probabilities

9.9.1 Introduction

Section 9.2 presented a pragmatic diagram showing semantic and causal relations among five 'factors' involved in a decision situation, (1) a DM's beliefs, (2) his wants, (3) his actions, (4) the facts, and (5) the results of his actions, and section 9.3 formulated a rough, qualitative pragmatic principle that relates them. *Results of actions correspond to what DMs want if the beliefs they act on correspond to the facts, and not if they don't; or, more briefly, actions succeed if the beliefs they are based on are true, and not if they are false.* This connects pure and practical reason by explaining why reasoners or DMs should want to reason in accord with principles that can lead them to conclusions that are true in the sense that they correspond to the facts. Now we want to speculate briefly and unsystematically on how this approach can be extended in such a way as to explain why DMs should want to reason in accord with principles that lead to conclusions about probabilities that correspond to the facts of probability. However, this will oblige us to say something about what the facts of probability are, what conclusions about them are, and what the practical advantage of being right in these conclusions is.

Rather than discussing the questions that concern us in generality, we will focus mainly on two examples that are extremes of a kind in a 'continuum of incertitude': (1) Acting on the belief that there is a probability π that a coin that is about to be tossed will fall heads and (2) Acting on the belief that there is a probability π that *if* the coin is tossed it will fall heads. What are these 'beliefs', what facts do they correspond to, and what is the advantage of being right about them? We will begin by tentatively identifying these 'probability beliefs' with having degrees of confidence π in 'simple beliefs' to the effect that the coin will fall heads, or that if it is tossed it will fall heads.[30] This allows us to apply the theories of decision making under uncertainty described in the previous sections of this chapter to 'probability beliefs', and thereby to fit them into our 'pragmatic scheme'.

[30]There are many reasons for questioning this, and, as already hinted, we shall do that ourselves later. One thing we will not do, however, is to treat degrees of confidence as 'subjective probabilities' that stand in no relation to objective probabilities, as many subjective probability theorists seem to do.

Later we shall be forced to modify this and the pragmatic picture itself, in order to deal with a 'paradox' to which the identification gives rise, but this will be postponed to the more detailed discussion of the unconditional case in the following subsection. The conditional case will be taken up in section 9.9.3.

9.9.2 The Unconditional Case

In this case a pragmatic diagram analogous to diagram 9.1 might go as follows, assuming that believing that the coin has probability π of falling heads is 'pragmatically equivalent' to having confidence π that it will fall heads, and with boxes corresponding to wants, actions, and results remaining to be filled in:

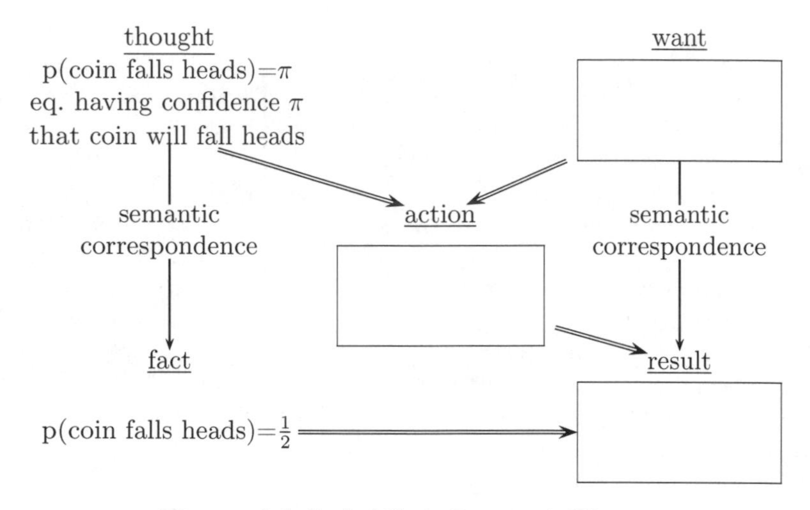

Diagram 9.5 Probabilistic Pragmatic Diagram

Postponing the question of how the empty boxes are to be filled in, let us note right away our assumption that the 'true probability' of the coin falling heads is $\frac{1}{2}$, and the 'paradox' to which this leads when it is combined with our identification of probability beliefs with degrees of confidence, and with pragmatic principles already stated. If the true probability is $\frac{1}{2}$, then the belief that p(the coin will fall heads) $= \pi$ can only be true if $\pi = \frac{1}{2}$. On the other hand, the degree of confidence pragmatic principle stated in section 9.4 asserts that actions based on degrees of confidence have the best results if these degrees correspond to the facts, and in this case the facts are whether or not the coin actually falls heads. I.e., if the coin is actually going to fall heads then from the practical point of view the best degree of confidence to have in its falling heads is 1, and if it is going to fall tails the best degree of confidence to have in its falling heads is 0. In sum, the judgment that p(the coin will fall heads) $= \pi$ is right if $\pi = \frac{1}{2}$, but the best degree of confidence to have in the coin falling heads is either 1 or 0. And, if that is so then it seems either that right probability judgments are not 'good' or 'useful' judgments, or else they are not 'pragmatically equivalent' to degrees of confidence.

Before accepting the above, however, we should note that there is an intuitive measure of truth to the idea, both that judging that p(the coin will fall heads) $= 1$ and acting on this would have desirable consequences if the coin actually falls heads, *and* that judging that p(the coin will fall heads) $= \frac{1}{2}$ is right and it is practically useful to know and act on this. We now suggest that the way to reconcile these things is to look more closely at the practical advantage that derives from knowing and acting on the belief that the coin has a 50-50 chance of falling heads. This will require us to say something about the how the blanks in diagram 9.5 are to be filled in, and that will require changes in the diagram.

The key idea is that the advantage of knowing that coins have a 50-50 chance of falling heads when they are tossed is that it is a good long-run *policy* to act on this, for instance by taking bets of $1 that they will fall heads, that pay off more than $1 if they win. The details are complicated, but the general idea is that, while persons who adopt this policy can't expect to win a high proportion of the time, they can expect to win at least as much as they lose and possibly more. These changes are incorporated in the following *policy decision diagram*, whose features will be commented on below:

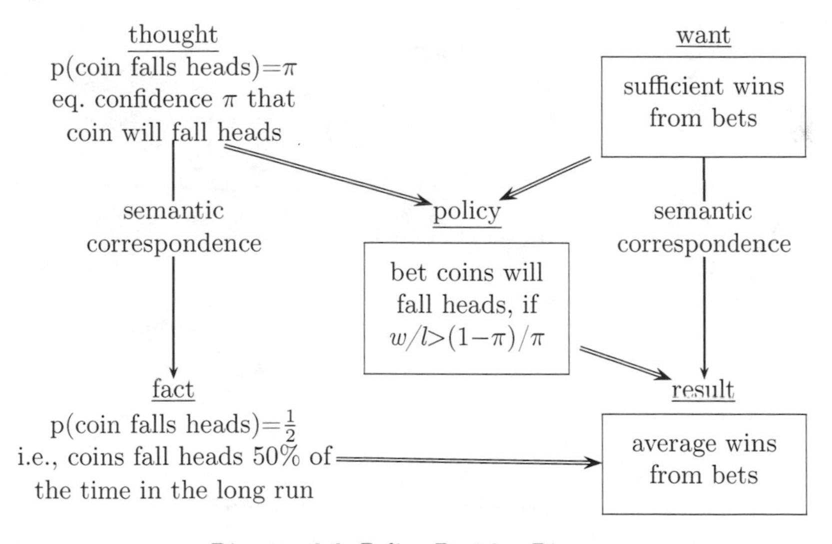

Diagram 9.6 Policy Decision Diagram

Here are brief comments on the new diagram.

First, the most important change is that instead of applying to a single action it applies to a *policy* for acting, namely, under circumstances described by the condition $w/l > (1-\pi)/\pi$, to bet that coins will fall heads. What that condition is will be returned to below. But given that it may obtain in many circumstances, both the thought and the fact are *general*, and apply to all occasions when the conditions obtain, rather than to one occasion when a coin is going to be tossed.

Second, the thought still combines the judgment that the probability of coins falling heads is π with having confidence π that they will fall heads, but the identification is more complicated than it was in simple cases like that represented in diagram 9.5. Very roughly, while the probability judgment applies to coin-tossing in general, the degrees of confidence they are identified with apply to coins falling heads in particular cases. Another way to put it is that the policy is that when a coin is going to be tossed, the DM is to act as though he had confidence π that it will fall heads.[31]

Third, the fact that the thought that p(coins fall heads) $= \pi$ corresponds to is itself general, and pertains to the frequency with which coins fall heads. Here we have supposed that, as a matter of empirical fact, they fall heads 50% of the time 'in the long run', and therefore the thought is true in the sense of corresponding to the facts if $\pi = \frac{1}{2}$. This is the relative frequency theory of probability (cf. Skyrms, section VII.3), but now connected with decision making under uncertainty.[32] *However:*

[31]That is, the DM should act as though he were maximizing expected utilities, as in the following matrix:

possible acts	expected utility	**possibilities**: the coin falls heads — thought confidence π	fact actual	it doesn't fall heads — thought confidence $1-\pi$	fact not actual
take bet	$w \times \pi - l(1-\pi)$	utitity w **win bet**		utility $-l$ lose bet	
don't take bet?	0	utility 0 neither win nor lose		utility 0 neither win nor lose	

According to this, the DM should make the bet if $w \times \pi - l \times (1-\pi)$ is greater than 0, which is equivalent to $w/l > (1-\pi)/\pi$. For instance, if he thought that the coin was biassed against heads, with a chance $\pi = 1/3$ of its falling heads and a chance $1-\pi = 2/3$ of its falling tails, so that $(1-\pi)/\pi = 2$, he should bet if he stands to win at least twice as much as he would lose if he lost the bet.

[32]Another complication that will be discussed briefly at the end of this subsection has to do with the fact that the coin-tossing case is an extreme one, in which a uniform policy of always having the same degree of confidence that coins will fall heads is as

Fourth, while the thought that p(coins fall heads) = π corresponds to is the actual frequency with which coins fall heads, which we have assumed is $\frac{1}{2}$, the correspondence doesn't have to be exact. Roughly, it is a matter of degree, and how well it corresponds can be measured by the difference between π and $\frac{1}{2}$. If we measure 'degrees of correspondence' as degrees of truth, we can even generalize the simple pragmatic principle stated earlier, as outlined in point 8, below.[33]

Fifth, the ratio w/l in the condition $w/l > (1-\pi)/\pi$ is between the amount, w, that the bettor stands to win on a bet that he might make that the coin will fall heads on a given occasion, and the loss, l, that he would suffer if he made the bet and lost because the coin fell tails on that occasion. What the whole condition says is that the ratio of the possible win to the possible loss should be greater than the ratio of the DM's confidence in losing to his confidence in winning.

Sixth, neither the 'want' nor the 'result' to which it is related are 'all or nothing quantities'. A DM adopting the policy doesn't aim to win on a single occasion, or even on most occasions. He only wants to win 'enough of the time', but how much is enough depends on him. For this reason the result, of winning on some proportion of the bets that the DM accepts, cannot be described simply as 'wanted' or 'unwanted', but rather as 'more or less satisfactory'.

Seventh, the result, i.e., the proportion of times that the DM wins on bets that he accepts, depends both on the policy that determines the bets he accepts, and on 'the facts', as to the occasions on which the coin actually falls heads.

Eighth, combining the foregoing remarks, and assuming that the exactnesses of correspondences with the facts themselves correspond to degrees of truth, suggests a very rough pragmatic principle:

Pragmatic Policy Principle The more closely the thoughts on which general policies of action are based correspond to the facts, i.e., the truer they are, the more satisfactory results of actions in accord with them are likely to be.

The most important thing about this principle is that, rough as it is, it suggests a motive for DMs to want their judgments concerning probabilities

good as any humanly applicable policy can be in the long run.

[33]For instance, we might take the quantity $1 - \left|\pi - \frac{1}{2}\right|$ as a measure of the degree of truth of p(coins fall heads) = π. Assuming this, if $\pi = \frac{1}{2}$ then the judgment that p(coins fall heads) = π would have truth 1—perfect truth. However, while it seems intuitively plausible to assign *degrees* of truth instead of 'all or none' truth values to numerical judgments, the 'right' way to do this is a controversial matter. Cf. Goguen (1968–9) as well as much subsequent writing on fuzzy logic. Appendix 9 considers the properties of a degree of truth measure that applies to approximate generalizations like "Birds fly," which is highly true because most birds do fly, although it has exceptions.

to correspond to the facts, not necessarily exactly, but 'closely enough' that the policies that they base on these judgments can be expected to be 'satisfactory enough'. This is part of the practical side of probability, which we said was ignored in pure probability theory.

The final points in this subsection have to do with the question of why we have shifted the focus of our inquiry from decisions whose results are described simply as 'wanted' or unwanted', to the choice of policies that lead to more or less satisfactory results when they are acted on over a period of time. In the case of the coins, the basic fact is that we know from long experience that while it would be 'ideal' to be certain and have confidence 1 that one would fall heads when it was tossed and fell heads and confidence 0 that it would fall heads when it was tossed and fell tails,[34] human beings cannot consistently attain this ideal. As F.P. Ramsey wrote: "...the highest ideal would be always to have a true opinion and be certain of it, but this ideal is better suited to God than to man."[35] We can go even further, and add that not only can't we always have true opinions as to how coins that are about to be tossed will fall, we can only be right in these opinions 50% of the time.[36]

But there are things that we can be right about far more than 50% of the time. One is that about 50% of the tosses in a series of 100 coin-tosses will fall heads. More exactly, it follows from the *law of large numbers* that if persons can only be right 50% of the time predicting how coins will fall on individual tosses, they must be right at least 95% of the time predicting that coins will fall heads between 40% and 60% of the time, when they are tossed at least 100 times.[37] Therefore, roughly, if they adopt policies of betting, based on the estimate that the probability of coins falling heads is between 40% and 60%, they will be right and the results of acting on this policy 'long enough' will be satisfactory at least 95% of the time.

The above points to a reason for considering policies that 'pay off' over

---

[34]Thus, to this extent the degree of confidence pragmatic principle is right.

[35]See pages 95–6 of Ramsey (1926). The earlier sections of this paper develop basic ideas of the utility theory that was outlined in sections 9.5–9.7⋆ of this chapter, but its final two sections, which comment on matters closely related to those discussed in this section, deserve much more attention than they have received so far. Ramsey's contribution to the latter subject is discussed in Adams (1988a), which follows up on the discussion in chapter III of Adams (1975).

[36]This is what Hans Reichenbach called *psychological randomness*, or indeterminism (Reichenbach, 1949: 150), which concerns the limits of human predictive capabilities, and which should be distinguished from the physical indeterminism that figures importantly in quantum physical theory.

[37]Cf. section 4 of chapter VI of Feller (1957). The application of this law to the present case is not perfectly rigorous, because the mathematical law assumes that the coin-tosses are statistically independent and we have only assumed that DMs cannot predict their results correctly more than 50% of the time. The assumptions needed to insure that this implies statistical independence should be investigated.

a period of time, as against decisions such as whether to bet on the outcome of a single coin-toss, which can only win or lose. These actions can only succeed 50% of the time, and there are strong practical reasons to consider 'behaviors' with higher success rates, albeit success is not winning '*simpliciter*'.[38] It is certainly the case that smart gamblers adopt such policies, and many books that set forth odds at cards, dice and other gambling devices provide the kind of permanent information that serves as the basis of the gambling policies of serious bettors.

But, to conclude this section, we must stress that while much and perhaps most empirical phenomena manifest a degree of uncertainty and unpredictability, very little of it is as unpredictable as the result of coin-tossing. For instance, the fall of a coin is much less predictable than the day's weather, although that often cannot be predicted with certainty. Related to this, also unlike coin-tossing, the chance of rain varies from day to day, even from hour to hour, and while sometimes it is highly uncertain, at other times it can be practically certain, say, that it is going to rain. And, deciding what to do when the weather is uncertain on a particular occasion, people are likely to mix general policies with considerations that apply in the individual circumstances.[39] They may make it a rule to take their raincoats when the skies are overcast in the morning, but decide on more idiosyncratic grounds whether to carry their umbrellas. In any case, laws of large numbers do not apply to such phenomena, and there are no published and generally agreed on rules that can guide decision making in such circumstances.[40]

But let us return to the ideal case, and comment very briefly on the probability that coins will fall heads *if* they are tossed.

---

[38]Behaviors of this kind are like the kind of 'habit' to which 'fixation of belief' leads, according to C.S. Peirce in his great paper of 1878, "How to Make Our Ideas Clear," (e.g., as explained on p. 29 of Buchler (1955)). Add that it is usually very much easier to 'fix one's beliefs' and follow a general rule of action than it is, say, to make up one's mind whether to bet each time a coin is about to be tossed.

[39]Two comments may be made here. One is that even 'idiosyncratic decisions' may involve the application of policies, but the circumstances may be so special that it is difficult to decide how to apply them. The other point relates to a distinction that is sometimes made in decision theory, between risk and uncertainty. Betting on the fall of a coin involves the former, because there is an objective probability of $\frac{1}{2}$ that it will fall heads, which the DM is likely to know. Deciding whether to carry an umbrella seems to involve the latter, because even if there is an objective probability of rain the DM may not know what it is.

[40]It might be suggested that persons should make it a rule to be guided by the probabilities published by professional weather forecasters, but at best that would provide good guidance in very limited circumstances. It would be foolish for a person to be guided by a forecast like "10% chance of rain by afternoon," if he saw unmistakable signs of approaching rain at noon. There is also controversy as to how the 'rightness' of probabilistic weather forecasts should be measured. This is the so-called *calibration problem*.

9.9.3 The Conditional Case

Actually, the entries in diagram 9.6 only need to be slightly modified to yield a 'conditionalized diagram':

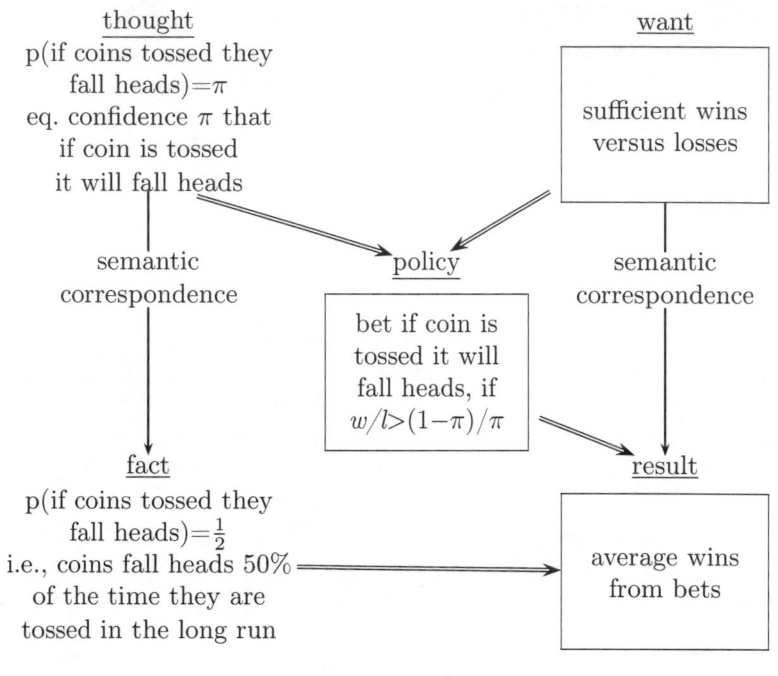

Diagram 9.7

The following comments primarily concern the differences between the foregoing and diagram 9.6.

Note first the probability, p(if coins tossed they fall heads), that enters into both the thought and the fact that it corresponds to. In this case it is explicitly defined as a *conditional frequency*, namely of coins falling heads in the class of instances in which they are tossed, as against the unconditional frequency of coins falling heads *'simpliciter'* that enters into diagram 9.6. Of course these frequencies are the same, because a coin is only counted as falling heads *when it is tossed*—simply discovering a coin lying face-up on a surface doesn't count as a case of its falling heads. Thus, diagram 9.7 only makes explicit something that is implicitly assumed in diagram 9.6. Similar points apply to the policy, want, and result that enter into diagram 9.7.[41]

[41]This is not to suggest that all judgments about the probabilities of conditionals only make explicit what is implicit in judgments about the probabilities of uncondi-

As to the correspondence itself, we may still say that the thought is true if it corresponds to the fact, i.e, if $\pi = \frac{1}{2}$. More generally, the more closely the thought corresponds to the fact, or the closer π is to $\frac{1}{2}$, the 'truer' it is. This sort of 'closeness to the truth' enters into the pragmatic policy principle stated earlier, which generalizes directly to 'conditional policies', as will be noted below.

The 'conditional' policy in diagram 9.7 is, in circumstances in which $w/l > (1-\pi)/\pi$ (the ratio of the possible win to the possible loss is greater than the ratio of the estimated chance of losing to the chance winning), to make a conditional bet that wins $\$w$ if a coin is tossed and falls heads, loses $\$l$ if it is tossed and falls tails, and neither wins nor loses if the coin isn't tossed. This contrasts with the policy of making unconditional bets that enters into diagram 9.6, but, again, the latter presupposes that the coins are tossed.

The entries in the 'want' and 'result' boxes in diagram 9.7 are also conditionalized versions of the corresponding entries in diagram 9.6. In effect, what the DM now aims for are sufficient wins among the bets that he makes that are either won or lost. And, the relevant result is the 'average win' that would result from making these bets.[42]

Given the above, the pragmatic policy principle still holds: the more closely the thought that p(if coins are tossed they fall heads) $= \pi$ corresponds to the fact that p(if coins are tossed they fall heads) $= \frac{1}{2}$, i.e., the closer π is to $\frac{1}{2}$, the more satisfactory the result of acting in accord with the policy that is based on it will be in the long run. More briefly, the truer the thought is, the more satisfactory the results of these actions are likely to be.

But now we must stress a major difference between the conditional probability fact that p(if coins are tossed they fall heads) $= \frac{1}{2}$, and the unconditional fact that p(coins fall heads) $= \frac{1}{2}$. p(coins fall heads) is the frequency with which statements of the form "The coin will fall heads," said of coins that are about to be tossed, prove to be true over a period of time. That is because "The coin will fall heads" is true if and only if the coin in question falls heads.[43] But p(if coins are tossed they fall heads) is not the frequency with which statements of the form "If the coin is tossed it will fall heads" are

tional propositions. For instance, telling someone "It is probable that if you eat those mushrooms you will be poisoned" wouldn't make explicit what would be implicit in saying "It is probable that you will be poisoned." Our coin examples are thus not representative of probability claims in general, but the negative conclusions that we draw from them here still stand, particularly as concerns the fact that the probabilities of conditionals are not necessarily equal to the probabilities of their being true.

[42]But note that this depends not only on how the coins fall when they are tossed, but on which coin-tosses the DM bets on, as in diagram 9.4.

[43]This is essentially in accord with the semantic theory of truth commented on in section 8.6⋆.

true. Rather, it is the frequency with which the consequents of these statements are true, out of those cases in which their antecedents are true.[44] And, this is the practical basis of the conditional probability thesis discussed in section 8.5, which, according to the triviality results, implies that while unconditional probabilities are probabilities of truth, and satisfy the Kolmogorov axioms, conditional probabilities are not probabilities of truth and do not satisfy these axioms.

9.10 Glossary

Critical Degrees of Confidence Degrees of confidence that, combined with result utilities, generate utilities of actions, two or more of which are equal.

Decision Maker ('DM') A person who has to choose between possible acts, on the basis of the consequences that may follow from them and how much they are desired.

Decision Matrix A matrix whose rows correspond to possible actions, whose columns correspond to possible 'facts', and each cell of which is filled in with the result to which the act in the cell's row and the fact in its column lead. In qualitative decision matrices possible facts are qualified as to whether the DM believes them to be the case, and whether they *are* the case, and the results are qualified as to whether or not the DM wants them. In quantitative decision matrices possible facts are quantified as to how confident the DM is in her beliefs concerning them, as well as whether they are the case, and the results are quantified by how much the DM wants them.

Degrees of Confidence A DM's degree of confidence in a possible fact measures how confident she is that it is the actual fact. These quantities are 'measured' on the assumption that they conform to the laws of probability, and that the utilities of actions are computed from them according to the expected utility rule.

Dominance Act A dominates act B (in a DM's judgment) if she would prefer the result that would result from act A to the one that would result from act B, no matter what the facts are.

Expected Utilities of Actions: Expected Value Rule The rule that the expected value or utility of an action is the sum of the utilities of its possible results, weighted by the DM's degrees of confidence that the action would have these results.

[44]Nor do we want to say that it is the frequency with which statements of the form "If the coin had been tossed it would have fallen heads" are true. Experience has taught us how often coins have fallen heads on occasions when they have been tossed, but not how often they would have fallen heads if they had been tossed. Appendix 6 comments further on counterfactual conditionals.

Implicit Definitions (of Degrees of Confidence and Value) These define a DM's degrees of confidence, and the utility to her of the results that could follow possible actions, not by directly measuring her subjective 'feelings', but by inferring them from her preferences, on the assumption that these are calculated from her degrees of confidence and utilities according to the expected value rule.

Practical Inference An 'inference' whose 'premises' are a DM's beliefs and wants, and whose conclusion is an action.

Pragmatic Diagram A diagram representing relations between five 'factors' that are involved in leading a DM to an action, and in the result of making the decision. The factors are: (1) the DM's beliefs that certain things are the case; (2) what she wants to attain as a result of her action; (3) the action she takes to attain this result; (4) 'the facts' that, together with the action, lead to the result; and (5) the result they lead to. The diagram involves two kinds of connections among the factors and two connections of each kind: (1) causal connections, (a) that lead from the belief and the want to the action, and (b) that lead from the action and the facts to the result; and (2) semantic connections, (a) between the belief and the facts, roughly determining whether the beliefs are true, and (b) between what the DM wants, and the result of the action, roughly determining whether the action 'succeeds'.

Pragmatic Principles

1. *Simple qualitative principle:* The results of actions correspond to what DMs want if and only if the beliefs they are based on correspond to the facts. More roughly: actions succeed if and only if the beliefs they are based on are true.
2. *Degree of Confidence Pragmatic Principle:* Other things being equal, decisions based on degrees of confidence that correspond to the facts, in the sense that things that DMs have high degrees of confidence in are actually the case, have at least as good or better results than ones that do not.
3. *Pragmatic Policy Principle:* The more closely the thoughts on which general policies of action are based correspond to the facts, i.e., the truer they are, the more satisfactory results of actions in accord with them are likely to be.

Result Utility How much a person values or wants something, generally the result of an action.

Semantic Correspondences There are two of these in a **Pragmatic Diagram**: (a) between a DM's beliefs and 'the facts', roughly determining whether the beliefs are true, and (b) between what the DM wants, and the result of the action, roughly determining whether the action 'succeeds'.

Appendix 1

Coherent Degrees of Confidence: Axiomatics and Rational Betting Odds

1.1 Introduction

The view on the nature of probability that was sketched in section 9.9⋆⋆ was *objectivist* in the sense that what makes statements about probabilities right is that they correspond to a special kind of fact, and it was *pragmatic* in the sense that it is to a DM's practical advantage to have degrees of confidence that correspond to them. Very often, however, it is simply held that there are fundamentally different kinds of probability, among which 'objective' (or possibly 'physical') are thought to be one kind, and 'subjective', or 'epistemic' are thought to be another.[1] The latter are allied to the degrees of confidence that chapter 9 was concerned with, but the theory of this kind of probability is usually developed intrinsically, without reference to the 'external facts' that make them 'right' or 'wrong' (or, perhaps 'better' or 'worse'). This appendix will sketch the basic ideas of two such theories, both of which are widely held, and both of which have the following objectives: (1) To show how it is possible to measure, or attach numerical values to degrees of confidence that are 'in the mind', and (2) To show that these numerical values can consistently be assumed to satisfy the basic laws of probability—the Kolmogorov axioms. Except for brief remarks in section 9.7⋆, these things were taken for granted in chapter 9, but theories that underlie them have been very influential. The first theory is axiomatic, starting with 'self-evident principles' that coherent or 'consistent' qualitative degrees of confidence should satisfy, and then proving that if they conform to these principles they must be numerically measurable in a way that conforms to the Kolmogorov axioms. The other approach considers the way that degrees of confidence influence the odds on bets that people will accept, as in section 9.9⋆⋆, but it only argues that it is

[1] Cf. Carnap 1945.

to bettors' practical advantage to accept bets at odds based on degrees of confidence that are measurable by probabilities that satisfy the axioms, and it does not try to prove that any one set of probabilities that satisfy the axioms is any more 'rational' than another. The axiomatic approach will be outlined in section 1.2 of what follows, and the betting odds approach will be outlined in section 1.3. As it happens, both approaches are quite complicated, and only their basic concepts and central results can be sketched here. This appendix will conclude with a brief critique of the two approaches.

1.2 The Axiomatic Approach

Section 9.7⋆ outlined aspects of a theory of the measurement of degrees of confidence and subjective value jointly, but the theories to be commented on here just focus on the former—the measurement of degrees of confidence independent of subjective value. Their basic assumption, usually, is that while a DM may not be directly aware of the numerical value of his degree of confidence in an 'event', say L, that Jane will take logic, he can at least compare this with his confidence in another event, say E, that Jane will take ethics. For purposes of conciseness, if the DM is more confident that Jane will take logic than he is that she will take ethics, we will write L $\gg$ E. The axioms of axiomatic approaches state laws that a rational DM's 'confidence ordering relation', $\gg$, should satisfy. Three that are widely regarded as 'self-evident' are as follows:[2]

A1. $\gg$ *orders* the field of events; i.e., for any events ϕ and ψ, not both $\phi \gg \psi$ and $\psi \gg \phi$, and for any events ϕ, ψ, and η, if $\phi \gg \psi$ then either $\phi \gg \eta$ or $\eta \gg \psi$.[3]

A2. If T is a logical truth and F is a logical falsehood, then T $\gg$ F, and for any event ϕ, it is not the case that F $\gg \phi$.[4]

A3. If η is logically inconsistent with ϕ and with ψ then $(\phi \vee \eta) \gg (\psi \vee \eta)$ if and only if $\phi \gg \psi$.

[2]There is a long history of axiomatizations of this kind, and axioms A1–A3 below are essentially due to de Finetti (1937). Chapter 5 of Krantz et al. (1971) gives a succinct summary of this history, as well as a precise and sophisticated exposition of the theory of qualitative probability representations. The full set of axioms to follow are very close to ones stated in section 5.2 of Krantz et al., op. cit.

[3]This ordering axiom is less familiar than more commonly stated ordering conditions that include transitivity, i.e., that if $\phi \gg \psi$ and $\psi \gg \eta$ then $\phi \gg \eta$. However, the latter is easily seen to follow from axiom A1. A field of 'events' may be considered to be a class of propositions, ϕ, ψ, etc., where if ϕ and ψ are in the class then so are $\sim \phi$, $\phi \& \psi$, and $\phi \vee \psi$, and these are assumed to satisfy the usual logical laws.

[4]Note the dependence of the qualitative degree of confidence axioms on underlying deductive concepts. That the DM is assumed to be able to recognize logical truths and falsehoods is one of the idealizations involved in the theories that we are concerned with.

These axioms are necessary for the existence of a *probability representation*, i.e., they are necessary to guarantee the existence of a probability function, p( ), that satisfies the condition that $p(\phi) > p(\psi)$ should hold if and only if $\phi \gg \psi$. They are commonly postulated without argument, since it is assumed to be self-evident that any rational DM's degrees of confidence should conform to them.[5] However Kraft, Pratt, and Seidenberg (1959) showed that the three axioms aren't sufficient by themselves to guarantee the existence of a probability representation,[6] and subsequent theoretical work has tended to supplement A1–A3 with other conditions, which are not strictly necessary for probabilistic representability, but which together with A1–A3 are sufficient for it.[7] Two axioms of that kind are given in section 5.2.3 of Krantz et al. (1971), rough and simplified versions of which are as follows:

A4. The field of events is *Archimedean*, i.e., if $\phi \gg$ F then there cannot be an infinite number of disjoint 'replicas' of ϕ, say $\phi_1, \phi_2, \ldots$, each of which is equivalent to ϕ in the sense that for any $i = 1, 2, \ldots$ it is not the case that $\phi \gg \phi_i$ or $\phi_i \gg \phi$.

A5. For any $\phi \gg \psi$, there exists some ψ' that logically entails ϕ, which is equivalent to ψ in the above sense.

Given A1–A5, we can state:

Theorem 1. If a field of events satisfies axioms A1–A5 then there exists a probability function $p(\phi)$ defined for all ϕ in the field such that $\phi \gg \psi$ holds if and only if $p(\phi) > p(\psi)$.[8]

It remains to comment briefly on the status of axioms A4 and A5, which are much less self-evident than A1–A3. Adams (1992) has discussed this question in detail, especially as concerns the question of whether axioms A4 and A5 have testable 'empirical content' in the sense that there might be 'data' of the form $\phi \gg \psi$ that are consistent with the existence of a

[5]But some writers have tried to justify certain of them. E.g., Davidson and Suppes (1956) have given an interesting argument for the transitivity of subjective orderings. Other authors, e.g., Chihara (1987), have worried about the idealization involved in assuming that DMs can recognize deductive properties and relations.

[6]See Krantz et al., op. cit., section 5.2.2. General results of Scott and Suppes (1958) are easily adapted to show that no purely universal axioms (roughly, axioms that don't postulate the *existence* of events of special kinds) can be necessary and sufficient for 'probabilistic representability'.

[7]Scott (1964) is an exception, since it presents an infinite set of purely universal axioms that are jointly sufficient to guarantee the existence of a probability representation for any finite field. These axioms are essentially the same as ones worked out and discussed independently in Adams (1965).

[8]Cf. Theorem 2 on p. 208 of Krantz et al. (1971). Section 5.6 of that work sets forth the results of R.D. Luce on the representation of qualitative conditional probability orderings.

probability representation, but which conflict with these axioms. Trivially, the combined axioms A4 and A5 have empirical content, but it is shown that an Archimedean condition akin to A4 does not. In a sense it is 'purely technical': it can be assumed 'for technical reasons' in the proof of a 'probability representation theorem', and we are assured *a priori* that no data that are consistent with the existence of such a representation will be inconsistent with this axiom.[9] But the justification of axioms like A5 raises deep questions of scientific method that cannot be pursued here.

Now we turn to the betting odds approach to the 'logical laws' of degrees of confidence.

1.3 Rational Betting Odds and Dutch Books

The discussion to follow will concentrate mainly on concrete examples of the behavior that theories of rational betting apply to, and it will enter into very few mathematical details. For an excellent elementary exposition of the mathematical theory the reader is referred to sections VI.4–VI.7 of Skyrms (1986).

Consider the following situation. A *bookie*, i.e., a person who takes bets at stated odds on the outcomes of horse races, posts the following odds in bets of different amounts on horses X, Y, and Z *placing*, i.e., he announces that he will accept bets at those amounts on their coming in first or second, in a three-horse race between them. Specifically, he will accept $8 bets that pay $10 if horse X places in the race, $6 bets that pay $10 if Y places, and $5 bets that pay $10 if Z places.[10]

It would not be unreasonable for a bookie to post odds like the ones stated above on any one of the three 'events' of horses X, Y, and Z placing in the race. For instance, if he knew that horse Z was a very poor runner in comparison to X and Y, it would be quite reasonable for him to post odds of less than 50% that Z would come in first or second in the race. But now we come to a crucial question: admitting that situations can be imagined in which it would be reasonable for the bookie to post the odds described above for bets on any one of the three events, does it follow that

[9]However, Manders (1979) examines this and points out the possibility of inductively confirming the existence of an infinite set of data that does this.

[10]Odds are more commonly given, e.g., as "odds of 4:1" that a given horse will win a race, which would imply that a bettor who placed a $1 bet that X would win, which pays off $5 if X does win, would come out a net $4 richer in that case. 4:1 is the ratio between the bettor's net gain of $4 if he wins on the bet and his net loss of $1 if he loses. Characterizing odds as is done here, in terms of $2 that a bettor must place on a bet that wins $10, is done for two reasons. (1) It makes explicit the amounts of money involved in the transactions both of the bettor paying the $2 to place the bet, and of his being paid $10 if he wins. (2) The ratio of these two amounts, namely $2/$10 = .2, is more directly related to probabilities than the four-to-one ratio involved in the more common way of stating odds.

there are situations in which it would be reasonable to post these odds on all three of them? If he did then a bettor could place bets on the three events simultaneously, or, as it is sometimes said, she could *make a book*, and what we are asking is whether it would be rational for the bookie to allow this. In fact, there is a simple argument that this would be highly irrational.

If the bettor made three bets simultaneously, that horse X would place, that horse Y would place, and that horse Z would place, he would pay a total of $\$8 + \$6 + \$5 = \$19$ making them. But clearly she would have to win two of the three bets, since in a race just between three horses only the horse finishing last wouldn't place. Therefore no matter which horses placed, the bettor would win a total of $2 \times \$10 = \$20$. Hence she would end up winning $\$20 - \$19 = \$1$ more than she paid out, and the bookie would end up \$1 worse off than he would have been if he had never accepted the bets.

A combination of bets in which a bettor necessarily ends up with an overall net gain and a bookie ends with a overall net loss, no matter what the results of the individual bets are, is called a *Dutch Book*,[11] and it seems irrational for bookies to post odds that would allow bettors to make such books. This leads us to inquire into the laws of 'coherent' betting odds, i.e., ones that would not allow bettors to make Dutch Books. A rough and ready argument to be given below relates these laws to laws of rational degrees of confidence, but first we will see that the laws of rational betting odds are themselves related to the laws of probability in a very simple way.

The Dutch Book described above is made possible by the following curious fact: The sum of the odds of 8 in 10, 6 in 10, and 5 in 10, namely $.8 + .6 + .5$, equals 1.9, and this is less than the sum of any possible probabilities of X placing, of Y placing, and of Z placing. This is made clear in the following table, which lists the bets on X, Y, and Z placing (as well as three other bets which will be returned to later), and which enumerates all of the possible orders in which X, Y, and Z might finish the race, with their probabilities. Thus, the probability of the horses finishing with X first, Y second, and Z last, abbreviated as $X > Y > Z$, is given as 'a', the probability of their finishing in the order X first, Z second, and Y last, abbreviated as $X > Z > Y$, is given as 'b', and so on. Then the probability

[11]A variant, called semi-Dutch Book in Shimony (1970) is a system of bets on which a bettor cannot suffer a net loss, no matter what the outcomes of the events bet on are, but in which some possible outcomes yield a net gain. For example, bets in which the bettor pays \$0 and wins \$10 if horse X places cannot lose, but it can win \$10 if X actually does place. This appendix focuses on 'strong Dutch Books', but Adams (1962) also presents necessary and sufficient conditions for systems of betting odds not to allow semi-Dutch Books, i.e., which exclude systems that are called weakly rational in the paper.

of X placing is the sum of the probabilities of the horses finishing in any of the orders X > Y > Z, X > Z > Y, Y > X > Z, and Z > X > Y, which is a + b + c + e. Similarly, you can see from the table that the probability of Y placing is a+c+d+f and the probability of Z placing is b + d + e + f:

		events					
possible finishes	distri-bution	X places	Y places	Z places	X wins	X last	if X doesn't win he places
X>Y>Z	a	yes	yes	no	yes	no	no bet
X>Z>Y	b	yes	no	yes	yes	no	no bet
Y>X>Z	c	yes	yes	no	no	no	bet wins
Y>Z>X	d	no	yes	yes	no	yes	bet loses
Z>X>Y	e	yes	no	yes	no	no	bet wins
Z>Y>X	f	no	yes	yes	no	yes	bet loses
probabilities→		a+b+c+e	a+c+d+f	b+d+e+f	a+b	d+f	(c+e)/(c+d+e+f)

Given the above, it follows that

$$\begin{aligned} &p(\text{X places}) + p(\text{Y places}) + p(\text{Z places}) \\ &\qquad = (a+b+c+e) + (a+c+d+f) + (b+d+e+f) \\ &\qquad = 2(a+b+c+d+e+f) = 2, \end{aligned}$$

and, as stated previously, that is greater than 1.9, which is the sum of the odds. This implies in turn that at least one of the probabilities of X placing, Y placing, and Z placing must be greater than the corresponding odds. The following theorem shows the importance of this:

Theorem 2. If in any possible system of probabilities that a set of 'events' might have, there is at least one event whose probability is greater than the odds posted on a bet on it, then it is possible to make a Dutch Book betting at these odds.[12]

The converse also holds:

Theorem 3. If the odds posted for bets on events in some system of events are no greater than some probabilities that these events might have, then it is not possible to make a Dutch Book betting at these odds.

For example, suppose that instead of posting odds of $8, $6, and $5,

[12]'Classical' Dutch Book theorems are due to de Finneti (1937), Kemeny (1955), Lehman (1955), and Shimony (1955; 1970)), and others. Chapter IV of Vickers (1976) gives very elegant proofs of the central classical results. More exact versions of the theorems stated here are given in Adams (1962) and they generalize the classical theorems in two important ways: (1) They allow the bookie to restrict the bets she will accept, e.g., she isn't necessarily 'fair', and (2) They consider the possibility of justifying the countable additivity axiom by reference to infinite sets of bets. Recent unpublished papers of Stinchcombe and of Spielman examine this axiom from the same point of view. It should be noted that the present theorems hold only provided that the bettor is allowed to place any number of bets on any of the events at the posted odds. An example that follows shows that it is sometimes necessary to place more than one bet on a single event in order to make a Dutch Book.

respectively, to win $10 on horses X, Y, and Z placing, the bookie had posted ones of $8, $6, and $6 or .8, .6, and .6, respectively, to win $10 on these events. These odds actually equal the probabilities that the events in question would have if they were generated by a probability distribution in which $a = b = c = e = .2$ and $d = f = .1$, hence according to theorem 2 the bettor couldn't make a Dutch Book betting at these odds. Similarly, if the odds were $9, $7, and $7 to win $10, or .9, .7, and .7, on X, Y, and Z placing, they would actually be greater than the probabilities of .8, .6, and .6 generated by the above distribution, and therefore the bettor couldn't make a Dutch Book betting at these odds either.

It follows from theorems 1 and 2 that if a bookie is 'rational' then the betting odds that he posts on events like the results of a horse race should be at least as great as the probabilities that these events could have, assuming that these probabilities satisfy the axioms of probability. Of course, this doesn't prove that the odds should *equal* probabilities that the events might have, and clearly odds on X's, Y's, and Z's placing of .9, .7, and .6, as above, *couldn't* equal any possible probabilities. In fact a bookie who depended on his winnings to make a livelihood would be foolish to post odds that did equal probabilities, because if he did, his income in the long run couldn't be expected to exceed the amounts that he would have to pay out to persons who won their bets.[13]

One way of forcing the bookie to post odds that equal probabilities that has often been suggested is to require the odds to be 'fair', in the sense that he should be willing to bet at either 'end' of the odds.[14] For instance, if he posts odds that horse X will place that pay $10 on an $8 bet, he should also be willing to make a bet of $8 to win $10 if X places. If fairness is required it is easy to see that the only way the bookie can avoid a Dutch Book is to post odds that actually equal some possible probabilities that the events in question might have. These odds can be related in turn to degrees of confidence.

Sometimes it is assumed that fair betting odds *are* rational degrees of confidence, and assuming this the Dutch Book theorems would guarantee that rational degrees of confidence should satisfy the laws of probability. But this is too simple. At best we can argue that a bookie's degrees of

[13]This is related to the fact that in parimutuel race track betting the state skims a percentage 'off the top', so that the 'bookie' always pays out a little less than he would if he posted 'fair' odds.

[14]Skyrms (1986: 186–9) defines fair bets differently. Adams (1962: 28) presents an argument that odds should be close to 'fair probabilities', as those are defined here. If there are several bookies and one of them posts odds that are significantly higher than such probabilities, then rivals can attract the first bookie's 'customers' away, by posting odds that are closer to fair probabilities, while still avoiding the possibility of a Dutch Book.

confidence should *influence* the odds that he posts, not that they should determine them. Clearly there are other influences as well, such as his need to gain a livelihood. In fact, as will be commented on in more detail in the concluding section, deciding to post betting odds and choosing which odds to post can be viewed as decision problems of the kind discussed in chapter 9, and when viewed this way, many complications emerge which can 'distort' the relation between the odds and degrees of confidence. But there is a plausible argument that the distortion can be eliminated by insisting on fairness. The argument goes as follows.

Suppose we assume that a bookie's betting odds are determined jointly by his degrees of confidence and his 'mercenary' need to gain a livelihood. The latter requires him to take in more money 'on the average' on bets that he accepts than he pays out on bets that win, and that requires him to post unfair odds. Conversely, it is suggested that the odds he would post if he were required to be fair would be independent of this kind of mercenary motive, and they should only depend on his degrees of confidence. If so, they should *measure* his degrees of confidence, i.e., the greater his confidence in the occurrence of an event is, the higher the odds he posts on it should be. And this is just what we wanted to show, because we have argued that fair odds must satisfy the laws of probability. A generalization of this has important implications.

Suppose that, rather than posting odds on X, Y, and Z's placing, the bookie posted odds as follows: (1) \$5 to win \$10 on X actually winning the race, (2) \$2 to win \$10 on X finishing last, and (3), \$5 to win \$10 on the contingent possibility of X at least placing, even if he doesn't win. In bet (3) it is assumed that if X does win the bettor's money is refunded, so that, in effect, bet (3) is a *conditional bet*.[15] Would these odds be rational? In fact, no, because the bettor could make a Dutch Book by simultaneously placing one bet of \$5 that X would win, two bets of \$2 that X would finish last, and two bets of \$5 that if X didn't win he would place. Then the total amount that the bettor would bet would be $1 \times \$5 + 2 \times \$2 + 2 \times \$5 = \$19$, but her 'net bet' would only be \$9 if X won the race, because in that case she would get back the amount she put down on the two bets of \$5 that if X *didn't* win he would place.

Now consider what the result would be in the three cases of X finishing first, finishing second, and finishing last. In the first case, in which the bettor's net bet would be \$9, she would win \$10 on the bet that X *would* win the race, so she would end up \$1 ahead. In the second case her net bet would be \$19, but now she would win \$10 on each of the two bets that she made that if X didn't win he would at least place, so she would end

[15]These are similar to the bets that if coins are tossed they will fall heads, that were commented on in the final part of section 9.9★★.

up with a $1 net win. In the last case her net bet would also be $19, but in this case she would win $10 on each of her bets that X would finish last, and again she would end up with a $1 net gain. In other words the bettor would make a Dutch Book, because she would come out $1 ahead no matter what the outcome of the race was. That this is possible follows from the next theorem, which combines and generalizes theorems 1 and 2:

Theorem 4. Suppose that odds of o(e) are posted on bets on events e and/or odds o(e|c) are posted on e, contingent on c being the case. Then it is possible to make a Dutch Book betting at these odds if and only if for every system of probabilities that these events might have there is at least one event e such that $p(e) > o(e)$, or one event e contingent on c being the case such that $p(e|c) > o(e|c)$.[16]

Thus, rational odds, o(e|c), on 'conditional events' are related to conditional probabilities, p(e|c), in the same way that odds on unconditional events, o(e), are related to the unconditional probabilities, p(e). And, assuming fairness, we can argue that these odds should measure the degrees of confidence that a bookie has in propositions that describe conditional events, such as the ones discussed in section 9.8, that are involved in decision making in nonindependence cases.

Similar considerations have been advanced to justify the assumption that degrees of confidence should change over time in accordance with Bayes' principle.

1.4 Dynamical Dutch Books

Consider the following variant of the conditional betting example. A coin will be tossed twice, and a bookie takes the following three actions: (1) He posts odds of $5 to pay $10 if the coin falls heads on the first toss, (2) He posts odds of $2 to pay $10 if the coin doesn't fall heads on either toss, and (3) He promises that if the coin doesn't fall heads on the first toss, he will post odds of $5 to win $10 if the coin falls heads on the second toss. Now, a variant of the bettor's earlier strategy allows her to make a 'dynamic Dutch Book': (1a) Before the coin is tossed the first time, she places one bet of $5 at the bookie's odds (1), that the coin will fall heads on the first toss, (2a) Also before the coin is tossed the first time, she places two bets of $2 at the bookie's odds (2), that the coin won't fall heads on either toss, and (3a) She decides that if the coin doesn't fall heads on the first toss, she will place two bets that it will fall heads on the second toss, at the odds (3), that the bookie has promised he would post if the coin doesn't fall

[16] This is also proved in Adams (1962), subject to the proviso that the bettor can place any number of bets at the given odds, but with the condition that $p(e|h) > o(e|h)$ replaced by $p(e\&c) > o(e|c) \times p(c)$. This sidesteps the problem of defining p(e|h) when $p(h) = 0$.

heads on the first toss. Readers can work it out for themselves that if the bettor adheres to this strategy and the bookie abides by his promise that if the coin doesn't fall heads on the first toss, he will post odds (3), then the bettor must end with a net gain of \$1, no matter what the results of the individual tosses are. Thus, the bettor has made a Dutch Book, which is 'dynamical' because it involves actions taken at two different times: odds posted and bets made before the coin is tossed the first time, and, if it doesn't fall heads the first time, odds offered and bets placed before it is tossed the second time.

The following diagram will help to 'diagnose' the problem above:

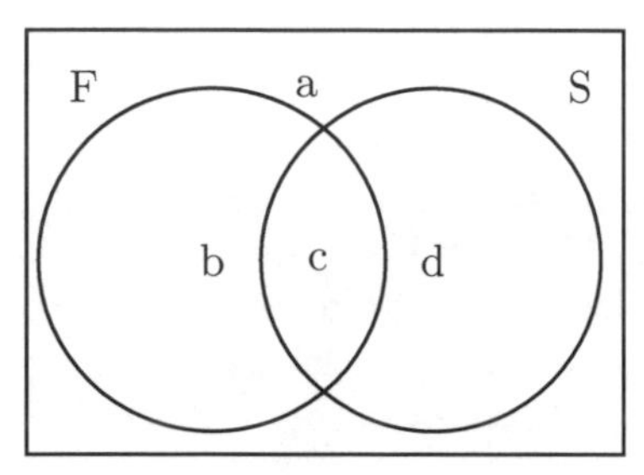

F = the coin falls heads on the 1st toss
S = the coin falls heads on the 2nd toss

$o(F) = .5 \geq p_0(F) = b + c$

$o(\sim(F \vee S)) = .2 \geq p_0(\sim(F \vee S)) = a$

$a/d \leq .2/.3$

$p_0(S|\sim F) = \frac{1}{1+(a/d)} \geq \frac{1}{1+(.2/.3)} = .6$

The probabilities given above, $p_0(F), p_0(\sim F \vee S))$, and the conditional probability $p_0(S|\sim F)$, are ones that obtain prior to the first coin-toss, and the first two are assumed to be 'rational' in the sense that they are not greater than the corresponding odds, $o(F) = .5$ and $o(\sim(F \vee S)) = .2$. These assumptions entail that $p_0(S|\sim F) \geq .6$, as shown. Now, if Bayes' principle is assumed then $p_0(S|\sim F)$ should equal $p_1(S)$, the unconditional probability of S *posterior* to learning that the coin didn't fall heads on the first toss. Thus, $p_1(S)$ should be at least .6. But that *shouldn't* be greater than the odds of .5 that the bookie promised to post on the coin's falling heads on the second toss, after it is learned that it didn't fall heads on the first toss. Generalizing, it seems that to avoid the possibility of a dynamic Dutch Book, a bookie should post odds that are at least as high as probabilities that conform to Bayes' principle, and to be fair he should post ones that equal probabilities conforming to the principle.

Thus Dutch Book considerations not only seem to justify the conclusion that degrees of confidence should satisfy the 'static' laws of probability, but they should also evolve in accord with Bayes' principle. However, this is not quite decisive.

1.5 Critique

One objection to the Dutch Book argument for Bayes' principle is that it assumes that the bookie promises in advance that if certain things happen he will then accept bets at predetermined odds, e.g., if the coin doesn't fall heads on the first toss then he will accept bets of $5 that pay $10 if it falls heads on the second one. But for the bookie to bind himself *a priori* in this way is in effect to post odds *a priori*, on the conditional bet that if the coin doesn't fall heads on the first toss it will fall heads on the second. All the argument shows, therefore, is that it would be incoherent for the bookie to post odds on three bets, including a conditional one, *a priori*. But theorem 3 already told us this, and there is no necessary connection between that and changes in betting odds. What we must ask is whether the bookie should be bound to accept bets placed *a posteriori*, at the odds that he posts on conditional bets placed *a priori*. That seems doubtful.

It is obvious that whatever odds a bookie posts, he only commits himself to accept bets at those odds during a limited period of time. For instance, it would be irrational for him to accept bets on a race that are placed after the race is over—even more so than to post odds that permitted a Dutch Book. In fact, this illustrates a point made earlier, that the odds that a bookie posts and the bets that he accepts are generally influenced by other things besides his degrees of confidence. That he wouldn't accept bets placed 'after the fact' reflects the fact that he is influenced by what he thinks bettors know. In particular, he knows that he ought to know at least as much about the events he posts odds on as bettors do when they place bets at those odds, and the reason he doesn't accept bets placed after the fact is because he knows that the persons who placed them then would know more than he did when he posted the odds.

The same considerations even call into question the thesis that 'static' betting odds ought to conform to the probability axioms. If no restrictions were placed on the times at which a bettor could make bets at a bookie's odds, say on the outcome of a horse race, she could always make sure of winning by waiting until the race was over before placing her bets. If the bookie had to accept such bets, the only way that he could avoid a sure loss would be to post odds of $10 to win $10 on all events, which would be inconsistent with the laws of probability, and would be in effect to accept no bets at all! Moreover, if he were compelled to be 'fair', and post odds that satisfied the laws of probability, he would have to offer ones that differed markedly from his degrees of confidence in order to minimize his losses. For instance, if he were 10% confident that horse X would win the race and 90% confident that he wouldn't, he should post odds of $2.50 to win $10 that X would win the race, and $7.50 to win $10 that he wouldn't win. If he did this and allowed bettors to wait until the race was over before placing their bets, his expected loss would be $6 per bet. While, if he posted odds equalling his degrees of confidence, his expected loss would be $10 per bet.[17]

[17]This and other 'oddities' of the relation between degrees of confidence and betting

But, to conclude, approaching the subject of consistency or coherence in degrees of confidence either axiomatically or via rational betting odds seems very dubious to the writer. What DMs and bookies *want* are results that are beneficial to themselves, and consistency should be valued primarily to the extent that it tends to such results. As F.P. Ramsey wrote "It is better to be inconsistent and right some of the time than to be consistent and wrong all of the time" ("Truth and Probability," p. 191).[18] Assuming this, if a DM's degrees of confidence conform to the laws of probability, this is most likely to be because he attempts to bring them into correspondence with objective probabilities such as those discussed in section 9.9⋆⋆, and these are what necessarily satisfy the axioms of probability.

---

odds that come out when the choice of odds is viewed as a decision problem in the manner of chapter 9 are discussed in section 2 of Adams and Rosenkrantz (1980). The special precautions that must be taken to assure that bettors are no more knowledgeable than bookies are discussed in Hacking (1967).

[18] Adams (1988a) points out a secondary virtue of consistency: namely to facilitate communication. Blatant contradiction generally produces confusion in a hearer, thus defeating a speakers' purposes.

Appendix 2

Infinitesimal Probabilities and Popper Functions

2.1 Introduction

Infinitesimal probabilities and Popper functions will be discussed here primarily because of their bearing on the characterization of conditional probabilities $p(\psi|\phi)$ in the case in which $p(\phi) = 0$. Recall that $p(\psi|\phi)$ cannot be defined as a ratio, $p(\phi\&\psi)/p(\phi)$, when $p(\phi) = 0$, and the stipulation that in this case $p(\psi|\phi) = 1$ was largely arbitrary. It was clearly not justified by practical considerations like those set forth in section 9.9⋆⋆, and the only practical justification for it is that it seems to be 'pragmatically harmless'. What difference does it make what we say the probability is that, if half of 5 is 3 then one third of 10 is 4, if it is certain that half of 5 isn't 3?[1] And, the stipulation has the merit of simplicity. Even if $p(\psi|\phi)$ is not always the ratio of $p(\psi\&\phi)$ and $p(\phi)$, at least it is uniquely determined by them; i.e., it is a function of these probabilities. But objections to the stipulation can be raised, and we will note two of them before turning to infinitesimals and Popper functions, which provide a way of meeting these objections.

2.2 Two Objections to the Stipulation

One objection is that if we assume that $p(\psi|\phi) = 1$ when $p(\phi) = 0$, then laws of probability, such that $p(\psi|\phi)+p(\sim\psi|\phi) = 1$, have exceptions when $p(\phi) = 0$. Clearly, in that case, $p(\psi|\phi) + p(\sim\psi|\phi) = 2$. This example might suggest that it would be better to stipulate that $p(\psi|\phi) = \frac{1}{2}$ when $p(\phi) = 0$, since that would entail that $p(\psi|\phi) + p(\sim\psi|\phi)$ equalled 1 even in the case in which $p(\phi) = 0$. But the more general law that if $\psi_1,\ldots,\psi_n$ are mutually exclusive and exhaustive then $p(\psi_1|\phi)+\ldots+p(\psi_n|\phi)$ should equal 1 would still have exceptions when $p(\phi) = 0$ and $n \neq 2$, since in that

[1] This is at the core of the so called 'Tartaglia's riddle', after the Italian Renaissance mathematician Nicolo Tartaglia, who asked "If half 5 were 3, what would a third of 10 be?"

case $p(\psi_1|\phi)+\ldots+p(\psi_n|\phi)$ would equal $\frac{n}{2} \neq 1$. In fact, no stipulation that $p(\psi|\phi)$ should have a constant value, 1, $\frac{1}{2}$, or whatever, can be consistent with the law $p(\psi_1|\phi)+\ldots+p(\psi_n|\phi) = 1$ for all n and all mutually exclusive and exhaustive $\psi_1,\ldots,\psi_n$.

A less formal objection to the stipulation is that it seems counterintuitive in many cases. For example, suppose that a point is picked at random inside the rectangle below, and consider the probability of its lying inside the circle P, given that it happens to be one of the two points π_1 or π_2, as shown.

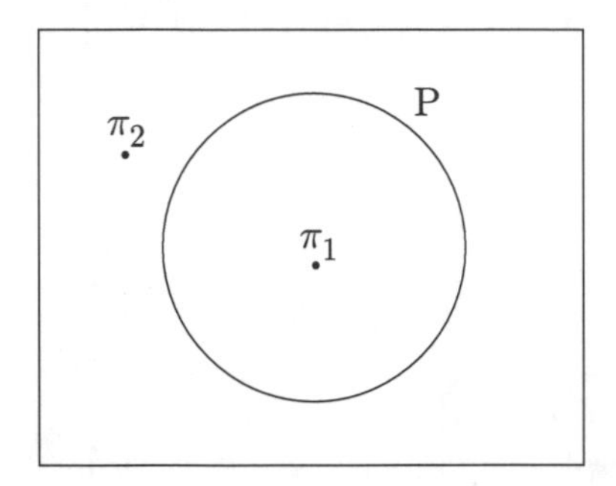

Intuitively, this probability ought to be $\frac{1}{2}$, given that π_1 lies inside P and π_2 lies outside it, and the point we are concerned with is chosen at random. But, if the probability we are concerned with is the conditional probability

$$p(\text{the point picked is in P}|\text{ it is either } \pi_1 \text{ or } \pi_2),$$

and $p(\text{the point is either } \pi_1 \text{ or } \pi_2) = 0$ because there is 0 chance that a point picked at random will be one of two out of an infinite number of points, then this conditional probability should be 1, according to our stipulation.

But the idea that the probability that the point is either π_1 or π_2 is 'absolutely' equal to 0 because it is chosen from an infinite number of points is counterintuitive, since we feel intuitively that there is *some* chance of its being one of π_1 or π_2. And, it is to allow for this that the idea of an *infinitesimal probability*, ι, is introduced, together with an 'infinitesimal arithmetic', according to which ι is greater than 0 but smaller than any 'standard number' like $\frac{1}{2}$, $\frac{1}{3}$, $\frac{1}{4}$, and so on. This is what we will now sketch, and show that assuming that

$$p(\text{the point is either } \pi_1 \text{ or } \pi_2) = 2\iota,$$

which is infinitesimal because ι is infinitesimal, we can still assume that

$$p(\text{the point picked is in P}|\text{it is either } \pi_1 \text{ or } \pi_2) = \tfrac{1}{2}$$

because it is equal to the ratio:

$$\frac{p(\text{the point is } \pi_1)}{p(\text{the point is either } \pi_1 \text{ or } \pi_2)} = \frac{\iota}{2\iota}$$

Before starting, though, it should be noted that the assumption that infinitesimal numbers exist was long a controversial matter in the foundations of mathematics. It was assumed by Isaac Newton in his 'invention' of the calculus, but it was criticized severely because it led to 'paradoxes of the infinite'. For instance, the noted philosopher George Berkeley called Newton's infinitesimals "ghosts of departed entities" in his famous work, *The Analyst*, 1734, and during most of the 19th and 20th centuries they have been banished from mathematics and replaced by *limits* of series of standard numbers. Abraham Robinson resurrected them in the 1960s and 1970s, by showing how the paradoxes can be avoided,[2] but this requires important revisions in ideas that are now orthodox in the foundations of mathematics. These cannot be entered into in detail here, and we will only give an intuitive sketch of the application of these ideas to the problem of defining $p(\psi|\phi)$ when $p(\phi) = 0$.

2.3 Intuitive Infinitesimals and Non-standard Probabilities

Symbolizing "The point picked is in P" as P, "the point picked is π_1" as Π_1, and "the point picked is π_2" as Π_2, the idea is that $p(\Pi_1)$ and $p(\Pi_2)$ can be assumed to have an infinitesimal value, ι.[3] Given this and the fact that $(\Pi_1 \vee \Pi_2)\&P$ is equivalent to Π_1 (because point π_1 is inside P and π_2 is outside of it), the probabilities of $(\Pi_1 \vee \Pi_2)\&P$ and of $\Pi_1 \vee \Pi_2$ must be ι and $\iota + \iota = 2\iota$, respectively, hence the conditional probability $p(P|\Pi_1 \vee \Pi_2)$ can be defined as the ratio $\iota/2\iota$. But what are these 'infinitesimal quantities', and what laws do they satisfy? As said, rigorous answers to these questions are provided by the modern theory of non-standard analysis, and here we will only sketch key ideas of the theory informally.

For our purposes, the two most important facts about infinitesimals like ι and 2ι are: (1) They are smaller than any 'finite' number like 1, .1. .01, .001, etc., and (2) They obey many of the same laws as finite numbers such as that $\iota+(\iota+\iota) = (\iota+\iota)+\iota$ (associativity of addition), and $\iota \times (\iota+\iota) = \iota \times \iota + \iota \times \iota$ (distributivity of multiplication).[4] They can also be combined with finite numbers, e.g., $1+\iota$, which is a non-standard real number. Most

[2]The modern theory of infinitesimals is largely Robinson's creation. His book, *Non-standard Analysis* (1996) sets forth the basic ideas of the theory, and chapter X gives a brief history of the controversies about them. Aspects of non-standard probability theory are set forth in Hoover (1982).

[3]The infinitesimal number ι mustn't be confused with the 'new information' that section 5.5 symbolizes as ι, and which will also be used in this sense in appendices 7, 8, and 9.

[4]More generally and more exactly, non-standard real numbers possess every property of the real number system that can be expressed in the 'first-order theory' of real numbers.

importantly for our purposes, infinitesimals and other non-standard real numbers can be divided, e.g., $\iota/2\iota = 1\iota/2\iota$, which is equal to 1/2 because you can cancel the quantity ι that appears in both the numerator and the denominator of the fraction $\iota/2\iota$.[5]

Now it is clear that if $p(P\&(\Pi_1 \vee \Pi_2)) = \iota$ and $p(\Pi_1 \vee \Pi_2) = 2\iota$, i.e., if the probability of picking a point at random that is both in the circle P and is one of the points π_1 or π_2 is ι, and the probability that it is one of π_1 or π_2 is 2ι, then $p(P|\Pi_1 \vee \Pi_2) = \iota/2\iota = 1/2$. I.e., the finite probability $p(P|\Pi_1 \vee \Pi_2)$ is still a ratio of unconditional probabilities, though the latter are themselves infinitesimals. In fact, but for one exception this gets around the problem of defining $p(\psi|\phi)$ when $p(\phi)$ is infinitely small. The exception is when ϕ is not only a 'practical impossibility', like picking a point at random that happens to be one of the two points π_1 or π_2, but it is a 'total impossibility' like picking a point that is both inside and outside circle P. In this case we still stipulate arbitrarily that $p(P|P\&{\sim}P) = 1$, because it cannot be defined as $p(P\&P\&{\sim}P)/p(P\&{\sim}P)$ if $P\&{\sim}P$ has a probability equal to absolute 0, which is smaller than any smallness, even infinitesimal.

These ideas are illustrated in the table below, which is like table 6.4 in chapter 6, except that it begins with a *non-standard probability distribution* over SDs formed from the atomic formulas P, Π_1, and Π_2, and then it gives the probabilities generated by the distribution for the three formulas $\Pi_1 \vee \Pi_2$, $(\Pi_1 \vee \Pi_2)\&P$, and $(\Pi_1 \vee \Pi_2) \Rightarrow P$. Note incidentally that four of the eight SDs in the table are assigned probabilities that equal absolute 0. E.g., SD #1, in which all three of P, Π_1, and Π_2 are true, has a probability of absolute 0 because it is logically impossible for a point picked at random to be in circle P, and at point π_1, and at point π_2. This is in contrast to the infinitesimal probability ι assigned to SD #2, which is greater than absolute 0 but less than any finite value. The so-called *Popper distribution*, to the left of the non-standard one, which 'conflates' the infinitesimal value ι with absolute 0 because that is the 'nearest standard real value' to ι, will

[5]But division doesn't always yield a finite number. For instance, $1/\iota$ is an infinite number, and therefore it is not close to any finite number, contrary to what will be assumed below. However, the ratios of the non-standard probabilities that we will be concerned with are always of the form $p(\phi\&\psi)/p(\phi)$, which are finite since even if $p(\phi\&\psi)$ and $p(\phi)$ are non-standard numbers, nevertheless $0 \leq p(\phi\&\psi) \leq p(\phi) \leq 1$. In fact, non-standard probabilities are assumed to satisfy all of the Kolmogorov axioms, extended to non-standard values. Thus, it is still the case that $0 \leq p(\phi) \leq 1$; if ψ is a logical consequence of ϕ then $p(\phi) \leq p(\psi)$, and if ϕ and ψ are logically inconsistent then $p(\phi \vee \psi) = p(\phi) + p(\psi)$.

The fact that in other cases ratios aren't always finite, shows that it cannot be taken for granted that non-standard numbers and infinitesimals in particular satisfy all of the laws that finite numbers do. Perhaps the most conspicuous difference between them and finite numbers is that finite numbers have numerical representations like 1, .1, .01, and so on, but 'decimal expansions' of infinitesimals like ι have to have a non-standard number of 'places'.

be commented on in the following section.

	distributions		formulas					
	Popper	non-standard	P	Π_1	Π_2	$\Pi_1 \vee \Pi_2$	$(\Pi_1 \vee \Pi_2)\&P$	$(\Pi_1 \vee \Pi_2) \Rightarrow P$
1	0	0	T	T	T	T	T	*T*
2	0	ι	T	T	F	T	T	*T*
3	0	0	T	F	T	T	T	*T*
4	.4	$.4-\iota$	T	F	F	F	F	......
5	0	0	F	T	T	T	F	*F*
6	0	0	F	T	F	T	F	*F*
7	0	ι	F	F	T	T	F	*F*
8	.6	$.6-\iota$	F	F	F	F	F	......
non-standard probabilities			.4	ι	ι	2ι	ι	$\iota/2\iota = .5$
Popper function values			.4	0	0	0	0	.5

Note that for the purpose of illustration we have assumed that the probability of P is .4, hence if the probability of picking point π_1 is ι, then the probability of P&$\sim\Pi_1$, that is of picking a point in P different from π_1, is $.4-\iota$. The probabilities of the formulas in the table are also computed according to the same rules that apply to standard numbers, but extended to non-standard values. Thus, the probability of $\Pi_1 \vee \Pi_2$ is the sum of the probabilities of the SDs in which it is true, and since it is true in lines 1, 2, 3, 5, 6, and 7, it is the sum of the probabilities of these lines; i.e., $p(\Pi_1 \vee \Pi_2) = 0 + \iota + 0 + 0 + 0 + \iota = 2\iota$. Similarly, the probability of the conditional $(\Pi_1 \vee \Pi_2) \Rightarrow P$ is the sum of the probabilities of the lines in which it has the *ersatz* value *T*, divided by the sum of the probabilities of the lines in which it has either the value *T* or the value *F* (cf. section 4.4⋆). Therefore, $p((\Pi_1 \vee \Pi_2) \Rightarrow P) = (0+\iota+0)/(0+\iota+0+0+0+\iota) = \iota/2\iota = 1/2$.

For present purposes the most important fact about non-standard probabilities is that they not only satisfy most of the laws of standard probabilities, but most of the theorems of probability logic stated in chapters 6 and 7⋆ are also true of them. Thus, all of the clauses of the second equivalence theorem stated in section 7.6⋆⋆ continue to hold. For instance, if a subset of the premises of an inference yields its conclusion then the uncertainty sum condition applies, so the uncertainty of the conclusion cannot exceed the sum of the uncertainties of the premises, while if no such subset exists then its premises can be arbitrarily highly probable while its conclusion has 0 probability. But now, since probabilities and uncertainties are non-standard, 'arbitrarily highly probable' can be infinitesimally close to 1.[6] For instance, the reader can easily verify that the non-standard distribution in the table generates a probability of $1-(.4-\iota)/.4 = 1-2.5\iota$ for the premise

[6]These and many other important properties of the reformulated 'logic of conditionals' are stated in section 3 of McGee (1994).

of the p-invalid inference $P \Rightarrow \sim\Pi_1 \therefore \Pi_1 \Rightarrow \sim P$ (contraposition),[7] while the conclusion has probability 0. Exercises at the end of this appendix ask the reader to construct non-standard probability counterexamples to other p-invalid inferences.

But there is a drawback to the infinitesimal characterization. That is that infinitesimals are not 'measurable', and it is not clear what sort of argument might be given, analogous to the one given in section 9.9★★, to show that an infinitesimal probability like the postulated value $p(\Pi_1 \vee \Pi_2) = 2\iota$ is the probability of a point picked at random being one of π_1 or π_2. There is no obvious sense in which this can be interpreted as the frequency with which one of π_1 or π_2 is picked in a series of 'trials', however long. So, while setting $p(P|\Pi_1 \vee \Pi_2) = \iota/2\iota$ solves the problem of defining $p(P|\Pi_1 \vee \Pi_2)$, the solution is a purely formal one. Fortunately, however, McGee (1994) has shown that the infinitesimals solution to the problem is equivalent to a solution in terms of *Popper functions*, that only involve finite probabilities.

2.4 Popper Functions[8]

These are functions that take ordinary real numbers as values, but which conform to slightly modified principles of conditional probability, including especially the laws:

1. If $\phi_1,\ldots,\phi_n$ are mutually exclusive then
$$p(\phi_1 \vee \ldots \vee \phi_n) = p(\phi_1) + \ldots + p(\phi_n);$$
2. $p(\phi\&\psi|\eta) = p(\phi|\eta) \times p(\psi|\phi\&\eta)$ for all ϕ, ψ, and η;
3. if $p_\phi(\psi) = p(\psi|\phi)$ then $p_\phi(\psi)$ satisfies the Kolmogorov axioms, unless $p_\phi(\psi)=1$ for all ψ; and
4. if t is a logical truth then $p(\psi|\phi) = p_t(\psi\&\phi)/p_t(\phi)$, unless $p_t(\phi) = 0$.

(1)–(4) are in fact laws of standard probability, now applied to Popper functions. But the extended laws aren't the same as the unextended ones in all respects. The most important difference is that when $p(\psi|\phi)$ isn't a ratio

[7]This is established by a little non-standard arithmetic:
$$(.4 - \iota)/.4 = 1 - (\iota/.4) = 1 - (\iota \times (1/.4)) = 1 - \iota \times 2.5.$$
That $1 - 2.5\iota$ is infinitesimally smaller than 1 follows because
$$1 - (1 - 2.5\iota) = 2.5\iota,$$
and 2.5ι can't be larger than any standard number, .1, .01, .001, ... For instance, if 2.5ι were larger than .001 then ι would be larger than $.001/2.5 = .004$, and ι wouldn't be infinitesimal.

[8]These are named after the philosopher Karl Popper, who described them in new appendices *ii–*v, in Popper (1959). A recent paper of J. Hawthorne (Hawthorne, 1996) applies Popper functions to analyze the logic of a range of conditionals, including ones of the form "If C then almost certainly B" and "If C then just possibly B," which are more general than the 'simple conditionals', "If C then B," with which this text has been concerned.

of unconditional probabilities because $p_t(\phi) = 0$, as in (4), then: (1) It isn't stipulated to equal 1, and (2) It is not necessarily a function of $p_t(\phi)$ and $p_t(\psi\&\phi)$. However, it should be noted that while Popper functions don't necessarily satisfy the condition that $p(\psi|\phi) = 1$ when $p_t(\phi) = 0$, they *can* satisfy it, and in that case they satisfy all of the laws set forth in chapter 4, without modification. But the example of Popper function values given in the table shows that $p(\psi|\phi)$ can be less than 1 when $p_t(\phi) = 0$. In that case the Popper function values in the table for both $\Pi_1 \vee \Pi_2$ and $(\Pi_1 \vee \Pi_2)\&P$ are equal to 0, but the Popper function value of P, given $\Pi_1 \vee \Pi_2$, is .5. This accords with an important theorem due to McGee (1994) roughly as follows.[9]

To begin with, note the correspondence between the non-standard probabilities in the table and the standard real numbers to the left of them. For instance, the standard real number 0 in line #2 corresponds to the infinitesimal number ι to the right of it, because 0 differs infinitesimally from ι, and the standard real number .4 in line #4 corresponds to the non-standard real number $.4-\iota$ to the right of it, because .4 is a standard real number which differs infinitesimally from $.4-\iota$.[10]

Now McGee's first theorem says the following: *The standard real values of non-standard probability functions form Popper functions; moreover, every Popper function is generated by the standard real values of some non-standard probability function.* Thus, there is a many-one correspondence between non-standard probability functions and Popper functions that permits us to 'translate' propositions about non-standard probabilities, including infinitesimals, into facts about ordinary real-valued functions, albeit ones that don't prefectly satisfy the laws set forth in chapter 4.

The practical import of this has to do with the fact that for the most part propositions about infinitesimals are more 'intuitive' than ones about Popper functions, and therefore it is easier to conceptualize chances, e.g., that a point picked at random should be one of points π_1 or π_2, in terms of infinitesimals than it is to conceptualize them in terms of Popper functions. The key point is that McGee's theorem 1 justifies this intuitive procedure, by telling us that everything expressed in terms of non-standard probabilities can be equivalently expressed in terms of Popper function probabilities.

---

[9]Theorem 1 on pages 181–2 of McGee (1994). An exact statement of this theorem requires a more exact characterization of infinitesimal probabilities, which is not entered into here.

[10]That there is the unique standard real number that differs only infinitesimally from any non-standard real number isn't self-evident. Of course, it is obvious in the cases ι and .4-ι, which differ only infinitesimally from 0 and .4, but it is false in the case of the 'infinite' non-standard real, $1/\iota$. That all finite non-standard reals are infinitely close to standard reals is, in fact, a fundamental and nonself-evident theorem of the theory of non-standard numbers.

The theoretical import of McGee's theorem 1 has to do with the fact that infinitesimal probabilities 'project' into the Popper function value of absolute 0. If this is coupled with the fact that if an inference is p-invalid then it is possible for its premises to have probabilities that are only infinitesimally smaller than 1 while its conclusion has a probability equal to absolute 0, we get McGee's theorem 3 (McGee, 1994: 190). That is: (1) *If an inference is probabilistically valid then the fact that its premises have Popper probabilities equal to 1 guarantees that its conclusion has this value, and (2) If it is p-invalid then it is possible for its premises to have Popper probabilities equal to 1 while its conclusion has Popper probability 0.* If we treat Popper probabilities of 1 and 0 as truth and falsehood, respectively, in a sense this 'reduces' p-validity to classical validity. However, as McGee himself points out on page 194 of his paper, the question of whether this is a 'true' reduction of the concept of probabilistic validity to the standard truth-conditional concept is a matter that is still in doubt.

Exercises

1. Calculate the probabilities that are generated by the non-standard distribution in section 2.3 for the following formulas:
   a. $\Pi_1 \& {\sim}\Pi_2$
   b. $\mathrm{P} \vee \Pi_1$
   c. $\mathrm{P} \Rightarrow \Pi_2$
   d. $(\mathrm{P} \vee \Pi_1) \Rightarrow \mathrm{P}$
   e. $(\mathrm{P} \leftrightarrow \Pi_1) \Rightarrow \mathrm{P}$
2. Consider the diagram below, and let Π_1–Π_3 be the propositions that a point chosen at random inside the outer rectangle is π_1–π_3, respectively, and let B and C be the propositions that it is inside circles B and C, respectively, and suppose that $p(\Pi_1)$, $p(\Pi_2)$, and $p(\Pi_3)$, all equal an infinitesimal value ι, and p(B) and p(C) have standard finite values b and c, respectively. Then determine the following non-standard probabilities:

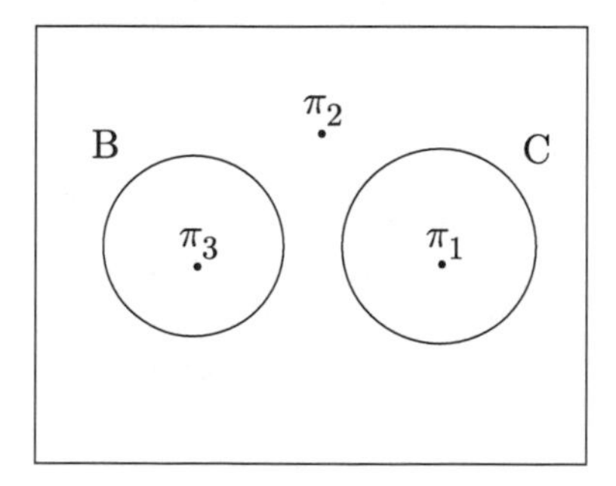

a. $p(B \vee \Pi_2 \vee \Pi_3)$

b. $p(\sim(B \vee \Pi_1))$

c. $p(\Pi_1|\Pi_1 \vee \Pi_2)$

d. $p(C|\Pi_1 \vee \Pi_2)$

e. $p(B \vee \pi_1|C \vee \pi_2)$

3. In each of the p-invalid inferences below, construct a non-standard probability counterexample in which its conclusions have probability 0 and whose premises have probabilities that differ at most infinitesimally from 1.

a. $B \mathrel{.\,.} A \Rightarrow B$

b. $A \vee B \mathrel{.\,.} \sim A \Rightarrow B$

c. $\sim(A\&B) \mathrel{.\,.} A \Rightarrow \sim B$

d. $(A \vee B) \Rightarrow C \mathrel{.\,.} A \Rightarrow C$

e. $\{A \Rightarrow B, B \Rightarrow C\} \mathrel{.\,.} A \Rightarrow C$

Appendix 3

David Lewis's Triviality Results

3.1 Introduction

Lewis's Triviality Results first appeared in a paper "Probabilities of Conditionals and Conditional Probabilities," (Lewis, 1976). From their first appearance they were regarded as a 'bombshell', since they established to most persons' satisfaction that the probability conditional theory is incompatible with assumptions about probability that are usually taken for granted by persons who employ probability concepts in their research. They tend to the same conclusion as the results of Alan Hájek that were briefly described in section 8.5, but they are arrived at in a way that is significantly different from Hájek's, and, as will be seen below, their presuppositions are different. This appendix will first describe Lewis's results (section 3.2), then an apparent 'way out',[1] and comparison with Hájek's results (section 3.3), and finally it will comment briefly on 'who is right' (section 3.4).

3.2 The Results

We assume as before that we have a language of ordinary sentential logic with a special nonmaterial conditional symbol $\Rightarrow$ added, and that probabilities attach to formulas of the language in the way that was assumed in chapters 6 and 7$\star$. However, Lewis's argument assumes that nonmaterial conditionals can be embedded in larger contexts like $\sim(A \Rightarrow B)$ or $B\&(A \Rightarrow B)$, and moreover that thus embedded they satisfy all of the standard laws of probability functions. E.g., the argument assumes that $p(\sim(A \Rightarrow B)) = 1 - p(A \Rightarrow B)$ and if $p(B) > 0$ then $p(A \Rightarrow B|B) = p(B\&(A \Rightarrow B))/p(B)$; i.e., $p(A \Rightarrow B|B)$ is a conditional probability. Let us see where this leads.[2]

Suppose that $p_0(A \Rightarrow B)$ and $p_1(A \Rightarrow B)$ are, respectively, the *a priori* and the *a posteriori* probabilities of $A \Rightarrow B$ before and after it is learned that B is the case, and assume that $p_0(A\&B)$ is positive, and therefore

[1] Which is similar to but less general than one originally due to van Fraassen (1976).

[2] There are slight differences between the present proofs of the Triviality Results and Lewis's own, but these are inessential.

so is $p_0(B)$. Given this and assuming Bayes' principle, $p_1(A)$ should equal $p_0(A|B)$ and that should equal $p_0(A\&B)/p_0(B)$, since $p_0(B)$ is positive. Moreover, $p_1(A \Rightarrow B)$ should equal $p_1(A\&B)/p_1(A)$. And, still assuming Bayes' principle, it should be the case that $p_1(A\&B) = p_0(A\&B|B)$, $p_1(A) = p_0(A|B)$, and, especially significantly, that

(1) $p_1(A \Rightarrow B) = p_0(A \Rightarrow B|B)$.

Now we can write

(2) $p_1(A \Rightarrow B) = \dfrac{p_1(A\&B)}{p_1(A)} = \dfrac{p_0(A\&B|B)}{p_0(A|B)} = \dfrac{p_0(A|B)}{p_0(A|B)} = 1$

and combining this with (1) gives:

(3) $p_0(B\&(A \Rightarrow B)) = p_0(B) \times p_0(A \Rightarrow B|B) = p_0(B) \times p_1(A \Rightarrow B) = p_0(B)$.

Assuming that $p_0(A\&{\sim}B)$ is positive, a similar argument leads to the conclusion that

(4) $p_0({\sim}B\&(A \Rightarrow B)) = 0$.[3]

Expanding $p_0(A \Rightarrow B)$ according to the laws of probability we get

(5) $p_0(A \Rightarrow B) = p_0(B\&(A \Rightarrow B)) \times p_0(B) + p_0({\sim}B\&(A \Rightarrow B)) \times p_0({\sim}B)$,

and substituting the values given by (3) and (4) for $p_0(B\&(A \Rightarrow B))$ and $p_0({\sim}B\&(A \Rightarrow B))$, respectively, in (5) yields:

(6) $p_0(A \Rightarrow B) = p_0(B) \times 1 + p_0({\sim}B) \times 0 = p_0(B)$,

assuming that $p_0(A\&B)$ and $p_0(A\&{\sim}B)$ are positive. This is an intuitive absurdity, since it implies that B and A are independent *a priori* (cf. section 4.3).

The foregoing leads to Lewis's Second Triviality Result, namely that *if the assumptions of the foregoing argument are correct then* $p_0(\ )$ *can have at most 4 distinct values.* An intermediate step in the proof of this is to show that there could not be three mutually exclusive propositions, P, Q, and R, each of which has positive probability, as in diagram 1 below:

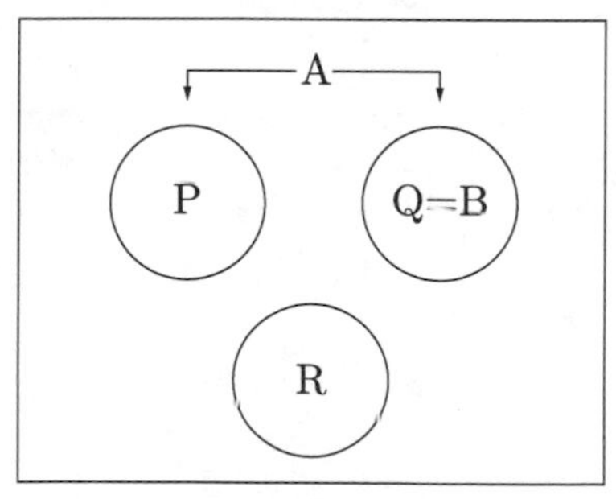

Diagram 1

[3]Exercise 1 at the end of this appendix asks the reader to prove this.

If P, Q, and R were mutually exclusive then setting $A = p \vee Q$ and $B = Q$, it would follow that both A&B and A&~B had positive probability, since A&B would be equivalent to P and A&~B would be equivalent to Q. But $p_0(B) = p_0(Q)$ would be less than $p_0(A \Rightarrow B) = p_0(Q)/p_0(P \vee Q)$ because $p_0(P \vee Q) \leq 1 - p_0(R)$ and $p_0(R)$ is positive, and this would be inconsistent with equation (6).

Now we can see that $p_0(\ )$ could not have five distinct values. If it did then at least three of them, say $p_0(P)$, $p_0(Q)$, and $p_0(R)$, would have to lie strictly between 0 and 1, and in that case we could picture P, Q, and R and their probabilities in a Venn diagram, as in diagram 2 below.

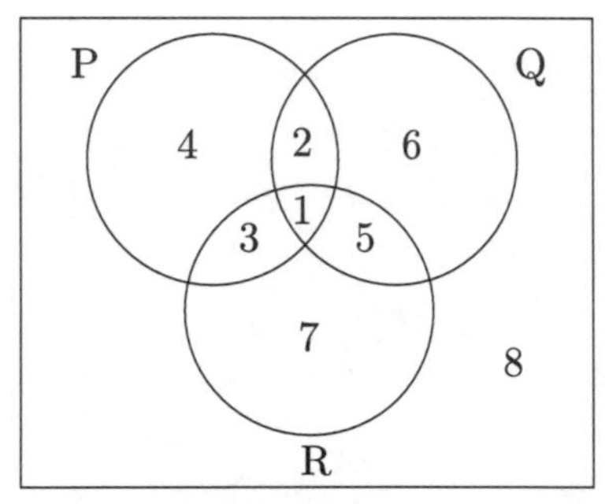

Diagram 2

But for $p_0(P)$, $p_0(Q)$, and $p_0(R)$ all to be different, at least three of the atomic regions 1–8 in the diagram would have to have positive probabilities, since if only two of them had them they would have to sum to 1 and they could only generate two distinct values strictly between 0 and 1, hence $p_0(P)$, $p_0(Q)$, and $p_0(R)$ couldn't all be distinct. But the formulas corresponding to the atomic regions are mutually exclusive, and at least three of them having positive probabilities would be inconsistent with what was proved just above.

But now we will consider the 'way out' that was mentioned earlier.

3.3 A Way Out

We may begin by noting one circumstance in which the assumptions of Lewis's argument cannot be satisfied, and in which the Triviality Results don't hold. That is one in which there is just one atomic formula, A, whose probability is strictly between 0 and 1. Then all factual formulas are equivalent to one of: (1) T (a tautology), (2) A, (3) ~A, or (4) F (a contradiction), and all conditional formulas, $\phi \Rightarrow \psi$, are equivalent to ones in which ϕ and ψ are among the formulas (1)–(4), hence $\phi \Rightarrow \psi$ itself is equivalent to one of these formulas, as shown in the following table.

			ψ		
		T	A	∼A	F
	T	T	A	∼A	F
ϕ	A	T	T	F	F
	∼A	T	F	T	F
	F	T	T	T	T

$\phi \Rightarrow \psi$

Table 1

Then there are at most four distinct probability values, namely those of formulas (1)–(4), there are not three mutually exclusive formulas, each of which has positive probability, and $p(\phi \Rightarrow \psi) = p(\psi|\phi)$ for all ϕ and ψ in the 'language', which is 'almost trivial.[4]

Two other points are to be noted as well. One is that because in this 'model' all conditionals are equivalent to factual formulas, they can be embedded in larger formulas in the same way as the formulas they are equivalent to. For instance, because A ⇒ A and A ⇒ ∼A are equivalent to T and F, respectively, ∼(A ⇒ A) is equivalent to ∼T = F, and (A ⇒ A) ∨ (A ⇒ ∼A) is equivalent to T ∨ F, which is equivalent to T.

The other point is that because probabilities satisfying the Kolmogorov axioms attach to formulas (1)–(4), and these are factual formulas with standard truth-conditions whose probabilities can be interpreted as their probabilities of being true, it follows that the probabilities of the conditionals to which they are equivalent can also be interpreted as probabilities of truth. Therefore, as in the purely factual case, the third validity theorem of section 7.5 applies, and p-validity is equivalent to truth-conditional validity. In sum, in 'almost trivial' languages conditionals act like factual formulas, and can be treated in all ways as though they were themselves factual. The interesting thing is that the construction to be described next seems to suggest that the foregoing can be extended to less trivial languages.

The idea is that, given formulas ϕ and ψ that correspond to regions in a Venn diagram, it is always possible to associate the conditional $\phi \Rightarrow \psi$ with a region in the diagram in such a way that its area is the proportion of the area corresponding to ϕ that lies inside the area corresponding to ψ, and 'truth conditional compounds' involving $\phi \Rightarrow \psi$ correspond to regions whose areas satisfy the Kolmogorov axioms. To illustrate, suppose that our language has two atomic formulas, A and B, which are represented by rectangles in diagram 3 below, whose areas have the values shown, and we want to associate the conditional A ⇒ B with a region whose area equals the proportion of A that lies in B.
Then we can add a rectangle labelled A ⇒ B, as in diagram 4, and dis-

[4]It would be completely trivial if it only contained tautologies and contradictions, or, equivalently, if the only probabilities were the 'truth values' 1 and 0.

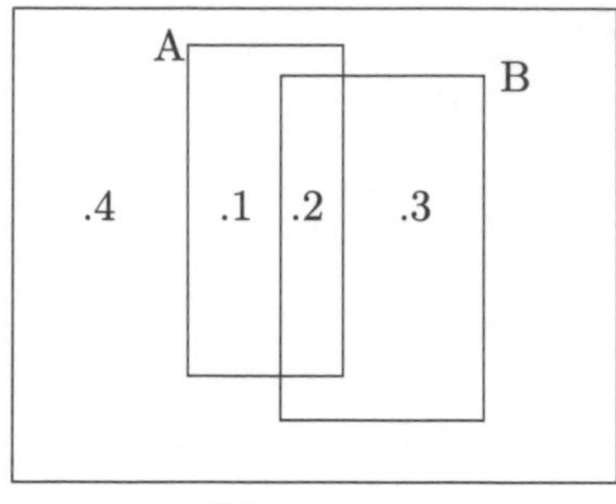

Diagram 3

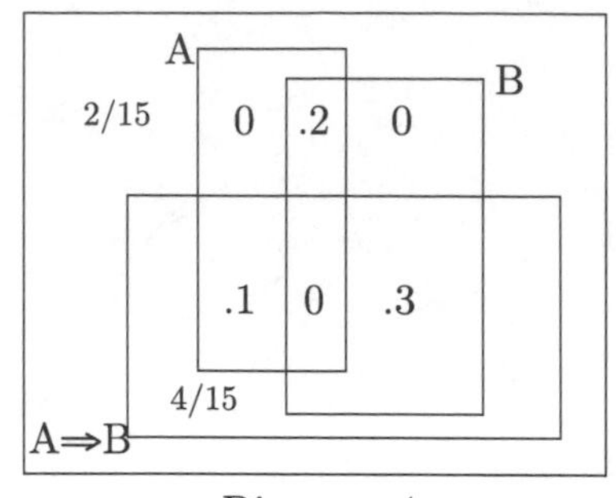

Diagram 4

tribute the probabilities in the atomic regions in diagram 3 into the subregions into which they have been divided in diagram 4. The sole requirement of this distribution is that the portions of the regions of diagram 3 that go inside the A $\Rightarrow$ B rectangle should add up to an amount equal to the proportion of region A that lies inside region B in diagram 3. It is easily verified that this condition is satisfied, since the total probability inside the A $\Rightarrow$ B region in diagram 4 is $.1+.3+\frac{4}{15} = \frac{2}{3}$, and that is the proportion of the A region that lies inside B in diagram 3. Now the diagram 4 probabilities of A and B and their truth-conditional compounds, ~A, A $\vee$ B, etc., are the same as the diagram 3 probabilities of these formulas, but there is a region in diagram 4 that corresponds to A $\Rightarrow$ B, whose area equals the conditional probability of B given A. In other words, a conditional now corresponds to a *region*, as in the Stalnaker theory (cf. section 8.3), but it avoids the objection that was raised against that theory in section 8.4, namely that the corresponding probability is not a conditional probability.

The construction method just described can be applied over and over again, each time adding a region corresponding to another conditional formula. E.g., exercise 2 at the end of this appendix asks the reader to apply the method so as to add a region corresponding to B $\Rightarrow$ (A $\Rightarrow$ B), whose area equals the proportion of B that lies in A $\Rightarrow$ B, and such that compounds containing B $\Rightarrow$ (A $\Rightarrow$ B) like ~(B $\Rightarrow$ (B $\Rightarrow$ A)) correspond to regions whose areas equal probabilities that satisfy the Kolmogorov axioms. So long as the 'base language' has only a finite number of atomic formulas like A and B, the process can be 'carried to infinity', so that if ϕ and ψ are any formulas of the 'extended language' there is a region corresponding to $\phi \Rightarrow \psi$ whose area equals the $p(\psi|\phi)$, and compounds like $\sim(\phi \Rightarrow \psi)$ and $\phi\&(\phi \Rightarrow \psi)$ correspond to regions whose areas equal probabilities that satisfy the Kolmogorov axioms.[5] Thus, any base language can be extended to a language which, for any formulas ϕ and ψ in it, contains

[5]The approach of van Fraassen (1976) seems to be able to show that the same thing holds for languages that don't have finite bases.

the conditional $\phi \Rightarrow \psi$ whose probability is the conditional probability of ψ given ϕ, and compounds of which have probabilities that satisfy the axioms. And given this, the concept of p-validity and laws involving it that were stated in chapter 7⋆ generalize directly to the inferences of the extended language.[6] In other words, we appear to have escaped the dilemma arrived at in section 8.4, by defining a truth-conditional and therefore embeddable conditional like Stalnaker's, but agreeing with the probability conditional theory concerning probabilities. But this seems to conflict with the Triviality Results, and to resolve the conflict we must reexamine the results and the arguments that lead to them. This is commented on very briefly in the concluding section.

3.4 Discussion

The key observation is that Lewis's and Hájek's Triviality Arguments depend on *assumptions*, and whatever conflicts with the results must conflict with the assumptions from which they are derived. Now, our version of Hájek's theory assumed that there were only a finite number of possible worlds (cf. section 8.5), and that can be seen to conflict with construction of our extended language. That is because it requires adding regions in a diagram *ad infinitum*, each time a probability has to be attached to a conditional $\phi \Rightarrow \psi$, as in diagram 4. But why assume that there are only finitely many possible worlds? After all, if the outer rectangle in the diagram is a real spatial region, doesn't it contain infinitely many subregions?

Lewis's argument didn't assume that there are a finite number of possible worlds, but it did assume Bayes' principle in writing $p_1(A \Rightarrow B) = p_0(A \Rightarrow B|B)$ (equation (1)). But we argued in section 4.7⋆⋆ that Bayes' principle is at best a default assumption, and the construction of the extended language avoids the Triviality Results by giving that up, at least in application to conditionals, as in equation (1).[7] Of course, whether the theory is 'right' is another matter, and we will end with a very inconclusive comment on

[6]That is not to say that all of the metatheorems stated in sections 7.6⋆⋆ and 7.7 continue to hold. For instance, 'true conjunctions' do exist in the extended languages, but they are no longer partitioned between factual and conditional formulas (what is $B\&(A \Rightarrow B)$?), and therefore the third validity and the factuality-conditionality theorems that depend on the partition no longer hold. So far as the author knows, the metatheory of such extended languages has not yet been studied in detail.

[7]In fact, distinct posterior and prior probabilities like $p_1(A \Rightarrow B)$ and $p_0(A \Rightarrow B|B)$ don't enter into the theory, which makes no assumptions about the ways in which probabilities change.

In interpreting the extended theory as giving up the Bayesian default specifically in its application to conditionals, recall that it was argued in section 6.5⋆ that special problems arise in extending the theory of probability dynamics based on Bayes' principle to conditionals. But also note that giving up the default in its conditional application could be giving it up in all applications, because there may not be any factual propositions in the languages of the theory of van Fraassen referred to in footnotes 1 and 5.

this, and on the contrast between this theory's resolution of the Triviality dilemma and the 'way out' offered by the theory outlined in chapters 6 and 7⋆.

As said, our construction method resolves the dilemma by giving up the 'Bayesian default', at least in application to conditionals, while the theory developed in chapters 6 and 7⋆ avoids the dilemma by not embedding conditionals in formulas like B&(A ⇒ B), but maintaining at least *compatibility* with the default. But this views the difference in the approaches in an overly formal way. Formulas like B&(A ⇒ B) are excluded in the latter theory, not because statements like "Bob will study, and if he stays up late he will study" aren't made in everyday speech, but because probabilities don't apply to them in a straightforward way, and the theory restricts itself to formulas that symbolize statements that probabilities apply to straightforwardly. The question we *should* ask is: Do probabilities satisfying this or that set of laws apply to statements like "Bob will study, and if he stays up late he will study," and, if so, how?

Recall David Lewis's comment quoted in section 8.6⋆, that conditional probabilities are 'probabilities in name only', because they are not probabilities of truth and they seem not to satisfy the standard laws of probability. Now the construction described in the previous section seems to suggest that they can consistently be assumed to be probabilities of truth and satisfy those laws.[8] However, the discussion in section 8.6⋆ suggests that, as with the simple conditional "If Bob stays up late he will study," the fundamental question in the case of statements like "Bob will study, and if he stays up late he will study" shouldn't be about whether their probabilities can be assumed to be consistent with such and such formal laws. Rather we should ask how, if at all, DMs are better off having degrees of confidence that correspond to these probabilities. And, when we approach things in this way, we may find that, as correspondence with 'the truth' isn't the pragmatically desirable thing in the case of simple conditionals, so correspondence with a single numerical probability isn't the pragmatically desirable thing in the case of compounds containing them. This may be the real lesson of Triviality.

The following appendix will take up again the problem of how to deal with compound sentences that have conditionals embedded in them like

[8]But there is a curious anomaly that the reader may wish to consider. While the inference A&B ∴ A ⇒ B is p-valid and presumably valid in the extended language, the region corresponding to A&B in diagram 4 isn't a subregion of the region corresponding to A ⇒ B. Therefore, if points in the diagram represent possible worlds, there are possible worlds in which A&B is true but A ⇒ B is false, and the inference A&B ∴ A ⇒ B seems to be represented as invalid in the classical sense, and this conflicts with the First Classical Validity Theorem, stated in section 7.1, which says that classical validity is a necessary condition for p-validity.

"Bob will study, and if he goes to bed early he will study," but no longer aiming, as our extension-construction does, at accounting for all constructions of this kind within the framework of a single theory.

Exercises

1. Give an argument that section 3.2 holds is 'similar' to the argument that leads to the conclusion that $p_0(B\&(A \Rightarrow B)) = p_0(B)$, to prove that $p_0(\sim B\&(A \Rightarrow B)) = 0$.
2. Distribute the probability in each atomic region of diagram 4 into the atomic regions into which it is subdivided in the diagram below, in such a way that the total probability of the $B \Rightarrow (A \Rightarrow B)$ region, which is new, is equal to the proportion of the B region that lies inside the $A \Rightarrow B$ region.

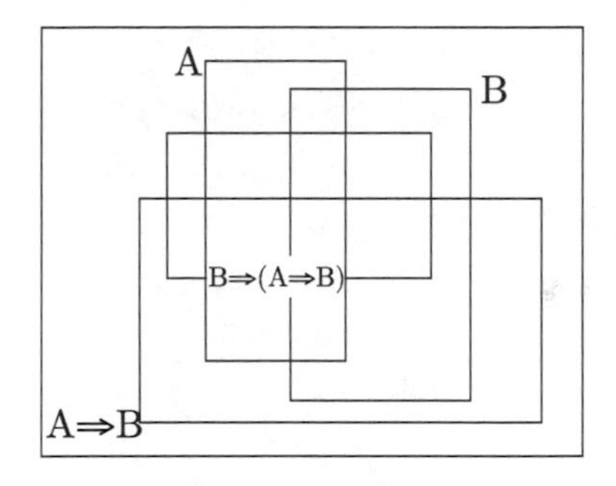

Appendix 4

The Problem of Embedded Conditionals

4.1 Introduction

The following sentences are easily intelligible:

S_1 It won't rain, but if it does rain it won't pour.

S_2 It is not the case that if it rains it will pour.

S_3 Either if it rains it will pour, or if it rains it won't pour.

S_4 If you leave your umbrella behind you'll get soaked if it rains.

S_5 If, if it rains you'll get soaked, then you'll get soaked.

S_1–S_5 might be symbolized as $\sim A \& (A \Rightarrow B)$, $\sim(A \Rightarrow B)$, $(A \Rightarrow B) \vee (A \Rightarrow \sim B)$, $C \Rightarrow (A \Rightarrow B)$, and $(A \Rightarrow B) \Rightarrow B$, respectively, each of which contains $A \Rightarrow B$ embedded in a larger context—a conjunction in S_1, a negation in S_2, a disjunction in S_3, and larger conditionals in S_4 and S_5. But our formal language doesn't contain any of these formulas, and therefore it cannot deal with reasoning involving them. This is a serious limitation, because sentences like S_1–S_5 occur in everyday reasoning, as illustrated below.[1] Of course, this is a problem for ordinary truth-conditional logic as well, because, given the fact that the material conditional doesn't deal adequately with reasoning involving simple conditionals, it doesn't with reasoning involving compounds that contain them either. In fact, these compounds give rise to new fallacies, such as the following:

> It won't rain. Therefore, it won't rain, but if it does rain it won't pour.
>
> It is not the case that if it rains it will pour. Therefore, it will rain.[2]

[1] But Cooper (1978) cites data to the effect that sentences containing conditionals embedded in them are statistically much rarer than simple conditionals.

[2] Note that this is the inverse of the simple fallacy of material implication "It won't rain: therefore, if it rains it will pour."

> If, if it rains you'll get soaked, then you'll get soaked. Therefore, if it doesn't rain you'll get soaked.

Try symbolizing these, using $\rightarrow$ for 'if–then', and use truth-tables to determine their validity!

A more controversial 'fallacy' due to McGee (1985), involving a statement of the form "if C, then if A then B," deserves special mention because it seems to call *modus ponens* into question. McGee's own examples involve rather complicated 'stories', but it is hoped that the reader will get the idea from the following. A person spies what looks like a dog in the distance, and says

> That is a dog. But if that *is* a dog, then, if it weighs 500 lbs., it is a 500 lb. dog.

But the person doesn't conclude by *modus ponens*

> If that weighs 500 lbs., it is a 500 lb. dog

because she thinks that whatever 'it' is, if it weighs 500 lbs. it is probably a larger animal, perhaps a bear. As said, McGee's counterexamples (actually, he presented three examples of this form) are controversial, but whether or not the reader is convinced by them they point up something important. That is that compounds with conditionals embedded in them present new and difficult problems, and nothing about them can be taken for granted, including *modus ponens* itself. And, *modus ponens* is taken for granted in almost all theories that apply to compounds of the form $C \Rightarrow (A \Rightarrow B)$, including the extended theory described in appendix 3, and therefore all of these theories are brought in question by examples like McGee's.

In what follows we will briefly sketch certain 'partial theories' that do not try to do what Stalnaker's theory and the extended theory of appendix 3 did, and account for all compounds with conditionals embedded in them, but rather to account for special cases of reasoning involving one or another of these compounds. Following that, we will make some general comments on the problems that are raised by these embeddings.

4.2 Some Partial Theories

It is generally agreed that conjunctions, e.g., "It won't rain, but if it does rain it won't pour," present no difficulties so long as their conjuncts are properly analyzed.[3] Thus, the inference

> It won't rain. Therefore, it won't rain, but if it does rain it won't pour.

[3]But the reader is reminded of the result concerning the undefinability of true conjunction that was stated in section 7.7. For example, there is no formula, $\phi \Rightarrow \psi$, in the language of chapters 6 and 7⋆ that is equivalent to the conjunction $\sim A \& (A \Rightarrow B)$ in the sense that $\phi \Rightarrow \psi$ p-entails both $\sim A$ and $A \Rightarrow B$, and $\{\sim A, A \Rightarrow B\}$ p-entails $\phi \Rightarrow \psi$.

should be valid if and only if both of the 'component inferences'

It won't rain. Therefore, it won't rain.

and

It won't rain. Therefore, if it does rain it won't pour.

are valid. Since the second is one of the fallacies of material implication, and it is invalid according to most 'nonorthodox' theories of conditionals, the whole original inference must be invalid according to those theories.

The negation of a conditional, e.g., "It is not the case that if it rains it will pour," is superficially simple to analyze, because it seems intuitively to be equivalent to the conditional denial, "If it rains it won't pour." In general, on this view $\sim(\phi \Rightarrow \psi)$ seems to be equivalent to $\phi \Rightarrow \sim\psi$. Assuming this, the inference

It is not the case that if it rains it will pour. Therefore, it will rain.

must be invalid because

If it rains it won't pour. Therefore, it will rain.

is invalid.

The intuition that $\sim(\phi \Rightarrow \psi)$ should be equivalent to $\phi \Rightarrow \sim\psi$ is supported by the observation that $p(\phi \Rightarrow \sim\psi) = 1 - p(\phi \Rightarrow \psi)$, which is exactly the relation in which factual negations stand to the things they negate. There is, of course, the possibility that $p(\phi) = 0$, in which case $p(\phi \Rightarrow \sim\psi) = p(\phi \Rightarrow \psi) = 1$, and $\phi \Rightarrow \sim\psi$ and $\phi \Rightarrow \psi$ appear to be compatible in that case. However, accepting the account of p-consistency given in section 7.8★★, $\phi \Rightarrow \sim\psi$ and $\phi \Rightarrow \psi$ are p-inconsistent, except in the extreme 'nonnormal case' in which ϕ itself is a logical contradiction, and theorists might be excused for overlooking that case.

However, doubts may still be raised about an ordinary language statement of the "It is not the case that if ϕ then ψ" form, because its function may be to contradict "If ϕ then ψ," not at the level of its 'correspondence with the facts', but rather at the level of the justification for asserting it. I.e., to assert "It is not the case that if ϕ then ψ" *can* mean that "If ϕ then ψ" isn't probable enough to be asserted. Or, slightly differently put, it may not mean that "If ϕ then ψ" is highly improbable, but rather that it is not highly probable—and something doesn't have to be highly *im*probable to be insufficiently probable to assert.[4]

[4]This point is made in Grice (1989: 80–1). Of course, the same point can be made about an ordinary factual denial expressed as "it is not the case that ϕ," which could either be a strong denial meaning "It is not the case that ϕ is true," or a weak denial meaning "It is not the case that ϕ is probable." Interestingly, for most formulas $\phi \Rightarrow \psi$ in the language of chapters 6 and 7★ there is no formula in the language that expresses the fact that $\phi \Rightarrow \psi$ is not probable—which is closely related to the undefinability of

Putting negation aside, though, let us turn to the more problematic case of disjunctions with conditional constituents, like S_3, "Either if it rains it will pour, or if it rains it won't pour."[5] The comments of two authors who have discussed disjunctions, including ones of the form $(A \Rightarrow B) \vee (C \Rightarrow D)$, may give the reader some idea of the nature of the considerations that have recently been brought to bear on them.

Frank Jackson (1987) discusses disjunctions of conditionals in appendix A.3 of *Conditionals*, holding that while not all ordinary language statements of the form $(A \Rightarrow B) \vee (C \Rightarrow D)$ are meaningful,[6] ones of the forms $(A \Rightarrow B) \vee (A \Rightarrow C)$ and $(A \Rightarrow C) \vee (B \Rightarrow C)$ are exceptions. Jackson suggests that $(A \Rightarrow B) \vee (A \Rightarrow C)$ is equivalent to $A \Rightarrow (B \vee C)$, and that $(A \Rightarrow C) \vee (B \Rightarrow C)$ is equivalent to $(A \vee B) \Rightarrow C$, an example being "If Mary comes, John will, or if Anne comes, John will," which is plausibly equivalent to "If Mary or Anne comes, John will" (Jackson, 1987: 137). However, R. E. Jennings in his recent book *The Genealogy of Disjunction* points out that:

> It is simply a fact of language that the form 'A if B or if C or if D' has, as it normally does, the reading as a succession of three conditionals: 'A if B; A if C; A if D'. No truth-functional account of 'or' or 'if' accounts for this phenomenon. (Jennings, 1994: 70)

Readers must decide for themselves which of these interpretations of ordinary language disjunctions of conditionals is the right one.

A final point that applies both to denials and to disjunctions of conditionals is that the language of the probability conditional theory set forth

---

true conjunction result mentioned in footnote 3. But, as will be described below, the language can be extended to include such formulas.

[5] S_3 is an instance of the so-called law of the conditional excluded middle. Note that the material conditional counterpart of S_3, symbolized as $(A \rightarrow B) \vee (A \rightarrow {\sim}B)$, is valid in classical logic. Moreover, symbolizing S_3 itself as $(A \Rightarrow B) \vee (A \Rightarrow {\sim}B)$, if $A \Rightarrow {\sim}B$ were equivalent to ${\sim}(A \Rightarrow B)$ then S_3 would reduce to the ordinary law of the excluded middle, because it would be equivalent to $(A \Rightarrow B) \vee {\sim}(A \Rightarrow B)$. But the validity of S_3 is more debatable if it is not assumed that $A \Rightarrow {\sim}B$ is equivalent to ${\sim}(A \Rightarrow B)$, and in particular it is valid in the theory of R. Stalnaker outlined in section 8.3 (cf. exercise $\star 1$ at the end of this appendix), but invalid in D. Lewis's variant of Stalnaker's theory, least in its counterfactual version. In this version the debate between Stalnaker and Lewis focussed on an example due to Quine (1950) involving the two counterfactual conditionals "If Bizet and Verdi had been compatriots then Bizet would have been Italian," and "If Verdi and Bizet had been compatriots then Verdi would have been French" (cf. the papers of Stalnaker and Lewis, reprinted in Part II of Harper, et al. (1981).

[6] Anthony Appiah has argued that assuming that the component conditionals $A \Rightarrow B$ and $C \Rightarrow D$ don't have truth conditions, it is hard to make sense of a disjunction $(A \Rightarrow B) \vee (C \Rightarrow D)$ because 'or' is a truth-functional compound (Appiah, 1985: 208). Nevertheless, the quotation from Jennings (1994) that is cited below points out that certain uses of 'or' connecting conditionals have the function of *conjoining* the conditionals they connect.

in chapters 6 and 7$\star$ can be extended to include not only simple conditionals but all 'boolean combinations' of them, such as $(A \Rightarrow B) \vee (A \Rightarrow \sim B)$ or $\sim[(A \Rightarrow B)\&(A \Rightarrow \sim B)]$. The idea, very roughly, is to define a formula like $(A \Rightarrow B) \vee (A \Rightarrow \sim B)$ to *hold* in an OMP-ordering (section 6.7$\star$) if and only if either $A \Rightarrow B$ or $A \Rightarrow \sim B$ holds in it, and to define $\sim[(A \Rightarrow B)\&(A \Rightarrow \sim B)]$ to hold in it if and only if not both $A \Rightarrow B$ and $A \Rightarrow \sim B$ hold in it.[7] Given this definition, it can be shown that $\sim[(A \Rightarrow B)\&(A \Rightarrow \sim B)]$ must hold in all OMP-orderings, hence it is valid in a sense, while $(A \Rightarrow B) \vee (A \Rightarrow \sim B)$ isn't valid because neither $A \Rightarrow B$ nor $A \Rightarrow \sim B$ hold in the ordering $1 \succ 2 \approx 3 \approx 4 \succ \emptyset$ that corresponds to the following diagram.

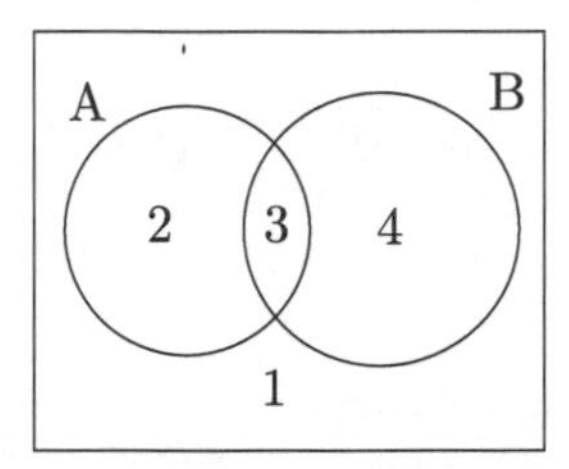

Thus, $A \Rightarrow B$ does not hold in the ordering because state-description #2, which is compatible with $A\&\sim B$, is not lower in the ordering than state-description #3, which is compatible with $A\&B$, and $A \Rightarrow \sim B$ does not hold for similar reasons.[8]

Despite the generality of the OMB-ordering approach it still doesn't apply to conditionals with other conditionals embedded in them, like ones of forms $C \Rightarrow (A \Rightarrow B)$ and $(A \Rightarrow B) \Rightarrow B$, and we will comment on the former in ending this section. *Prima facie*, the most natural interpretation of $C \Rightarrow (A \Rightarrow B)$ is that it is equivalent to $(C\&A) \Rightarrow B$. Thus, "If that is a dog, then if it weighs 500 lbs. then it is a 500 lb. dog" seems to be

[7] This approach is essentially equivalent to the theory of the logic of high probability developed in Adams (1986). It is noteworthy that while the approach applies to formulas like $(A \Rightarrow B) \vee (A \Rightarrow \sim B)$ and $\sim[(A \Rightarrow B)\&(A \Rightarrow \sim B)]$, it does not assume that these formulas have probabilities. In effect, what the fact that $(A \Rightarrow B) \vee (A \Rightarrow \sim B)$ doesn't hold but $\sim[(A \Rightarrow B)\&(A \Rightarrow \sim B)]$ does hold in $1 \succ 2 \approx 3 \approx 4 \succ \emptyset$ means is that this ordering represents a situation in which neither $A \Rightarrow B$ nor $A \Rightarrow \sim B$ are highly probable, and *ipso facto*, they are not both highly probable. But these are statements about the probabilities of $A \Rightarrow B$ and $A \Rightarrow \sim B$, and not about the probabilities of compounds containing them.

[8] This agrees with the David Lewis's thesis that the law of the conditional excluded middle shouldn't be valid in conditional logic. In fact, the OMP-ordering approach agrees completely with the so-called 'case VW' of Lewis's theory in its application to boolean combinations of simple conditionals, though it doesn't, as Lewis's theory does, extend to conditionals embedded in other conditionals. Cf. Metametatheorem 4, p. 269, of Adams (1986).

equivalent to "If that is a dog that weighs 500 lbs., then it is a 500 lb. dog." This equivalence combines the so-called rules of *importation* (that A can be imported into the antecedent of C $\Rightarrow$ (A $\Rightarrow$ B), to yield (C&A) $\Rightarrow$ B), and *exportation* (that A can be exported into the consequent of (C&A) $\Rightarrow$ B to yield C $\Rightarrow$ (A $\Rightarrow$ B)), and its seeming validity is what lies at the heart of the counterexample to *modus ponens* that was cited in section 1 of this appendix. In fact, it is easy to see that, in general, *modus ponens* in combination with importation–exportation must lead to a 'paradox' when they are applied to iterated conditionals.[9]

The argument is that on any theory of conditionals, (B&A) $\Rightarrow$ B is a logical validity. Moreover, if that is equivalent to B $\Rightarrow$ (A $\Rightarrow$ B), then B $\Rightarrow$ (A $\Rightarrow$ B) should also be a logical validity. Now, if *modus ponens* could be applied to B $\Rightarrow$ (A $\Rightarrow$ B) then $\{B, B \Rightarrow (A \Rightarrow B)\}$ $\therefore$ A $\Rightarrow$ B should be a valid inference. But if $\{B, B \Rightarrow (A \Rightarrow B)\}$ $\therefore$ A $\Rightarrow$ B is a valid inference and B $\Rightarrow$ (A $\Rightarrow$ B) is a logical validity, then the inference B $\therefore$ A $\Rightarrow$ B must also be valid.[10] But that is one of the fallacies of material implication. Thus, one of *modus ponens* or importation-exportation must be given up in application to iterated conditionals, on pain of leading to fallacies of material implication—as in the example of the 500 lb. dog and McGee's examples.

Three prominent writers on conditionals, Anthony Appiah, Frank Jackson, and Vann McGee, have interestingly different attitudes towards the dilemma posed by the foregoing argument. Appiah accepts *modus ponens*, but, based on a proposed characterization of the probability of an iterated conditional, C $\Rightarrow$ (A $\Rightarrow$ B), he argues that it is not equivalent to (C&A) $\Rightarrow$ B (Appiah, 1985: 241). Jackson accepts the equivalence of C $\Rightarrow$ (A $\Rightarrow$ B) and (C&A) $\Rightarrow$ B *and* the validity of *modus ponens*, but says "...I am, in effect, allowing that a valid inference pattern, to wit *modus ponens*, may fail to preserve assertibility in a fairly dramatic fashion" (appendix A.1 of Jackson 1987: 133).[11] McGee takes the counterexamples at face value, and concludes that inferences of the form $\{C, C \Rightarrow (A \Rightarrow B)\}$ $\therefore$ A $\Rightarrow$ B are not universally valid; i.e., *modus ponens* is not always valid in application to iterated conditionals. He builds a positive theory on the assumption that importation-exportation *is* always

[9]The argument to follow was originally given in Adams (1975) on page 33.

[10]This is in accord with a fundamental principle of all theories of deductive logical consequence, that if a conclusion $\mathcal{C}$ is a logical consequence of premises $\mathcal{P}$ and $\mathcal{P}_1, \ldots, \mathcal{P}_n$, and $\mathcal{P}$ is itself a logical validity, hence a consequence of $\mathcal{P}_1, \ldots, \mathcal{P}_n$, then $\mathcal{C}$ must be a logical consequence of $\mathcal{P}_1, \ldots, \mathcal{P}_n$.

[11]This reflects Jackson's (1987), as well as David Lewis's (1976), view that indicative conditionals have truth values. In fact they are those of the material conditional, but the probabilistic theory measures what they call *assertibility*, which in the case of conditionals differs from probability.

valid, though his theory coincides with the theory developed in chapters 6 and 7⋆ in its application to simple, uniterated conditionals (McGee, 1989).

We leave it to the reader to choose which, if any, of the above theories to accept. We will only point out that to date none of them is widely accepted, and 'the problem of iterated conditionals' is still very much an open one. However, we will make a general observation.

So far no one has come up with a pragmatics that corresponds to the truth-conditional or probabilistic semantics of the theories that they propose, which might justify these theories in the same way that the pragmatics outlined in section 9.9⋆⋆ attempts to justify the conditional probability theory of simple conditionals. One adjunct to such a pragmatic investigation ought to take account of the ways in which conclusions expressed by the sentences that the theories deal with, e.g., sentences involving embedded conditionals, influence reasoning and behavior. This must in turn take into account the fact that what such a sentence means is a matter of fact that cannot be 'legislated' *a priori*, e.g., we are not entitled to assume that *modus ponens* must be valid in application to iterated conditionals. And, to reiterate a point made at the end of appendix 3, it may be that when we approach things in this way, we will find that, as 'correspondence with the truth' isn't the pragmatically desirable thing in the case of simple conditionals, so correspondence with a single numerical probability isn't the pragmatically desirable thing in the case of compounds containing them. A good place to begin may be with the problematic compounds discussed in this appendix.

Exercises

⋆1. Apply the OMP-ordering theory to determine the validity of the following inferences; i.e., to determine whether it is possible for their premises to hold in an OMP-ordering without their conclusions holding in it.

a. $A \Rightarrow \sim B \therefore \sim(A \Rightarrow B)$

b. $\sim(A \Rightarrow B) \therefore A \Rightarrow \sim B$

c. $B \therefore (A \Rightarrow B) \vee \sim A$

d. $B \therefore (A \Rightarrow B) \vee (\sim A \Rightarrow B)$

e. $\{A \Rightarrow B, B \Rightarrow A\} \therefore (\sim A \Rightarrow \sim B) \vee (\sim B \Rightarrow \sim A\}$

⋆2. Show that for any uncertainty function, u( ), it must be the case that

$$u(A \Rightarrow B) \times u(\sim A) \leq u(B).$$

This implies that if $p(B) \geq .01$ then either $p(A \Rightarrow B) \geq .1$ or $p(\sim A) \geq .1$, which substantiates the claim that if B is highly probable then at least one of $A \Rightarrow B$ or $\sim A$ must be 'reasonably probable'.

Adams (1986) gives necessary and sufficient conditions for premises to 'disjunctively entail' conclusions in a set $\{\phi_1 \Rightarrow \psi_1, \ldots, \phi_n \Rightarrow \psi_n\}$, in the sense that the fact that the premisses are highly probable guarantees that at least one of $\phi_1 \Rightarrow \psi_1, \ldots, \phi_n \Rightarrow \psi_n$ is reasonably probable. It is shown that the only way this can happen when the conclusions are all factual is for the premises to logically entail at least one of them.

$\star$3. Can you depict a probabilistic state of affairs of the kind that was discussed in appendix 3, in which (A ⇒ B) ⇒ B is represented as probable but ~A ⇒ B is represented as improbable? The possibility of doing this suggests that the inference (A ⇒ B) ⇒ B ∴ ~A ⇒ B couldn't be valid in this theory.

Appendix 5

A Geometrical Argument for the 'Statistical Reasonableness' of Contraposition

Bamber (1997) gives a geometrical argument that while contraposition is not p-valid (section 6.1), nevertheless cases in which its premise is probable but its conclusion is improbable must be statistical rarities, and therefore the inference can be said to be 'statistically reasonable'.[1] To illustrate the argument we will consider contraposition in the form $A \Rightarrow \sim B \therefore B \Rightarrow \sim A$, and show that while it is possible for $A \Rightarrow \sim B$ to be probable while $B \Rightarrow \sim A$ is improbable, most of the time when $A \Rightarrow \sim B$ is probable, $B \Rightarrow \sim A$ is also probable.

Focusing on the atomic sentences A and B, there are four state-descriptions to consider, as in table 1 below:

				formulas	
regions	distribution	A	B	A⇒∼B	B⇒∼A
1	.90	T	F	*T*	....
2	.09	T	T	F	*F*
3	.01	F	T	....	*T*
4	.00	F	F	....	....
	↳	.99	.10	.91	.10

Table 1

The distribution in table 1 is a υ-distribution with $\upsilon = .9$, which is roughly reflected by the corresponding areas in diagram 1 (region 3 is made smaller than region 4 to simplify the picture, but this is not essential to what follows), which in turn determines the OMP-ordering $1 \succ 2 \succ 3 \succ 4 \approx$

[1] Bamber (1997) generalizes to other patterns of inference including $A \vee B \therefore \sim A \Rightarrow B$ and transitivity.

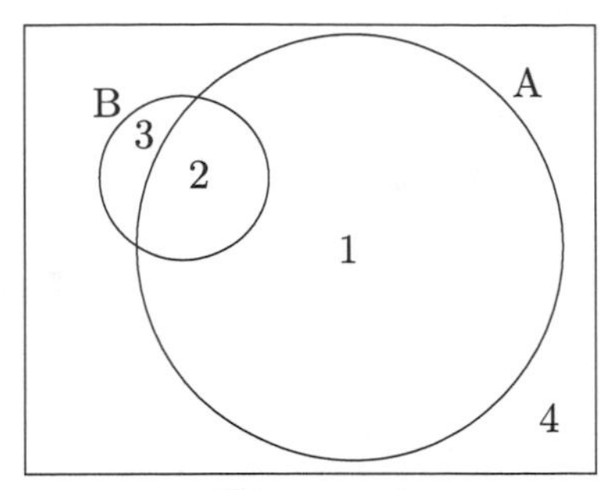

Diagram 1

$\emptyset$. Because A $\Rightarrow$ ~B holds in this ordering and B $\Rightarrow$ ~A does not, this is an OMP-ordering counterexample to A $\Rightarrow$ ~B $\therefore$ B $\Rightarrow$ ~A (section 6.5$\star$), and the probabilities of A $\Rightarrow$ ~B and B $\Rightarrow$ ~A generated by the υ-distribution constitute a numerical counterexample to the inference. Note too that it conforms to the rule noted in section 6.6$\star$, that in these cases the consequent of the premise, namely ~B, has to be independently probable. But this is intuitively unlikely, and Bamber's argument is designed to substantiate this impression quantitatively.

Assuming what is plausible, that we can neglect the possibility that s.d. #4 has positive probability, we can assume that p(A&~B) = p_1, p(A&B) = p_2, and p(B&~A) = p_3 sum to 1, and we can picture a distribution, (p_1, p_2, p_3), corresponding to a point, π, inside an equilateral triangle with altitude 1.[2] Given the theorem that the distances from a point π inside an equilateral triangle of altitude 1 to the sides must sum to 1, we can assume that they correspond to the values of p_1, p_2, and p_3 (see diagram 2, compare diagram 9.3 in chapter 9).

In this representation points that are located below a line $l(\delta)$ that passes through the lower right vertex of the triangle (vertex 3), slanting upwards to the left,[3] correspond to distributions (p_1, p_2, p_3) that generate probabilities p(A $\Rightarrow$ ~B) that are at least $1 - \delta$, for some quantity δ. Similarly, points that correspond to distributions that generate probabilities p(B $\Rightarrow$ ~A) that are at least $1 - \epsilon$ are located below a line $l(\epsilon)$, passing through the

[2]That the possibility that p(~A&~A) is positive can be neglected for the purpose of the present argument can itself be demonstrated by a rigorous geometrical argument. However, this requires a three-dimensional geometrical representation, in which the four probabilities p(A&~B) = p_1, p(A&B) = p_2, p(B&~A) = p_3, and p(~A&~B) = p_4 define a distribution (p_1, p_2, p_3, p_4) that corresponds to a point π inside a regular tetrahedron of altitude 1 (a three-sided pyramid with distance 1 from base to summit, whose faces are equilateral triangles), and p_1, p_2, p_3, and p_4 are the distances from π to its four faces.

[3]And intersecting the left leg of the triangle a proportion δ of the distance from vertex 1 to vertex 2. Similarly, $l(\epsilon)$ intersects the right leg of the triangle a proportion ϵ of the distance from vertex 3 to vertex 2.

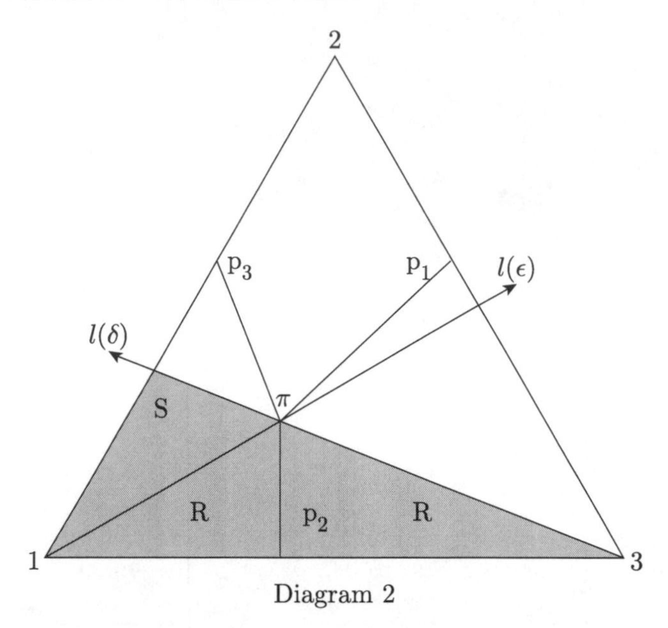

Diagram 2

lower left vertex (vertex 1), and slanting upwards to the right. Given this, the shaded region R consists of points that correspond to distributions that generate probabilities such that both $p(A \Rightarrow \sim B) \geq 1-\delta$ and $p(B \Rightarrow \sim A) \geq 1-\epsilon$, and the shaded region S consists of points that correspond to distributions that generate probabilities such that $p(A \Rightarrow B) \geq 1-\delta$ but $p(B \Rightarrow \sim A) < 1-\epsilon$.

Now, if we assume that the area of a region is proportional to the probability that a distribution corresponds to a point in it, the ratio of the area of R to the area of the union $R \cup S$ measures the conditional probability that $p(B \Rightarrow \sim A)$ is at least $1-\epsilon$, given that $p(A \Rightarrow \sim B)$ is at least $1-\delta$. Thus because diagram 2 represents the area of S as small in relation to that of R, it represents the conditional probability that $p(B \Rightarrow \sim A)$ is at least $1-\epsilon$, given that $p(A \Rightarrow \sim B)$ is at least $1-\delta$, as being high. Making this precise, it can be shown in general that, given that $p(A \Rightarrow \sim B) \geq 1-\delta$, the conditional probability that $p(B \Rightarrow \sim A) \geq 1-\epsilon$ is $1-[(1-\epsilon)\delta]/[\epsilon+(1-\epsilon)\delta]$. For instance, setting $\delta = .01$ and $\epsilon = .1$, $1-[(1-\epsilon)\delta]/[\epsilon+(1-\epsilon)\delta] = 1-.009/.109 \simeq .918$, hence if $p(A \Rightarrow \sim B)$ is .99 the chance that $p(B \Rightarrow \sim A)$ is at least .9 is approximately .918.

Now let us pass to the limit, and suppose that δ approaches 0 while ϵ has a fixed positive value. In the geometrical picture, when δ approaches 0 the line $l(\delta)$ rotates downwards around vertex 3, and its slope approaches 0, i.e., $l(\delta)$ approaches horizontal, and if ϵ is held fixed line $l(\epsilon)$ stays fixed. Given this, it is intuitively obvious that the ratio of the area of R to the area of R∪S has to approach 1. Furthermore, it is easy to see that in this case the quantity $1 - [(1-\epsilon)\delta]/[\epsilon + (1-\epsilon)\delta]$ must approach 1. I.e., the conditional probability of p(B ⇒ ∼A) being at least $1-\epsilon$ must approach 1 for any positive ϵ, given that p(A ⇒ ∼B) approaches 1.[4] Roughly, the more probable A ⇒ ∼B is the more likely it is that B ⇒ ∼A is probable, and this likelihood approaches certainty as the probability of A ⇒ ∼B approaches 1. This constitutes a 'statistical vindication' of contraposition, by showing that even if B ⇒ ∼A can be improbable while A ⇒ ∼B is probable, the likelihood of that approaches 0 as the probability of A ⇒ ∼B approaches 1.

Two points can be made in conclusion. One is that in speaking of the probability of p(B ⇒ ∼A) being at least $1 - \epsilon$, it is assumed that there exist probabilities that probabilities have certain properties, i.e., there exist *second order probabilities.* This may be granted, but it should be noted that these probabilities aren't necessarily proportional to areas in the way that has been assumed here. The assumption that they can be represented in this way is not justified by the logical laws of probability (i.e., it is not justified by the Kolmogorov axioms), and further study of the assumption is needed.[5]

The other point is that Bamber's is one of a number of recent studies of defaults that justify reasoning in accord with p-invalid principles in special circumstances. Adams' (1983) "Probabilistic Enthymemes" is a relatively early paper in this genre, Schurz (1994), and especially Pearl and Goldszmidt (1991), with its nonmonotonic System Z+, to which both Bambers' and Schurz's systems are closely related, are more recent contributions.

Exercise

1. When p(A&B) = p(A&∼B) = p(B&∼A) = 1/3, hence p(A ⇒ ∼B) = p(B ⇒ ∼A) = 1/2, the point π is at the centroid of the equilateral triangle in diagram 2, and lines from it to the sides of the triangle must divide it into three equal diamond-shaped regions. Using this fact, argue that if p(A ⇒ ∼B) = .5, the probability that p(B ⇒ ∼A) $\geq$ .5 must equal 2/3. Note that this equals $1-[(1-\epsilon)\delta]/[\epsilon+(1-\epsilon)\delta]$ when $\delta = \epsilon = .5$.

[4] What can you say about the chances that p(B ⇒ ∼A) $\geq 1 - \epsilon$ when $\epsilon = 0$?

[5] Skyrms (1990) uses second order probability representations in important recent studies of dynamical equilibria in the evolution of degrees of confidence. Bamber has pointed out in private correspondence that results very similar to those described here also follow if second-order probabilities are proportional not to areas but to other sufficiently uniform measures on the 'space' of possible first-order probability distributions.

Appendix 6

Counterfactual Conditionals

'Counterfactual' is the standard but poorly chosen term for such conditionals as "If Lee Harvey Oswald hadn't shot Kennedy then Lyndon Johnson wouldn't have become president", as contrasted with so called *indicative* conditionals like "If Lee Harvey Oswald didn't shoot Kennedy then Lyndon Johnson didn't become president".[1] The student should recognize an intuitive difference between these statements, since the first seems right, or at least probable, while the second seems absurd. In fact, its absurdity is explained by the probability conditional theory because, given that Lyndon Johnson certainly did become president, the probability that he didn't become president given that Oswald didn't shoot Kennedy is close to zero. But the difference between counterfactual and indicative conditionals was recognized long before the probability conditional theory was developed, when it was generally assumed that indicatives could but counterfactuals couldn't legitimately be symbolized as material conditionals.[2] Moreover, the problem of analyzing counterfactual conditionals was regarded as an important one, because it was thought that certain important scientific concepts depended on it. For instance, it was thought that a so called *dispositional* concept like *brittleness* should be analyzed counterfactually, since saying that an object was brittle seems to mean that it would have broken if it had been dropped or struck—not that it *was* broken if it was dropped or struck. And, granted the importance of counterfactual conditionals, there has been a very large literature on 'the problem of the counterfactual', which cannot be entered into here.[3] However, something may be said about how the problem looks from the probabilistic point of view.

[1]The example of Lee Harvey Oswald and John F. Kennedy was originally discussed in Adams (1970) and it has often been cited to show the logical difference between counterfactual conditionals and the indicative conditionals that correspond to them.

[2]Cf. Quine 1959: 14.

[3]Cf. Chapter 1 of Goodman (1955).

One thing is that while counterfactuals may differ from indicative conditionals, there are many situations in which their probabilities are just as clearly defined as those of indicatives. For instance, we may say of a person who didn't draw a card from a shuffled 52-card deck that the chances are 1 in 13 that she would have drawn an ace if she had drawn a card from the deck—but we wouldn't say that the chances are 1 in 13 that she drew an ace if she drew a card from the deck. Moreover, these 'counterfactual probabilities' obey many of the same laws as ordinary conditional probabilities. For instance, the chain rule still holds, so that the chance that both cards would have been aces if the person had drawn *two* cards from the deck equals the chance that the first card would have been an ace multiplied by the chance that the second card would have been an ace, given that the first card had been one.[4]

Because counterfactual probabilities conform to many of the laws of conditional probability, inferences involving counterfactual conditionals conform to many of the same laws as ones involving indicatives. In fact, some of the real life counterexamples given in section 6.3 change to counterexamples to counterfactual principles by simple changes of mood and tense. Thus, a counterexample to transitivity transforms to:

$B \Rightarrow C$	=	If the sun had risen no more than 5 minutes later yesterday then it would have risen less than 5 minutes later yesterday.
$A \Rightarrow B$	=	If the sun had risen exactly 5 minutes later yesterday it would have risen no more than 5 minutes later yesterday.
$A \Rightarrow C$	=	If the sun had risen exactly 5 minutes later yesterday it would have risen less than 5 minutes later yesterday.

There are two important things to note about this kind of transformation, the first of which is that it doesn't seem to extend to inferences that combine conditional and factual statements. For instance, the real life counterexample to the derivation of a conditional from a disjunction cited in section 6.3, "Either the sun will rise tomorrow or it will warm up during the day; therefore, if the sun doesn't rise tomorrow it will warm up during the day" has the factual premise "Either the sun will rise tomorrow or it will warm up during the day." But, while this has a simple past tense transform, "Either the sun rose yesterday or it warmed up during the day," which should be a practical certainty, there doesn't seem to be a counterfactual mood that corresponds to it.[5] This is related to the fact that while

[4]Note the subjunctive-counterfactual mood in 'given that the first card had been an ace'.

[5]Section IV.4 of Adams (1975) suggests that "Either the sun should have risen yesterday or it should have warmed up during the day" might stand logically to "Either the

there are counterfactual *conditional* probabilities, e.g., the 1 in 13 chance that the person would have drawn an ace if she had drawn a card, there don't seem to be any nonconditional counterfactual probabilities.

The other point about the grammatical difference between indicative and counterfactual conditionals is that it often seems to make no logical difference in relation to statements about the future, but they 'split apart' logically in the past. For instance, it seems to make no difference whether we say that there is a 1 in 13 chance that our person *will* draw an ace if she draws a card, or we say that this is the chance she would have if she *were* to draw a card. But, knowing that the person didn't draw a card, we would say that there is a 1 in 13 chance that she would have drawn an ace if she had drawn a card, but not that there is a 1 in 13 chance that she did draw an ace if she drew a card. In fact, learning that the person didn't draw a card seems to 'transform' the *a priori* indicative "There is a 1 in 13 chance that if she draws a card she will draw an ace" into the *a posteriori* counterfactual "There is a 1 in 13 chance that if she had drawn a card she would have drawn an ace,"[6] which has led some to argue that counterfactual conditionals are past tenses of indicative conditionals.[7] This is 'practically' implausible, however, since the *bets* that if the person draws a card then she will draw an ace and that if she drew a card then she drew an ace are settled in the same way, but there is no 'standard' way of settling a bet that if she had drawn a card it would have been an ace.[8]

To conclude, however, it is important to stress that some counterfactual conditionals play a role in practical reasoning—contrary to what we might think. Given that our person did *not* draw a card, we might be inclined to ask: who cares what would have happened if she had drawn one? It could seem to be purely speculative 'Monday morning quarterbacking' to say that if she had drawn one it would have been an ace. But the appearance of pure

---

sun rose yesterday or it warmed up during the day" as "If the sun hadn't risen yesterday it would have warmed up during the day" stands to "If the sun didn't rise yesterday it warmed up during the day." Cf. also section 10.1 of Edgington (1995).

[6]Chapter IV of Adams (1975) advances a general 'epistemic past tense hypothesis' concerning counterfactuals, roughly to the effect that their probabilities are equal to the probabilities that the corresponding indicative conditionals had or might of had under hypothetical and possibly nonactual circumstances. But section IV.8 of this work also cites a putative counterexample to the hypothesis, which is connected in turn with Skyrms' (1981) 'prior propensity' interpretation of counterfactuals. Cf. Section 10 of Edgington (1995) on this.

[7]Cf. Dudman (1994) and Bennett (1995).

[8]Except in appendix 1, betting considerations have been ignored in this work, but they can be very helpful in clarifying ideas about meaning and probability. It is assumed that a bet on an indicative conditional like "If she draws a card she will draw an ace" wins if the person draws a card and it is an ace, loses if she draws a card and it isn't an ace, and it neither wins nor loses if she doesn't draw a card. I.e., it is assumed to be a conditional bet. Cf. section 9.9⋆⋆ and appendix 1.

speculativeness is misleading at least in some cases. Here is an example.

Suppose that you are hoping to find your friend at home in her house during the evening, but, driving up, you see that all of the house lights are out. Given this, you reason "She is not at home, because if she were at home some lights would be on."[9] This seems quite reasonable, but the conditional premise "if she were at home some lights would be on" has to be counterfactual and not indicative, since seeing that the lights are out, it would make no sense to assert "if she is at home the lights are on."

The foregoing counterfactual reasoning is also supported by probabilistic analysis, as an *inverse probable inference* following the pattern of exercise ⋆4 in section 4.7⋆⋆:

$$\frac{p(\sim H|O)}{p(H|O)} = \frac{p(\sim H)}{p(H)} \times \frac{p(O|\sim H)}{p(O|H)}$$

where O constitutes the 'observed effect', namely that the house lights are out, and ~H and H are the hypotheses, respectively, that your friend isn't and that she is at home. The conditional probabilities on the right in the equation, p(O|~H) and p(O|H), are not plausibly interpreted as the probabilities of indicative conditionals, $\sim H \Rightarrow O$ and $H \Rightarrow O$, respectively. E.g., given that you see that the lights are out, the probabilities of the indicative conditionals "If she (your friend) is at home the lights are out," and "If she isn't at home the lights are out" must both be close to 1, and the posterior probability ratio on the right, p(~H|O)/p(H|O) will be almost the same as the prior probability ratio p(~H)/p(H). On the other hand, if p(O| ~ H) and p(O|H) are interpreted as the probabilities of the counterfactual conditionals "If she were not at home the lights would be out," which seems probable, and "If she were at home the lights would be out", which seems improbable, then posterior p(~H|O)/p(H|O) would be very high compared to the prior probability ratio.

Of course the foregoing argument is very hasty, but at least it strongly suggests that counterfactual conditionals are important not only in informal reasoning, but their probabilities figure centrally in a scientifically important kind of formal probabilistic inference, namely what we have called inverse probable inference 'from effects to causes'. Moreover, the probabilities that enter into this kind of inference require a special kind of analysis, because they cannot be accounted for straightforwardly within the 'pragmatic framework' of section 9.9⋆⋆.

[9]This kind of reasoning is what Charles Sanders Peirce (1903) called *abduction*, and which is now often called "inference to the best explanation."

Appendix 7

Probabilistic Predicate Logic[1]

7.1 Introduction

This appendix sketches how the theory of probability logic developed in this work can be generalized to apply to propositions that are normally formalized in the first-order predicate calculus. Because these propositions are bivalent, it will be seen that the theory can be applied to them with fairly simple modifications.[2]

The most important modification derives from the fact that because state-descriptions are no longer in the picture, neither are probability distributions over them. What takes their place in the first-order theory are *models*, or 'possible worlds', in which first-order sentences are true or false, and because there are infinitely many of these, distributions are *measures* over classes of them. Given this, the probability of a sentence is not generally a simple sum of the probabilities of the worlds in which it is true, but an *integral* over this set, which is calculated in a more complicated way—although when it comes to describing probability counterexamples we will see that it is not necessary to consider distributions over more than finitely many worlds and finite sums of their probabilities. But there are other interesting aspects of the application, and much of the important technical work that has been done in probability logic has to do with them.

7.2 Language, Worlds, and Probabilities

For illustration, and for simplicity, we will concentrate on a language with two monadic predicates, P and S, a single variable, x, and individual constants, b and c. Thus, we have 'factual' monadic formulas Pb $\rightarrow$ Sc,

[1]This and the following appendices presuppose a knowledge of elementary first-order logic, such as is usually attained in first-year courses in symbolic logic. The symbolism of the formal language that will illustrate basic concepts is fairly straightforward, but we use the terms *model* or *possible world* to refer to what are often called *interpretations*.

[2]It should be noted that important alternatives to the present approach to probabilities of universal statements are presented in Scott and Krauss (1966) and Vickers (1976), which there is not space to discuss here.

$(\forall x)(Px \rightarrow Sx)$, $(\forall x)Px \rightarrow Pc$, etc., as well as conditional formulas $\phi \Rightarrow \psi$, where ϕ and ψ are factual formulas, e.g., $Pc \Rightarrow Sc$ and $(\forall x)Px \Rightarrow Pc$.

As before, it is important to stress that probability conditionals cannot be embedded, hence expressions like $(\forall x)(Px \Rightarrow Sx)$ are not formulas of our language. Therefore statements like "Everyone at the party was a student" must be symbolized using material conditionals, e.g., as $(\forall x)(Px \rightarrow Sx)$, although "If Charles was at the party then he is a student" can still be symbolized as a probability conditional, $Pc \Rightarrow Sc$. But that isn't a particular 'instance' that can be inferred from $(\forall x)(Px \rightarrow Sx)$ by universal instantiation, and the relation between the two statements becomes more complicated, as will be discussed in section 4.

A model or *possible world* for the language is a 'system', $\Gamma = \langle D, P, S, b, c\rangle$, that assigns extensions to the predicates P and S, which are subsets of Γ's domain, D, which is itself a nonempty set, and assigns individual members of D to the individual constants b and c. A factual formula ϕ is defined to be true in a world Γ in the usual way. It will also be convenient to introduce the *truth function*, $\tau(\ )$, *of* Γ, defined as the function $\tau(\phi)$ that equals 1 when ϕ is true in Γ and equals 0 when ϕ is false in it. For instance, if Pc is true in Γ then $\tau(Pc) = 1$, and this holds if and only if the individual assigned to c belongs to the set assigned to P. Similarly, if $(\exists x)(Px\&Sx)$ is true in Γ then $\tau((\exists x)(Px\&Sx)) = 1$, and this holds if there exists an individual in D that belongs to both of the sets that are assigned to P and S. As would be expected, truth isn't defined for conditional formulas, hence $\tau(\phi \Rightarrow \psi)$ is not defined, though probabilities, $p(\phi \Rightarrow \psi)$, are defined.

Probabilities can be thought of intuitively as probabilities of being in possible worlds, and the probability of a factual formula can be thought of as the probability of being in a world in which it is true. If there are only a finite number of worlds, $\Gamma_1, \ldots, \Gamma_n$, then the probabilities of being in them constitute a distribution, $p(\Gamma_1), \ldots, p(\Gamma_n)$, and for any factual formula ϕ, $p(\phi)$ is the sum of those $p(\Gamma_i)$ for the worlds Γ_i in which ϕ is true. It follows that if $\tau_1, \ldots, \tau_n$ are the truth functions corresponding to $\Gamma_1, \ldots, \Gamma_n$, respectively, then $p(\phi)$ can be written as:

(1) $$p(\phi) = p(\Gamma_1)\tau_1(\phi) + \ldots + p(\Gamma_n)\tau_n(\phi).$$ [3]

As previously noted, first-order logic generally considers infinite classes of worlds, and in this case probabilities are generated from probability measures over those classes (cf. Halmos 1950), but, as also said, this complication will be avoided here.

In the finite case it is easy to see that the probabilities of factual formulas satisfy the Kolmogorov axioms, and it can be shown that this is also true of probabilities that are generated from measures over infinite classes of

[3] Generalizations of this formula and equation (2) to follow will be especially important to the theory developed in appendix 9.

worlds. Hence, if ϕ is a factual formula then $p(\sim\phi) = 1 - p(\phi)$, and if ϕ and ψ are mutually exclusive then $p(\phi \vee \psi) = p(\phi) + p(\psi)$.[4] More generally, p( ) must satisfy all of the laws of probability set forth in section 2.3.

Probabilities of conditional formulas $\phi \Rightarrow \psi$ are defined as before, namely as conditional probabilities, and they satisfy all of the laws of conditional probability such as the chain rule (section 4.3). Also, equation (1) generalizes to give the probability of $\phi \Rightarrow \psi$:

$$\text{(2)} \qquad p(\phi \Rightarrow \psi) = \frac{p(\Gamma_1)\tau_1(\phi\&\psi) + \ldots + p(\Gamma_n)\tau_n(\phi\&\psi)}{p(\Gamma_1)\tau_1(\phi) + \ldots + p(\Gamma_n)\tau_n(\phi)}$$

which says that $p(\phi \Rightarrow \psi)$ is the ratio of the probability of being in a world in which $\phi\&\psi$ is true to the probability of being in a world in which ϕ is true.

7.3 Probabilistic Validity

Since the probabilities of first-order formulas and of conditionals formed from them satisfy the same laws as the probabilities of sentential formulas and conditionals formed from them, p-validity can be defined in the same way for both classes of formulas, and it satisfies the same laws in both cases. Therefore, we can ask in a precise way which of the inferences below is p-valid:

$$\text{a.} \quad \frac{(\exists x)Px \Rightarrow (\forall x)Px}{(\forall x)(Px \rightarrow Sx)} \qquad\qquad \text{b.} \quad \frac{Pb \rightarrow Pc}{Pc \Rightarrow Sc}$$

and we can answer this question in the same way, using the same methods as were described in chapters 6 and 7⋆. In particular, all of the rules of p-valid inference involving probability conditionals that were stated in sections 7.2 and 7.3 still hold, and so does the rule stated in section 7.5 for determining p-validity in terms of 'yielding'. Specifically, a conditional $\phi \Rightarrow \psi$ can be said to be *verified* in a world Γ if ϕ and ψ are both true in it, to be *falsified* if ϕ is true while ψ is false, and to be neither verified nor falsified in it if ϕ is false. Then premises $\phi_1 \Rightarrow \psi_1, \ldots, \phi_n \Rightarrow \psi_n$ *yield* $\phi \Rightarrow \psi$ if: (1) $\phi \Rightarrow \psi$ is verified in every world in which at least one of $\phi_1 \Rightarrow \psi_1, \ldots, \phi_n \Rightarrow \psi_n$ is verified and none of them is falsified, and (2) it is falsified in every world in which at least one of $\phi_1 \Rightarrow \psi_1, \ldots, \phi_n \Rightarrow \psi_n$ is falsified. And, finally, $\{\phi_1 \Rightarrow \psi_1, \ldots, \phi_n \Rightarrow \psi_n\}$ p-entails $\phi \Rightarrow \psi$ if and only if some subset of $\{\phi_1 \Rightarrow \psi_1, \ldots, \phi_n \Rightarrow \psi_n\}$, possibly empty and possibly the entire set, yields $\phi \Rightarrow \psi$.[5] However, it is important to notice that this method for determining p-validity is no longer 'mechanical', since

[4] They actually satisfy an infinite summation law: if $\phi_1, \phi_2, \ldots$ is an infinite series of mutually exclusive formulas then $p(\phi_1 \vee \phi_2 \vee \ldots) = p(\phi_1) + p(\phi_2) + \ldots$.

[5] Because the p-entailment isn't a compact relation, problems arise in generalizing this to inferences with infinitely many premises, which are considered in more general first-order theories. Thus, conclusions can be p-entailed by infinite sets of premises that are not p-entailed by any finite subset of these premises, an example of which is given in

according to Church's theorem (Church, 1936a) there is no mechanical procedure for determining whether there exist worlds that verify or falsify given formulas.[6] But the two following examples show that the method still has considerable practical usefulness, as well as important theoretical consequences.

The third validity theorem stated in section 7.5 is still valid, and it implies that an inference with a factual conclusion is p-valid if and only if it is classically valid when the probability conditional is replaced by the corresponding material conditional. It follows that inference (a) above is p-valid, because it becomes classically valid when $\Rightarrow$ is replaced by $\rightarrow$.

The factuality-conditionality theorem stated in section 7.7 also continues to hold, and it applies to inference (b) above, since its premise is factual and its conclusion is conditional. The theorem says that if the premise is essential, it can only p-entail the conclusion if it logically entails both the antecedent and its consequent. It follows that inference (b) must be p-invalid, because its premise is obviously essential but it doesn't logically entail either the antecedent or the consequent of the conclusion; i.e., $(\forall x)(Px \rightarrow Sx)$ doesn't logically entail either Pc or Sc.

Let us now see how the method outlined in section 7.6⋆⋆ can be used to construct a probabilistic counterexample to inference (b). The 'yielding test' implies that if $(\forall x)(Px \rightarrow Sx)$ ∴ $Pc \Rightarrow Sc$ is p-invalid, then neither $(\forall x)(Px \rightarrow Sx)$ nor the empty set can yield $Pc \Rightarrow Sc$, and if this is the case there must exist worlds Γ_1 and Γ_2 such that: (1) $(\forall x)(Px \rightarrow Sx)$ is verified in Γ_1 but $Pc \Rightarrow Sc$ isn't, and (2) $Pc \Rightarrow Sc$ is falsified in Γ_2. Such worlds are depicted below, with domains represented by sets of points inside the outer rectangles, with P and S represented by sets of points in the inner circles, as shown, and with c represented as a point outside of both P and S in world Γ_1 and as a point inside P and outside S in world Γ_2 (the constant b is omitted, since it doesn't enter into this inference):

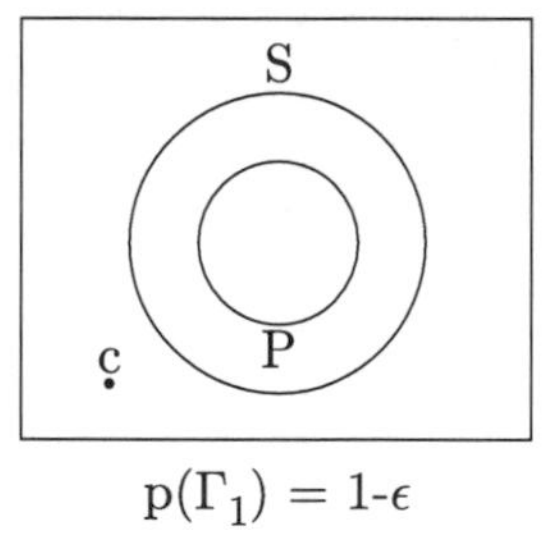

$p(\Gamma_1) = 1\text{-}\epsilon$

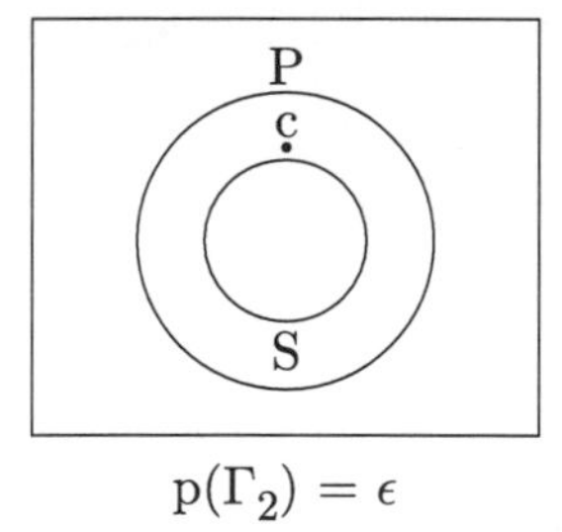

$p(\Gamma_2) = \epsilon$

Adams (1966: 276–7). For a related reason, the order-of-magnitude ordering method of determining p-validity described in section 6.7⋆ doesn't generalize directly to first-order inferences involving conditionals.

[6]This theorem is named for the great American logician Alonzo Church. Actually, in the special case of languages with only one-place predicates and formulas like $(\forall x)(Px \rightarrow Sx)$ and $Pc \rightarrow Sc$, there are mechanical procedures for determining whether there exist worlds in which these formulas are true or false. Cf. chapter IV of Ackermann (1954).

Note that $(\forall x)(Px \rightarrow Sx)$ is intuitively true in Γ_1 and false in Γ_2, since in Γ_1 everything in P is in S, but that is not the case in Γ_2. And, $Pc \Rightarrow Sc$ is neither verified nor falsified in Γ_1, since Pc is false in it, while it is falsified in Γ_2, because Pc is true but Sc is false in it.

Now, if we suppose that the probability of being in world Γ_1 is $1 - \epsilon$, and the probability of being in world Γ_2 is ϵ, then these must be the only 'probabilistically possible' possible worlds. Moreover, since $(\forall x)(Px \rightarrow Sx)$ is true in Γ_1 and false in Γ_2 its probability must equal $1 - \epsilon$, i.e., $p((\forall x)(Px \rightarrow Sx)) = 1 - \epsilon$ while $p(Pc \Rightarrow Sc)$ must equal 0 since Pc&Sc is false in both Γ_1 and Γ_2.[7] Since ϵ can be arbitrarily small, it follows that $(\forall x)(Px \rightarrow Sx)$ can be arbitrarily highly probable while $Pc \Rightarrow Sc$ has zero probability, and therefore, according to equivalence (4) of the first equivalence theorem of Section 7.1, inference (b) must be p-invalid.

Concluding this section, let us note that the formal result, that inference (b) is not p-valid, is confirmed by a real-life counterexample, which itself corresponds to some extent to the possible probabilistic state of affairs represented by the distribution $p(\Gamma_1) = 1 - \epsilon$ and $p(\Gamma_2) = \epsilon$ over worlds Γ_1 and Γ_2. Let us interpret

Px = x was at the party,
Sx = x is a student,
c = the Chancellor.

Given this, $(\forall x)(Px \rightarrow Sx)$ symbolizes "Everyone at the party was a student," and $Pc \Rightarrow Sc$ symbolizes "If the Chancellor was at the party then he is a student," which is obviously absurd. The following section considers this from the dynamical point of view.

7.4 Dynamics and Scope of Universal Statements: Confirmation

Now let us introduce *prior* and *posterior* probability functions, $p_0(\ )$ and $p_1(\ )$, and assume that the latter is derived from the former by conditionalizing on new information, ι, according to Bayes' principle. Section 5.5 states the following formula, which relates the prior and posterior uncertainties, $u_0(\phi)$ and $u_1(\phi)$, of a factual conclusion, ϕ, that might be deduced from a prior factual premise ϕ_1, after acquiring new information, ι:

$$u_1(\phi) \leq u_0(\phi_1|\iota).$$

Combining this with the hypothesis that $u_0(\phi_1|\iota)$ is the prior uncertainty of the conditional $\iota \Rightarrow \phi_1$, we get

$$u_1(\phi) \leq u_0(\iota \Rightarrow \phi_1),$$

[7] This follows from equations (1) and (2), assuming that $\tau_1(\ )$ and $\tau_2(\ )$ are the truth functions that correspond to Γ_1 and Γ_2, respectively.

which is equivalent to

$$p_1(\phi) \geq p_0(\iota \Rightarrow \phi_1).$$

I.e., the posterior probability of the conclusion ϕ must be at least as high as the prior probability of the 'old premise' ϕ_1 given the new information, ι. Now, apply this to the case in which $\phi_1 = (\forall x)(Px \rightarrow Sx)$, $\iota = Pc$, and $\phi = Sc$, i.e., in which the prior premise was "Everyone at the party was a student," the new information is "The Chancellor was at the party," and the conclusion is "The Chancellor is a student." Then we get:

$$p_1(Sc) \geq p_0(Pc \Rightarrow (\forall x)(Px \rightarrow Sx)).$$

I.e., the posterior probability of the Chancellor being a student must be at least as high as the prior probability of "If the Chancellor was at the party then everyone at the party was a student." But, while "Everybody at the party was a student" might have been highly probable *a priori*, "If the Chancellor was at the party then everybody at it was a student" would have been absurd—in fact, to infer it from "Everybody at the party was a student" would have been an instance of one of the fallacies of material implication (cf. section 6.2). Let us comment briefly on this.

The fact that it might be a fallacy to deduce "If the Chancellor was at the party then everybody at it was a student" from "Everybody at the party was a student" doesn't mean that the conditional couldn't have been independently probable *a priori*. However, whether it was probable or not would have been a matter of fact and not of form, depending on the circumstances and what we know about students, chancellors, parties, and so on. In short, as was pointed out in section 5.6⋆, it would depend on whether "The Chancellor was at the party" belonged to the *scope* of "Everyone at the party was a student," as uttered in the circumstances imagined.[8]

A final remark on probability dynamics relates to *confirmation*, and

[8]It is worth recalling the distinction that was made in section 5.6⋆, between two kinds of scope that a universal claim like "Everyone at the party was a student" can have: (1) a 'traditional scope' that is the class of individuals included in "Everyone" that the claim is meant to 'apply' to, and (2) the class of propositions, the learning of which would not lead to giving it up. Only universals have scopes of type (1) while all asserted propositions have scopes of type (2), but when they both exist there is a *prima facie* connection between them. For instance, that "The Chancellor was at the party" falls outside the propositional scope of "Everyone at the party was a student" suggests that the "Everyone" wasn't meant to include the Chancellor, and therefore he didn't belong to the traditional scope of this universal. This suggested a crude approximate rule, that it is likely that an individual belongs to the traditional scope of a universal if and only if he or she is the subject of a proposition that belongs to its propositional scope. This applies in the present case because the Chancellor, who doesn't belong to the traditional scope of "Everyone at the party was a student," who is the subject of "The Chancellor was at the party," lies outside the propositional scope of the universal.

A more general point that was made in section 5.6⋆ also applies to our example. That is that no proposition that is not a priori certain can have universal propositional scope, and therefore it is possible that it might have to be given up (this follows from rule

specifically to the question of whether learning a 'positive instance' like "Charles was at the party, and he is a student" ought to 'support' or 'confirm' a universal like "Everyone at the party was a student." In general, should propositions of the form Pc&Sc support universal statements of the form $(\forall x)(Px \rightarrow Sx)$? In fact, this is closely connected to the question of whether inferences of the form

$$\text{c.}\quad \frac{(\forall x)(Px \rightarrow Sx)}{(Pc\&Sc) \Rightarrow (\forall x)(Px \rightarrow Sx)}$$

are p-valid. As before, the factuality-conditionality theorem implies that this inference is p-invalid, since its premise is clearly essential but it does not logically entail the antecedent of the conclusion. That means that Pc&Sc doesn't necessarily belong to the scope of $(\forall x)(Px \rightarrow Sx)$, and therefore $(\forall x)(Px \rightarrow Sx)$ can be probable *a priori* but learning Pc&Sc not only may not confirm it, it can actually lead to giving it up. A slight modification of the counterexample to inference (b) suggests how this might happen:

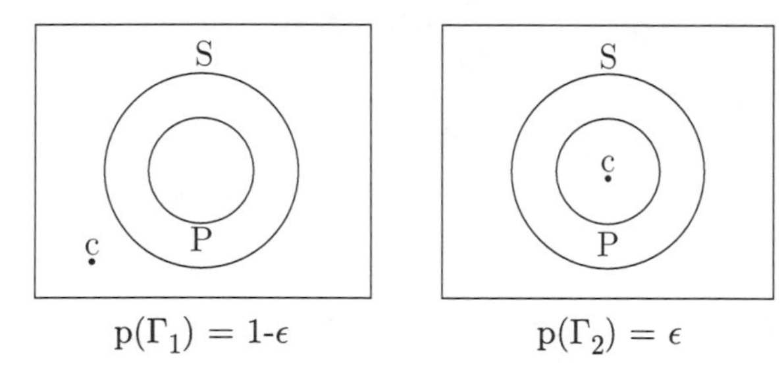

$p(\Gamma_1) = 1\text{-}\epsilon$ $\qquad$ $p(\Gamma_2) = \epsilon$

This shows $(\forall x)(Px \rightarrow Sx)$ to be true in Γ_1 and false in Γ_2, while Pc&Sc is false in Γ_1 and true in Γ_2. Since $p(\Gamma_1) = 1 - \epsilon$ and $p(\Gamma_2) = \epsilon$, it follows that $p((\forall x)(Px \rightarrow Sx)) = 1 - \epsilon$ and $p((Pc\&Sc) \Rightarrow (\forall x)(Px \rightarrow Sx)) = 0$. Given this, $(\forall x)(Px \rightarrow Sx)$ should be probable *a priori*, but learning Pc&Sc would make it certain *a posteriori* that one was in world Γ_2, and therefore that $(\forall x)(Px \rightarrow Sx)$ was false. The following possibly far-fetched example suggests that this could happen in real life.

Suppose that you think that everyone who was at a given party was a student, *and* that a certain Charles, who is himself a student, was not at the party. You also know that whenever Charles goes to parties he goes with his wife, who happens not to be a student. Now suppose you learn, to your surprise, that Charles was at the party: a positive instance, since Charles is a student. Under the circumstances, however, you are likely to

S3—that anything inconsistent with something that is not a certainty doesn't belong to its scope). This applies to "Everybody at the party was a student," which would obviously not be a certainty, and which would have to be given up if it was learned that the Chancellor was at the party.

abandon your earlier opinion, that everyone at the party was a student, because now that you know that Charles was at the party you are pretty sure that his wife was with him, and she isn't a student.

The foregoing is only an illustration of the potentialities for applying ideas of deductive dynamics to the confirmation of universal statements, and not an 'in depth' study. However, although a generalization to approximate generalizations will be commented on in appendix 9, this cannot be pursued further here. Prior to that, though, appendix 8 will discuss briefly the extension of probability logic to the logic of identity statements, which will bring out a special problem of Bayes' principle applied to reasoning involving formulas of predicate logic.

Exercises

1. Determine the p-validity of the following inferences, and for those that are p-invalid draw diagrammatic counterexamples like the pair of Venn diagrams the were given in section 3 as a counterexample to the inference $(\forall x)(Px \rightarrow Sx)$ ∴ $Pc \Rightarrow Sc$, and using these as a basis, construct real life counterexamples to these inferences.
   a. $(\exists x)Px \Rightarrow (\forall x)(Px \rightarrow Sx)$ ∴ $(\forall x)(Px \rightarrow Sx)$
   b. Pc ∴ $(\forall x)Px \Rightarrow (\exists x)Px$ (Be careful on this one!)
   c. $(\exists x)(Fx \vee Sx)$ ∴ $\sim(\exists x)Fx \Rightarrow (\forall x)Sx$
   d. $(\forall x)(Pc \rightarrow Px)$ ∴ $Pc \Rightarrow (\forall x)Px$
   e. $\sim(Fc\&Sc)$ ∴ $Fc \Rightarrow \sim Sc$

⋆2. *The bald man paradox*[9] Let "Bx" symbolize "x is bald", let "yHx" symbolize "y has one more hair on his head than x", and let $m_1, m_2, \ldots, m_n$ be man #1, man #2, ..., man #n. Then "$(Bm_1\&m_2Hm_1) \rightarrow Bm_2$" says that if man #1 is bald and man #2 has one more hair on his head than man #1 then man #2 is also bald, and so on. Then if we have n premises,

$$Bm_1, (Bm_1\&m_2Hm_1) \rightarrow Bm_2, (Bm_2\&m_3Hm_2) \rightarrow Bm_3, \ldots, (Bm_{n-1}\&m_nHm_{n-1}) \rightarrow Bm_n,$$

the valid conclusion is Bm_n. I.e., man #n is bald. But this is absurd if n is very large.

[9]This paradox, which is essentially the same as the Sorites paradox (the 'paradox of the heap') that was known to the ancient Greeks, has been widely discussed in recent philosophical literature, especially with the aim of applying fuzzy logic to it (cf. Goguen 1968–9). This application assumes that what 'accumulates' in the conclusion is not its degree of uncertainty, but something like its 'distance from the truth' (as the number of hairs on a man's head increases, the claim that he is bald becomes further and further from the truth). But Edgington (1997) suggests that distance from the truth might conform to the same laws as uncertainty.

★a. Apply theorems of chapter 3 to answer the following: If $p(B_1) = .99$, $p(m_{i+1}Hm_i) = 1$ for $i = 1, \ldots, n-1$, and $p(Bm_{i-1} \mathbin{\&} m_iHm_{i-1}) \rightarrow Bm_i) = .9999$ for $i = 2, \ldots, n$, how large does n have to be for it to be possible that $p(Bm_n)$ should equal 0?

★b. Can you answer this in the case in which the material conditionals above are replaced by probability conditionals?

★3. Consider the set of six relational premises $\{a < b, a < c, a < d, b < c, b < d, c < d\}$, where a, b, c, and d are the ages of four people, and the conclusion $a < b < c < d$ (the people are ordered in increasing age).

a. Assuming that the laws of numerical ordering are logically true, and using the laws of essentialness stated in section 3.3★, determine the degrees of essentialness of these premises.

b. Assuming that each premise has a probability of at least .9, what is the minimum probability of the conclusion?

c. Describe a possible probabilistic state of affairs in which each premise has a probability of at least .9, but the conclusion has the minimum probability calculated in your answer to part b.

★4. Adams (1988b) considers the so-called 'independence of pure instancehood' condition for the confirmation of an approximate generalization by a positive instance, namely that, provided that Pc by itself doesn't disconfirm an approximate generalization "Ps are Ss," a positive instance Pc&Sc confirms it, or at least it doesn't disconfirm it. For instance (looking ahead to appendix 8), provided that "Tweety is a bird" doesn't by itself disconfirm "Birds fly," the positive instance "Tweety is a bird that flies" confirms "Birds fly," or at least it doesn't disconfirm it.

Argue that the following analogue applies to exact generalizations $(\forall x)(Px \rightarrow Sx)$: Provided that its posterior probability just given Pc isn't lower than its prior probability, it follows that its posterior probability given Pc&Sc also isn't lower than its prior probability.[10] Put in terms of conditional probabilities, argue that if $p((\forall x)(Px \rightarrow Sx)|Pc) \geq p((\forall x)(Px \rightarrow Sx))$ then $p((\forall x)(Ps \rightarrow Sx)|Pc\&Sc) \geq p((\forall x)(Px \rightarrow Sx))$. Note that the problem in the example of Charles going to a party arose from the fact that merely learning that Charles went to the party without being told whether or not he was a student was enough to disconfirm "Everyone at the party was a student."

[10]There are connections between this example and the famous 'Ravens paradox' due to C.G. Hempel (cf. chapter 1 of Hempel 1965), which shows that two seemingly self-evident principles of confirmation have the paradoxical consequence that discovering a pink elephant should confirm the law "All ravens are black."

Appendix 8

Probabilistic Identity Logic

8.1 Generalities

From the formal point of view probabilistic identity logic is a very minor extension of probabilistic predicate logic. That is, it extends the language of the latter only by adding the formal identity predicate, =, which is assumed to satisfy the usual laws of the identity calculus.[1] As in appendix 7, for purposes of exposition we will assume that our language contains just the single monadic predicate A, individual constants t and q, and possibly others, with the binary identity relation symbol = added. Factual formulas are the usual formulas of the identity calculus such as $t = q$ and $(\exists x)Ax$, and conditionals $\phi \Rightarrow \psi$ are formed from factual ϕ and ψ, e.g., $(t = q) \Rightarrow (\exists x)Ax$.

Also as in appendix 7 we will assume that truth and probability apply to factual formulas, ϕ and ψ, beginning with possible worlds, $\Gamma = \langle D, A, t, q\rangle$, in which these formulas are defined to be true or false in the standard way. An identity like $t = q$ is defined to be true in Γ if and only if Γ assigns the same member of D to both t and q. The probability of ϕ, $p(\phi)$, is also conceived as the probability of being in a world in which it is true, and the probability of $\phi \Rightarrow \psi$ is defined as $p(\psi|\phi) = p(\phi\&\psi)/p(\phi)$ (or $p(\phi \Rightarrow \psi) = 1$ if $p(\phi) = 0$). P-validity is defined as before, the rules of conditional inference stated in chapter 7$\star$ continue to hold, and it continues to be the case that an inference is p-valid if and only if a subset of its premises yields its conclusion. For instance, inference (a) below is p-valid, and inference (b) is p-invalid:

$$\text{a.}\quad \frac{(b = c) \Rightarrow (b = d)}{(b = c) \Rightarrow (c = d)} \qquad\qquad \text{b.}\quad \frac{At}{(q = t) \Rightarrow Aq}$$

Inference (a) is valid because $(b = c) \Rightarrow (b = d)$ yields $(b = c) \Rightarrow (c = d)$. Thus, any world in which $(b = c) \Rightarrow (b = d)$ is verified must be one in which $b = c$ and $b = d$ are both true, hence $c = d$ must be true, and

[1] For example, see chapter 9 of Mates (1965).

therefore $(b = c) \Rightarrow (c = d)$ must be verified. And, any world in which $(b = c) \Rightarrow (c = d)$ is falsified must be one in which $b = c$ is true and $c = d$ false, hence $b = d$ must be false and therefore $(b = c) \Rightarrow (b = d)$ must be falsified. And, given that $(b = c) \Rightarrow (b = d)$ yields $(b = c) \Rightarrow (c = d)$, it follows from the uncertainty sum theorem that $u[(b = c) \Rightarrow (c = d)] \leq u[(b = c) \Rightarrow (b = d)]$, hence $p[(b = c) \Rightarrow (c = d)] \geq p[(b = c) \Rightarrow (b = d)]$, i.e., the conclusion of inference (a) must be at least as probable as its premise.

The factuality-conditionality theorem of section 7.7 applies to inference (b) because it has a factual premise and a conditional conclusion. The premise is obviously essential, hence the inference can only be p-valid if its premise logically entails both the antecedent and consequent of its conclusion, i.e., if At logically entails both $q = t$ and Aq, which it obviously does not.

As before, the yielding test can be used to construct probabilistic and real-life counterexamples that substantiate the claim that the inference is p-invalid, and it wouldn't always be rational in everyday life. That At doesn't yield $(q = t) \Rightarrow Aq$ is shown by the world Γ_1, diagrammed on the left below, in which At is true but $q = t$ is false, hence $(q = t) \Rightarrow Aq$ isn't verified:

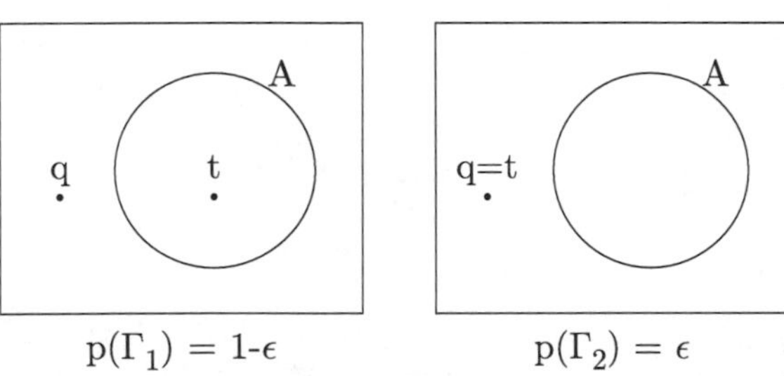

$p(\Gamma_1) = 1\text{-}\epsilon$ $\quad$ $p(\Gamma_2) = \epsilon$

That $(q = t) \Rightarrow Aq$ isn't yielded by the empty set is shown by world Γ_2, diagrammed on the right, since $q = t$ is true but Aq is false in it. Then, assuming that $p(\Gamma_1) = 1 - \epsilon$ and $p(\Gamma_2) = \epsilon$, it follows that $p(At) = 1 - \epsilon$ while $p((q = t) \Rightarrow Aq) = 0$, hence At can be arbitrarily highly probable while $(q = t) \Rightarrow Aq$ has zero probability.

To construct a real-life counterexample, suppose that, speaking of cards turned face down in a shuffled pack, we interpret:

t = the top card,
q = the queen of spades,
Ax = x is an ace.

Thus interpreted, inference (b) becomes "The top card is an ace; therefore, if the queen of spades is the top card then it is an ace." In certain circumstances it could be probable that the top card should be an ace, but it would be absurd to infer that if the queen of spades is the top card then it is an ace. In such circumstances the scope of the prior premise "The top card is an ace" wouldn't include its being the queen of spades, and learning that would result in giving up the premise rather than in inferring "If the queen of spades is the top card then it is an ace."

But bringing probability dynamics into the picture introduces certain complications in the analysis of reasoning related to inference (b), including connections with Frege's puzzle, which the following section comments on briefly.

8.2 Complications: Connections with Frege's Puzzle

The inference below is related to form (b):

$$\text{c.} \quad \frac{\mathrm{At}, \mathrm{q} = \mathrm{t}}{\mathrm{Aq}}.$$

This is a purely factual and classically valid inference, hence it should be p-valid, and therefore the uncertainty sum theorem implies that

(1) $u(Aq) \leq u(At) + u(q = t)$.

Therefore if each premise has an uncertainty of at most 1%, the conclusion's uncertainty can be at most 2%.

But this analysis neglects probability dynamics, and the possibility that learning the second premise might lower the probability of the first. To take this into account, let prior and posterior probabilities be given by the functions $p_0(\ )$ and $p_1(\ )$, respectively, and assume that the new information is $q = t$. Then, assuming Bayes' principle, the dynamical theory of two-premise inferences outlined in section 5.3 implies that

$$u_1(Aq) \leq u_0(At|q = t) = u_0((q = t) \Rightarrow At),$$

where $u_0(\ )$ and $u_1(\ )$ are the uncertainty functions corresponding to $p_0(\ )$ and $p_1(\ )$, respectively. Therefore

(2) $p_1(Aq) \geq p_0((q = t) \Rightarrow At)$.

Thus, Aq can only be expected to be probable *a posteriori*, after $q = t$ is learned, if the conditional $(q = t) \Rightarrow At$ was probable *a priori*. But that is exactly what isn't the case in the card example, since "The top card is an ace" may have been probable *a priori*, but the conditional "If the queen of spades is the top card then it is an ace" would have been absurd. As said, "The queen of spades is the top card" would almost certainly lie outside the scope of "The top card is an ace." But there is another complication.

The dynamical 'fallacy' of deducing "The queen of spades is an ace" from "The top card is an ace," given that the queen of spades is the top card, is in some ways like Frege's Morning Star–Evening Star puzzle.[2] In fact,

[2]Frege (1892). The literature on this puzzle is enormous. Salmon (1986) gives a very detailed account and analysis of it.

it seems that, given that the queen of spades actually is the top card, if this weren't known it could be probable that the card that was *actually* the queen of spades was an ace. This is due to what used to be called the 'referential opacity' of the expression "... that the top card is an ace," which does not permit "the top card" to be replaced by "the queen of spades" in "It is probable that the top card is an ace," even though the top card happens to be the queen of spades. The question to be asked is whether the fact that expressions like "... that the top card is an ace" may be opaque in contexts like "It is probable that the top card is an ace" further complicates the problem of accounting for the validity of reasoning such as that which was formalized as inference (c).

It seems plausible that the static analysis of the inference is unaffected by referential opacity considerations. Consider inequality (1), and assume that the quantities it involves are 'real uncertainties', i.e., they are derived from frequencies in psychologically random series, as discussed in section 9.9⋆⋆. If t and q are, as above, the top card and the queen of spades, respectively, in a shuffled standard 52-card pack, and Ax is the property that x is an ace, then, intuitively, the real probabilities of At, q = t, and Aq should be 1/13, 1/52, and 0, respectively, and the real uncertainties u(At), u(q = t), and u(Aq) should be 12/13, 51/52, and 1, respectively, which clearly satisfy inequality (1). Moreover this intuition is supported by the analysis of section 9.9⋆⋆.

Focus on the 'real probability' that q = t, that the queen of spades is the top card, which intuition says is 1 in 52.[3] According to the pragmatic theory of section 9.9⋆⋆, what saying that the probability that the queen of spades is the top card is 1/52 really means is that in situations in which persons predict that the queen of spades is the top card, they are right only 1 time in 52, and wrong 51 times in the long run. That is why they are best off in the long run if they act on degrees of confidence that correspond to this probability. And at least on superficial consideration, the proportion of times that persons are right in predictions that the queen of spades is the top card in shuffled packs of cards does not depend on how they refer to or otherwise single out the cards.[4]

On the other hand, it is not clear that the dynamical analysis is indepen-

[3]Given that every standard deck of cards has a queen of spades, one might question calling "the queen of spades" a definite description that refers to a unique individual. An alternative would be to say that "... is the queen of spades" designates a property that a card can have, but this complication can be left aside here because it doesn't affect the probabilisitic analysis.

[4]It is we, external observers, who determine by observation that persons make guesses about cards, but what allows us to say of an item of a person's behavior that it is a prediction that the queen of spades is the top card in a deck is a complex matter. It need not involve identifying an element of that behavior as 'referring' to something.

dent of referential opacity considerations. This analysis depends on Bayes' principle, and that has to do with the effects of learning, or acquiring information, say that the queen of spades is the top card in a shuffled deck. But what is it for a person to learn that the queen of spades is the top card? If the queen of spades *is* the top card, then isn't whatever the person learns about the top card equally something she learns about the queen of spades? And, if she learns, what she already knows, that the top card is the top card, doesn't she also learn, and in fact know, that the queen of spades is the top card? That's Frege's puzzle, which must remain unsolved in the present context.

The upshot is as follows. In spite of referential complications, it seems that the static analysis given by inequality (1) makes a contribution to our understanding of reasoning involving identities, because it makes their probabilities explicit, and 'real' probabilities are independent of reference. For instance, inequality (1) seems to justify inferring that the top card is probably a face card, given that it is probably the queen of spades and the queen of spades is a face card.[5] That is because the identity of the top card and the queen of spades is treated as a factual relation subject to probabilities, on a par with other factual relations like being of the same age. On the other hand, whether probability considerations can help to throw light on the special problems that are associated with referential opacity, which seem to be involved in the Bayesian theory of probability change,[6] is a matter for further inquiry.

Exercises

1. Use the yielding test to determine which of the inferences below are p-valid, and to construct probabilistic and real-life counterexamples to any that are not.
   a. $(\mathrm{b} = \mathrm{c}) \Rightarrow \mathrm{Ab} \therefore (\mathrm{c} = \mathrm{b}) \Rightarrow \mathrm{Ac}$
   b. $\mathrm{Ab} \vee \mathrm{Ac} \therefore (\mathrm{b} = \mathrm{c}) \Rightarrow \mathrm{Ac}$
   c. $(\forall \mathrm{x})(\mathrm{Ax} \rightarrow (\mathrm{x} = \mathrm{f})) \therefore \mathrm{Af} \Rightarrow \mathrm{As}.$
2. a. Following the discussion of section 3.3$\star$, determine the sizes of the minimal essential premise sets in the inference $\{\mathrm{a} = \mathrm{b}, \mathrm{b} = \mathrm{c}, \mathrm{a} = \mathrm{c}\} \therefore (\mathrm{a} = \mathrm{b})\&(\mathrm{b} = \mathrm{c})$ and the degrees of essentialness of each premise.

[5] The queen of spades might be recognizable in marked decks, and persons who could recognize their markings could be right most of the time in predictions they made of the form "The top card is the queen of spades."

[6] These problems do not only affect reasoning involving identities, but any 'dynamical', Bayesian reasoning about propositions that involve references to particular individuals.

b. What is the minimum probability of the conclusion of the inference above, if each premise has a probability of .9?

c. Give a probabilistic model in which the premises of the inference above have probability .9 and the conclusion has the minimum probability calculated in part b.

d. Can you describe a real-life situation in which the premises of the inference might have probability .9 and the conclusion have a minimum probability consistent with this?

⋆3. Consider the premises "a=b", "a=c", "a=d", "a=e", "b=c", "b=d", "b=e", "c=d", "c=e", and "d=e" (there are 5 individuals, a, b, c, d, and e, and for any two of them there is a premise that states that those two are equal), and the valid conclusion "a=b=c=d=e" (all five are equal). What is the smallest essential set of premises that contains "a=b", and what is its degree of essentialness? If each premise has probability .9, what is the minimum probability of the conclusion? Describe a probabilistic model in which all of the premises have probability .9 and the conclusion has this minimum probability.

Appendix 9

Approximate Generalizations[1]

9.1 Introduction

Appendix 7 dealt with general statements like "All birds fly," symbolized using universal quantifiers, say as $(\forall x)(Bx \rightarrow Fx)$, which were evaluated simply as true or false in possible worlds or models. It follows that if there is at least one bird, say Tweety, symbolized as t, for whom the corresponding instance, $Bt \rightarrow Ft$, is false, then the general statement is false in the world. Therefore, since there are nonflying birds in the real world, namely penguins, "All birds fly" must be false in it.[2] So, it seems to follow that saying "All birds fly" is just as wrong as saying "No birds fly." Both are false, and one is no more false than the other. Or is it?

This appendix will explore the idea that "All birds fly" can be interpreted as an *approximate generalization*, more properly expressed as "Birds fly."[3] This isn't strictly true, but it has practical utility and a *degree of truth* because, given that birds fly you know what to expect when you see a bird, even though the expectation may occasionally be mistaken. The important thing, however, is that it can be more useful to have guidance that is right most of the time than to have no guidance at all.[4] That is the 'pragmatic

[1]The approach that is outlined here to approximate generalizations is more fully developed in several papers including Adams (1974; 1986; 1988b), Adams and Carlstrom (1979), and Carlstrom (1975; 1990).

[2]The example of birds, flying, and Tweety has been used repeatedly in recent writings on Artificial Intelligence. Cf. Pearl (1988).

[3]The famous philosopher John Stuart Mill, 1806–73, pioneered in the study of these propositions. C.f. Chapter XXIII of Book III of Mill (1895 [1843]).

[4]Some would say that we should never make unqualified statements of the "All As are Bs" form when they have exceptions. But to restrict ourselves to exceptionless generalizations in practical matters would be to restrict ourselves almost exclusively to tautologies like "All birds are birds," which provide no useful guidance. We might insist that when there are exceptions we should say "Almost all As are Bs" or "Probably almost all As are Bs" instead of "All As are Bs", but to have to keep repeating these qualifications when they are obvious in context would be irksome. In any case, no matter how they are expressed in ordinary language, logic should be able to deal with them if it is to deal

dimension' of the approximate generalization.

Methods of formal logic can be applied to formalize and analyze reasoning involving generalizations like "Birds fly" (henceforth we will omit the word "approximate," and speak simply of "generalizations"). The basic idea is to measure the pragmatic value or degree of truth of "Birds fly" by the proportion of birds that fly, and given this we would expect "Birds fly" to be fairly highly true—much more so than "No birds fly", or "Birds don't fly", whose degree of truth is measured by the proportion of birds that don't fly.

From the formal point of view, the interesting thing about measuring the degrees of truth of generalizations as proportions is that proportions satisfy the same laws as the probabilities that our earlier chapters were concerned with do, and relative proportions, e.g., of birds that fly, satisfy the laws of conditional probability.[5] This suggests that reasoning involving the generalizations can be evaluated in much the same way as reasoning involving conditionals. In fact, generalizations can be treated formally as conditionals, although we will see in the final section of this appendix that it would be a serious mistake to suppose that they *are* conditionals. The next two sections outline this in more detail, beginning with a formalism for symbolizing the generalizations that we are concerned with. The final section examines the connection between generalizations and conditionals more closely.

9.2 Approximate Monadic Generalizations

The expressions that we will focus on in this section involve only unary predicates like Bx and Fx, symbolizing "x is a bird" and "x flies" with one free variable, which we can always suppose is 'x'. Let us call Bx, Fx, and compounds like Bx&Fx *indefinite sentences*,[6] and we will use the $\Rightarrow$ connective in the 'indefinite generalization' Bx $\Rightarrow$ Fx, which symbolizes "If x is a bird then x flies," or, equivalently, "Birds fly." Indefinite sentences, including generalizations, are in contrast to definite sentences like Bt, Ft, Bt&Ft, Bt $\Rightarrow$ Ft, etc., that involve individual constants like t, and which might symbolize "Tweety is a bird," "Tweety flies" and "Tweety is a bird that flies," and "If Tweety is a bird then Tweety flies." Aside from conditionals, definite sentences are simply true or false, or true or false in models or possible worlds; but indefinite sentences, including generalizations, will be supposed to have 'degrees of truth'. In the case of such a nonconditional indefinite sentence, as Bx&Fx, its degree of truth in a world will simply

with generalizations that are practically useful.

[5]There is more than a formal resemblance between proportions and probabilities, since the fact that a high proportion of birds fly is part of what makes it probable that if Tweety is a bird then he flies. But it will be seen in section 9.4⋆ of this appendix that the probability and the proportion should not be confused.

[6]These were often called *propositional functions* in more traditional logical parlance. Cf. section 04 of Church (1956).

be taken to be the proportion of values of 'x' that satisfy it in the world $\Gamma = \langle D, B, F, t \rangle$, and in the case of such a generalization as $Bx \Rightarrow Fx$, its degree of truth is taken to be the proportion of values of 'x' that satisfy Fx out of all values that satisfy Bx.[7] Proceeding in this way, the degrees of truth of Bx&Fx and of $Bx \Rightarrow Fx$ will be written $\tau(Bx\&Fx)$ and $\tau(Bx \Rightarrow Fx)$, where $\tau(\ )$ is the *degree of truth function* for the world Γ (this measure generalizes the truth functions that were introduced in appendix 7). This is in contrast to prefixing $Bx \rightarrow Fx$ with the universal quantifier $(\forall x)$, which applies to the strict universal generalization $(\forall x)(Bx \rightarrow Fx)$, which properly symbolizes the false statement "Every bird, without exception, flies." But "Birds fly" symbolized by $Bx \Rightarrow Fx$ isn't strict, and its degree of truth, $\tau(Bx \Rightarrow Fx)$, can be quite high.[8]

Turning to deductive relations among our generalizations, we will make use of the fact that their degrees of truth satisfy the same laws as probabilities, and they can be represented diagrammatically in the same way. For instance, adding another predicate and letting Px symbolize "x is a penguin", we can represent the populations of birds, of things that fly, and of penguins as regions in a Venn diagram, thus:

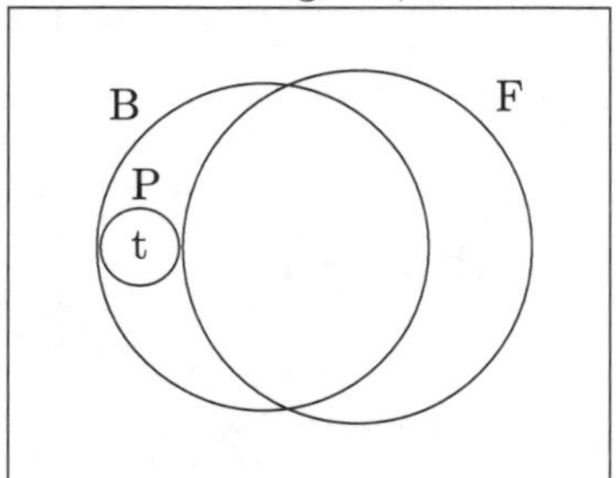

Proportionality Diagram 1

Here, birds and penguins are represented by the points in circles B and

[7]That a definite conditional like $Bt \Rightarrow Ft$ doesn't have a truth-value in a world, while its indefinite counterpart, $Bx \Rightarrow Fx$, has a degree of truth is explained by the fact that if a *constant*, t, has only one value, it doesn't make sense to talk of the proportion of its values that satisfy Ft out of all those that satisfy Bt.

For certain purposes it is sometimes necessary to consider the possibility that a population might be infinite, and that creates a problem when it comes to defining proportions. It is possible to deal with this, but it involves considerable mathematical complications which we will avoid here by assuming that all of the populations that we deal with are finite.

[8]There is a subtle distinction between a formula, say $Bx \Rightarrow Fx$, having the highest degree of truth in a world, namely $t(Bx \Rightarrow Fx) = 1$, and the 'perfect' or 'mathematical truth' of the corresponding universal statement $(\forall x)(Bx \rightarrow Fx)$ in the world. The former is a necessary but not a sufficient condition for the latter, since in worlds $\Gamma = \langle D, B, F, t \rangle$ in which D is infinite and $t(Bx \Rightarrow Fx)$ is not a simple proportion, it can happen that $t(Bx \Rightarrow Fx) = 1$, but $(\forall x)(Bx \rightarrow Fx)$ is false. This is related to the infinitesimal probabilities that were discussed in appendix 2.

P, respectively, flying things are represented by points in region F, and Tweety is represented by a point inside P. Thus, Tweety is represented as being a penguin, i.e., as a nonflying bird. On the other hand, the proportion of birds that fly corresponds to the proportion of region B that lies inside region F, which is fairly high, although it is less than 1. Therefore, the strict universal $(\forall x)(Bx \rightarrow Fx)$ is pictured as being false but the unstrict generalization "Birds fly", symbolized as $Bx \Rightarrow Fx$, is pictured as being highly true, though not perfectly true. On the other hand "Birds who are penguins fly," symbolized as $(Bx\&Px) \Rightarrow Fx$, is pictured as perfectly false, because none of the intersection of circles B and P lies inside of F. Adapting terminology introduced in section 6.2, we can say that being a penguin doesn't belong to the scope of the generalization "Birds fly", since birds that are penguins don't fly.

We also see that proportionality diagram 1 constitutes a diagrammatic counterexample to the inference $Bx \Rightarrow Fx \therefore (Bx\&Px) \Rightarrow Fx$, since it pictures its premise as highly true at the same time that its conclusion has no truth. This brings out the fact that we can define a concept of validity that applies to inferences involving approximate generalizations, in the same way that probabilistic validity applies to inferences involving probability conditionals. We will call this *measure-validity* (m-validity) to distinguish it from the probabilistic version, but it clearly obeys the same laws. In particular, we can use the probability diagram technique to show that inferences like $Bx \Rightarrow Fx \therefore (Bx\&Px) \Rightarrow Fx$ are m-invalid, and to construct real-life examples in which reasoning of a given pattern would be irrational. Furthermore, the principles of p-validity stated in section 7.2 translate into principles of m-validity for inferences involving approximate generalizations. In fact, the trivalent truth-table method of section 7.5 can be adapted to determine whether inferences like $Bx \Rightarrow Fx \therefore (Bx\&Px) \Rightarrow Fx$ are m-valid, and you can use the p-validity rules to deduce conclusions from the premises of inferences that are m-valid. Thus, the truth-table method shows this inference to be m-invalid, since making Bx and Fx both true and Px false verifies the premise but it doesn't verify the conclusion—so the premise doesn't *yield* the conclusion, as proportionality diagram 1 shows.[9]

But there is an important extension that doesn't directly correspond to the p-validity principles that were stated in section 7.2.

9.3 Multi-variate Approximate Generalizations

It is natural to use two variables to symbolize generalizations like "Husbands are taller than their wives", thus: $xHy \Rightarrow xTy$, where xHy symbolizes "x is

[9]Of course, Bx and Fx are true 'in name only', and Px is false in name only. Strictly speaking, these formulas have degrees of truth and not 'simple' truth values.

the husband of y" and xTy symbolizes "x is taller than y." And, it seems obvious that it would be unreasonable to infer from this that "Short husbands are taller than their wives", symbolized by (Sx&xHy) $\Rightarrow$ xTy, where Sx symbolizes "x is short." But we want to demonstrate this quantitatively, by showing that the fact that the premise has a high degree of truth is compatible with the conclusion having a low one.

We can define the degree of truth of "Husbands are taller than their wives" as the proportion of married couples in which the husband is taller than the wife; i.e., τ(xHy $\Rightarrow$ xTy) is the proportion of ordered pairs of individuals that satisfy xTy out of all those that satisfy xHy.[10] Similarly, τ((Sx&xHy) $\Rightarrow$ xTy) is the proportion of pairs that satisfy xTy out of all those that satisfy the conjunction Sx&xHy. Then we can use a variant of the proportionality diagram method to show that the fact that τ(xHy $\Rightarrow$ xTy) is high is compatible with τ((Sx&xHy) $\Rightarrow$ xTy) being low. Now we picture pairs of individuals, like John and his wife Mary, as points in a two-dimensional space, and we picture sets of pairs which satisfy formulas like xHy and xTy as regions in this space (see Proportionality diagram 2)

Here the husband-wife pair of John and Mary is represented by the point with coordinates ⟨John,Mary⟩, the set of husband-wife pairs is represented by the set of points in the lower left circle, and the set of pairs of persons in which the first is taller than the second is represented by the set of points occupying the upper right circle. Most of the first circle lies inside the second, hence we have pictured a situation in which in a high proportion of husband-wife pairs, the husband is taller than the wife. On the other hand, the region of persons who satisfy Sx is shown as the vertical column at the left,[11] so the class of husband-wife pairs in which the husband is short

[10]Note that in defining τ(xHy $\Rightarrow$ xTy) as the proportion of pairs $\langle x, y\rangle$ that satisfy xTy out of all pairs that satisfy xHy, we have introduced a 'quantifier' that applies to x and y simultaneously, which is something that we have not considered before. We can also regard this as the degree of truth of the multiply-general statement xHy $\Rightarrow$ xTy, symbolizing "if x is the husband of y then x is taller than y". But we should be careful to distinguish this from "For almost all x, for almost all y, if x is the husband of y then x is taller than y". For instance, while it is true that for almost all (in fact for all) positive integers x, for almost all positive integers y, x is less than y, but it is only 'half true' that for almost all positive integers x and y, x is less than y.

Curiously, we sometimes want to consider the degrees of truth of formulas like Sx $\Rightarrow$ Sy, in which one of the variables doesn't occur in the antecedent. Then τ(Sx $\Rightarrow$ Sy) is the proportion of pairs $\langle x, y\rangle$ that satisfy Sy out of all those that satisfy Sx, which isn't necessarily the same as τ(Sx $\Rightarrow$ Sx), which is equal to 1 since it is the proportion of individuals that satisfy Sx out of all that satisfy Sx. In fact, τ(Sx $\Rightarrow$ Sy) = τ(Sy), since the proportion of pairs that satisfy Sy out of all those that satisfy Sx necessarily equals the proportion of individuals that satisfy Sy (you can easily verify this for yourself if you think about it). This marks an important difference between degrees of truth and probabilities, since while we can give precise definitions of Prob(Sx $\Rightarrow$ Sy) and Prob(Sy) (cf. section 9.4$\star$), these quantities are not necessarily equal.

[11]Strictly, the vertical column marked Sx represents a set of *pairs* of people, namely

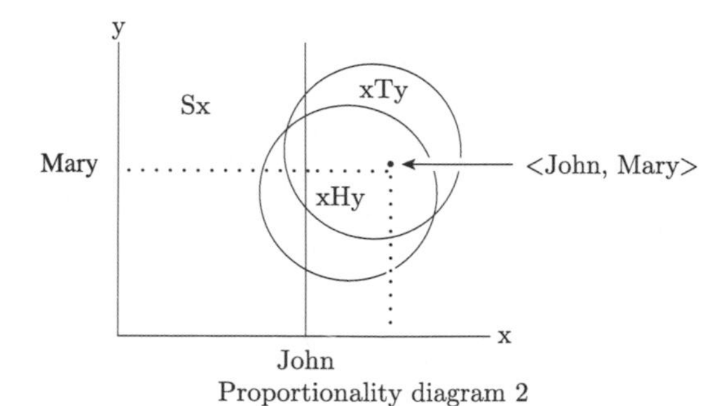

Proportionality diagram 2

corresponds to the intersection of the xHy circle with the vertical column. But the proportion of this intersection that lies inside the xTy circle is not very high. Hence, as we have pictured it, the proportion of husband-wife pairs in which the husband is short, but he is taller than his wife is not high. This is a quantitative demonstration of the fact that the inference xHy $\Rightarrow$ xTy $\therefore$ (xHy&Sx) $\Rightarrow$ xTy is not m-valid.

It would be nice to be able to give a general method for determining the m-validity of multi-variate inferences like the above, like the trivalent truth-table method that applies to uni-variate inferences, and also to give rules for deriving the conclusions of these inferences like the ones that were transferred over from the probabilistic rules stated in section 7.2. Unfortunately this theory has not yet been worked out in detail, and exercise 2 at the end of this appendix, which deals with these inferences, requires you to use your ingenuity.

In the final section of this appendix we will consider 'proper probabilities'. The reader is warned that this material is somewhat more technical than what has been discussed heretofore.

9.4⋆ Generalizations and Conditionals: Degrees of Truth and Probability

Let us return to the monadic case, and consider especially the relation between the generalization "Birds fly," symbolized as Bx $\Rightarrow$ Fx, and the particular conditional statement "If Tweety is a bird then he flies," symbolized as Bt $\Rightarrow$ Ft. How are these two statements related, and what are their similarities and differences?

The most obvious similarity is that we are apt to think that the probability p(Bt $\Rightarrow$ Ft) should equal the degree of truth τ(Bx $\Rightarrow$ Fx); i.e., we are apt to think that the probability that if Tweety is a bird then he flies ought to equal the proportion of birds that fly.

But there is a fundamental difference between p(Bt $\Rightarrow$ Ft) and τ(Bx $\Rightarrow$

those in which the first person of the pair is short.

Fx). $\tau(\text{Bx} \Rightarrow \text{Fx})$ is the degree of truth of Bx ⇒ Fx in a possible world, say $\Gamma = \langle \text{D}, \text{B}, \text{F}, \text{t} \rangle$, and it depends on the world and on the proportions of values of 'x' that satisfy Bx and Bx&Fx in it. In contrast, Bt⇒Ft has a probability that is independent of possible worlds, but not a degree of truth in a world. Therefore, if $\tau(\text{Bx} \Rightarrow \text{Fx})$ is different for different worlds Γ, while the probability p(Bt ⇒ Ft) is 'constant', independent of possible worlds, then p(Bt ⇒ Ft) cannot always equal $\tau(\text{Bx} \Rightarrow \text{Fx})$. However, we would like to understand this more clearly, in a way that permits us to explain why we are likely to think that the probability of "If Tweety is a bird then he flies" ought to equal the proportion of birds that fly. One approach to this is to define a probability not just for the definite proposition "If Tweety is a bird then he flies," but for the indefinite proposition "Birds fly" as well. The interpretation of probabilities of definite sentences that was described in appendix 7 suggests a way of doing this.

In appendix 7 we represented the probabilities of unconditional propositions like Bt&Ft and Bt as the probabilities of being in worlds in which they are true, and represented the probability of the conditional Bt⇒Ft as the ratio of these probabilities, i.e., as the ratio of the probabilities of being in worlds in which Bt&Ft and Bt are true. Now, this idea can be generalized to apply to the indefinite formulas Bx&Fx and Bx if we replace the probabilities of being in worlds in which they are *true* by their *expected truth values* in worlds, and then define the probability of Bx ⇒ Fx as the ratio of these expectations. This can be formalized as follows, assuming that there are only a finite number of possible worlds, $\Gamma_1, \ldots, \Gamma_n$, as in appendix 7.

Suppose that the probabilities of being in worlds $\Gamma_1, \ldots, \Gamma_n$ are $p(\Gamma_1)$, $\ldots, p(\Gamma_n)$, respectively, and the degree of truth functions that correspond to them are $\tau_1, \ldots, \tau_n$, respectively, and for a sentence like Bt&Ft, we set $\tau_i(\text{Bt\&Ft})$ equal to 1 or 0 according as Bt&Ft is true or false in Γ_i. Given this, as in appendix 7, the probability of being in a world in which Bt&Ft is true can be written as

$$(1) \qquad p(\text{Bt \& Ft}) = p(\Gamma_1)\tau_1(\text{Bt \& Ft}) + \ldots + p(\Gamma_n)\tau_n(\text{Bt \& Ft}).$$

This says that p(Bt&Ft) is the expected truth value Bt&Ft, and it allows us to define p(Bt ⇒ Ft) as a ratio of expected values:

$$(2) \quad p(\text{Bt} \Rightarrow \text{Ft}) = \frac{p(\text{Bt\&Ft})}{p(\text{Bt})} = \frac{p(\Gamma_1)\tau_1(\text{Bt\&Ft}) + \ldots + p(\Gamma_n)\tau_n(\text{Bt\&Ft})}{p(\Gamma_1)\tau_1(\text{Bt}) + \ldots + p(\Gamma_n)\tau_n(\text{Bt})}$$

But now, if we replace the individual constant t by the variable x, we immediately arrive at definitions of probability for the indefinite propositions Bx&Fx and Bx ⇒ Fx:

$$(3) \qquad p(\text{Bx \& Fx}) = p(\Gamma_1)\tau_1(\text{Bx \& Fx}) + \ldots + p(\Gamma_\text{n})\tau_\text{n}(\text{Bx \& Fx})$$

and

$$(4)\qquad p(Bx \Rightarrow Fx) = \frac{p(Bx \,\&\, Fx)}{p(Bx)}$$

$$= \frac{p(\Gamma_1)\tau_1(Bx \,\&\, Fx) + \ldots + p(\Gamma_n)\tau_n(Bx \,\&\, Fx)}{p(\Gamma_1)\tau_1(Bx) + \ldots + p(\Gamma_n)\tau_n(Bx)}$$

In effect, (3) says that p(Bx&Fx) is the expected degree of truth of Bx&Fx, which, unlike Bt&Ft, can have a value different from 1 or 0, and (4) says that p(Bx $\Rightarrow$ Fx) is the ratio of the expected degrees of truth of Bx&Fx and Bx. And, now that we have a precise definition of p(Bx $\Rightarrow$ Fx), we can consider its relation to p(Bt $\Rightarrow$ Ft) in detail.

The first thing we see is that there is no necessary connection between p(Bt $\Rightarrow$ Ft) and p(Bx $\Rightarrow$ Fx). Proportionality diagram Γ_1 makes this obvious, if it is interpreted as picturing the possible world that we are certain to be in; i.e., suppose that it pictures the world Γ for which $p(\Gamma) = 1$. Assuming this, p(Bx $\Rightarrow$ Fx) must equal τ(Bx&Fx)/τ(Bx), and this must have a high value because it is represented by the proportion of circle B that lies inside F, while p(Bt $\Rightarrow$ Ft) must equal 0 because Bt is certainly true and Bt&Ft is certainly false in this world.

So why should we think that p(Bt $\Rightarrow$ Ft) equals p(Bx $\Rightarrow$ Fx)? This is a somewhat complicated question, but part of the answer lies in the following fact: While we may know what the proportion of birds that fly is, or at least know that it is high, we may know nothing about Tweety. This becomes clearer if we picture possible worlds as equally probable and as having the same proportions of birds, penguins, flying things, and combinations in them, but picture Tweety at different places in them. A very simple example involves five equiprobable worlds, with one 'thing' in the minimal regions of each, but where which of those things Tweety is varies from world to world:[12]

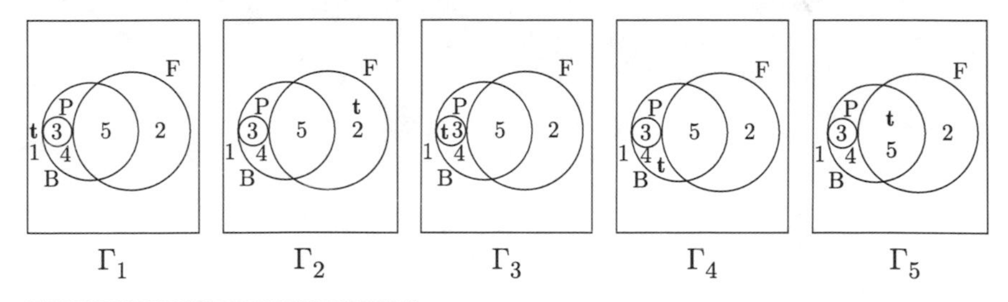

[12]In the parlance of modal logic, "Tweety" isn't a *rigid designator*. Cf. Kripke (1980).

What this pictures is a 'universe' consisting of 5 possible worlds, Γ_1, ..., Γ_5, each containing 5 'objects' labelled '1', '2', '3', '4', and '5', and the objects have the same properties in all of the worlds. E.g., object 3 is a penguin and therefore a nonflying bird in each of Γ_1–Γ_5. On the other hand, if we don't know which object Tweety is, we can assume that the name 'Tweety' labels object 1 in Γ_1, object 2 in Γ_2, and so on. Thus, Tweety would be a flying bird in world Γ_2, and a penguin, hence a nonflying bird, in Γ_3.

Now suppose that the probabilities of being in these worlds are all equal, hence $p(\Gamma_1) = p(\Gamma_2) = p(\Gamma_3) = p(\Gamma_4) = p(\Gamma_5) = .2$. Then the reader can easily verify that p(Bt) = .6 because Bt is true in worlds Γ_3, Γ_4, and Γ_5 while p(Bt&Ft) = .2 because Bt&Ft is only true in Γ_5. Therefore $p(Bt \Rightarrow Ft) = .2/.6 = 1/3$. As to degrees of truth, they equal .6 for Bx in any world because 3 out of 5 of the objects in any world satisfy Bx, and they equal .2 for Bx&Fx in any world because only 1 out of the 5 objects in any world satisfy Bx&Fx. Hence the expected degrees of truth p(Bx) and p(Bx&Fx) must equal .6 and .2, respectively, and therefore $p(Bx \Rightarrow Fx) = .2/.6$. And, this is equal to $p(Bt \Rightarrow Ft)$.

But it is clear that the equality of $p(Bt \Rightarrow Ft)$ and $p(Bx \Rightarrow Fx)$ depends critically on the assumptions made above, which don't necessarily hold. In fact, if the probabilities of being in the worlds had not been equal, $p(Bt \Rightarrow Ft)$ would not necessarily have equalled $p(Bx \Rightarrow Fx)$. For instance, if $p(\Gamma_3)$ had equalled 1 and $p(\Gamma_1)$, $p(\Gamma_2)$, $p(\Gamma_4)$, and $p(\Gamma_5)$ had all equalled 0, then $p(Bx \Rightarrow Fx)$ would have equalled 1/3 while $p(Bt \Rightarrow Ft)$ would have equalled 0 (this would have been like the situation depicted in Proportionality diagram 1).

Exercises

1. Use the trivalent truth-table method to determine which of the following inferences are m-valid, and in the m-invalid cases construct proportionality diagrams representing their premises as highly true while their conclusions are highly untrue.

   a. $\dfrac{(Px \vee Qx) \Rightarrow Rx}{Px \Rightarrow Rx}$

   b. $\dfrac{\sim(Px \,\&\, Qx)}{Px \Rightarrow \sim Qx}$

   c. $\dfrac{Px \leftrightarrow Qx}{Px \Rightarrow Qx}$

   d. $\dfrac{(Px \,\&\, \sim Qx) \Rightarrow Qx}{Px \Rightarrow \sim Qx}$

e. $\dfrac{\text{Px}, \text{Px} \rightarrow \text{Qx}}{\text{Qx} \Rightarrow \text{Px}}$

f. $\dfrac{\text{Px} \Rightarrow (\text{Qx} \vee \text{Rx}), \text{Qx} \Rightarrow \sim\text{Rx}}{(\text{Px} \mathbin{\&} \text{Qx}) \Rightarrow \sim\text{Rx}}$

g. $\dfrac{\text{Px} \Rightarrow (\text{Qx} \mathbin{\&} \text{Rx})}{(\text{Px} \mathbin{\&} \text{Qx}) \Rightarrow \text{Rx}}$

2. In each of the following inferences try to determine whether it is m-valid by either: (1) Giving an argument that if its premise is highly true its conclusion must also be, or (2) By drawing a proportionality diagram similar to diagram 2 which depicts its premise as highly true while its conclusion is much less true.

a. $\dfrac{\text{xHy}}{\text{yHx}}$

b. $\dfrac{\text{xHy}}{\text{xHx}}$

c. $\dfrac{\text{Px} \vee \text{Qx}}{\text{Px} \vee \text{Qy}}$

d. $\dfrac{\text{Px} \vee \text{Qy}}{\text{Px} \vee \text{Qx}}$

e. $\dfrac{\text{xHy} \Rightarrow \sim\text{yHx}}{\text{yHx} \Rightarrow \sim\text{xHy}}$

3. Which of either of the inferences below are m-valid? Give arguments to support your answers.

a. $\dfrac{\text{xHy}}{(\forall\text{x})\text{xHy}}$

b. $\dfrac{\text{xHy}}{(\exists\text{x})\text{xHy}}$

4. Can you describe a possible state of affairs in which the premise of the inference below would be highly true while its conclusion would be highly untrue? Note that such a state can't be pictured in a 2-dimensional graph because the inference involves four variables.

$$\frac{(\text{xHy} \mathbin{\&} \text{yHz}) \Rightarrow \text{xHz}}{(\text{xHy} \mathbin{\&} \text{yHz} \mathbin{\&} \text{zHw}) \Rightarrow \text{xHw}}$$

5. What can you say about a society in which almost all husbands are taller than their wives and almost all wives are taller than their husbands?

Answers to Selected Exercises

Chapter 2

Section 2.1

1a. 1,4 1c. 3 1e. 2,3,4 1g. none 1i. 1,2,3

2a. 1,2 2c. 1,2,7,8 2e. 1,3,4,5,6,7,8 2g. 2,6,7,8 2i. 1,2,3,6,7,8

Section 2.2

1.

	formula	probabilities in distributions #1	#2	#3	#4
a.	$E \leftrightarrow L$	.50	.3	0	a+d
c.	$\sim(L \rightarrow E)$	.25	.6	1	c
e.	$(E\&L) \rightarrow \sim E$	.75	.8	1	b+c+d
g.	$(E \leftrightarrow \sim E)$	0	0	0	0
i.	$(E \rightarrow L) \rightarrow L$	.75	.9	1	a+b+c

probabilities of formulas

2a.

distribution		case	truth values E	L	$E \vee L$	$\sim E$	L
0	a	1	T	T	T	F	T
.3	b	2	T	F	T	F	F
.5	c	3	F	T	T	T	T
.2	d	4	F	F	F	T	F
					.8	.7	.5

⋆2b. .5

3.

prob. distributions particular					truth values			formulas			
#1	#2	#3	#4 variable	region	A	B	C	A&B	$A \leftrightarrow C$	$A\&(B \vee C)$	$(A \rightarrow B) \vee C$
.125	.01	0	a	1	T	T	T	T	T	T	T
.125	.02	0	b	2	T	T	F	T	F	T	T
.125	.03	0	c	3	T	F	T	F	T	T	T
.125	.04	0	d	4	T	F	F	F	F	F	F
.125	.1	0	e	5	F	T	T	F	F	F	T
.125	.2	0	f	6	F	T	F	F	T	F	T
.125	.3	1	g	7	F	F	T	F	F	F	T
.125	.3	0	h	8	F	F	F	F	T	F	T

	A&B	$A \leftrightarrow C$	$A\&(B \vee C)$	$(A \rightarrow B) \vee C$
truth regions	1,2	1,3, 6,8	1,2 3	1,2, 3,5 6,7,8
Dist. #4 prob.	a+b	a+c +f+h	a+b +c	1−d
Dist. #3 prob.	0	0	0	1
Dist. #2 prob.	.03	.54	.06	.96
Dist. #1 prob.	.25	.5	.375	.875

Table 2.3

4a.

prob. dist. particular	prob. dist. variable	region	truth values A	B	C	formulas A	B	C	(A&B&C)
.7	a	1	T	T	T	T	T	T	T
.1	b	2	T	T	F	T	T	F	F
.1	c	3	T	F	T	T	F	T	F
0	d	4	T	F	F	T	F	F	F
.1	e	5	F	T	T	F	T	T	F
0	f	6	F	T	F	F	T	F	F
0	g	7	F	F	T	F	F	T	F
0	h	8	F	F	F	F	F	F	F
						.9	.9	.9	.7

⋆4b. .7 ⋆4c. .85

Section 2.4⋆

1a. 1. L logically entails E → L. By pure logic.
2. p(L) ≤ p(E → L). From step 1, by axiom K3. QED

1b. 1. A ∨ B is logically equivalent to A ∨ (B&∼A). By pure logic.
2. p(A ∨ B) = p(A ∨ (B&∼A)). From step 1 and Theorem 4.
3. A and B&∼A are logically inconsistent. By pure logic.
4. p(A ∨ (B&∼A)) = p(A) + p(B&∼A). By step 3 and Axiom K4.
5. p(A ∨ B) = p(A) + p(B&∼A). From steps 2 and 4 by algebra.
6. B is logically equivalent to (B&∼A) ∨ (B&A). By pure logic.
7. p(B) = p((B&∼A) ∨ (B&A)). From step 6, and Theorem 4.
8. B&∼A and A&B are logically inconsistent. By pure logic.
9. p((B&∼A) ∨ (A&B)) = p(B&∼A) + p(A&B). From step 8 and Axiom K4.
10. p(B) = p(B&∼A)+p(A&B). From steps 7 and 9 by algebra.
11. p(B&∼A) = p(B)− p(A&B). From step 10, by algebra.
12. p(A ∨ B) = p(A) + p(B)− p(A&B). From steps 5 and 11 by algebra.
13. p(A)+p(B) = p(A∨B)+p(A&B). From step 12 by algebra. QED

2. The following distribution gives counterexamples to parts a, b, and c of this exercise:

	E	L	B	E	L	E&B	L&B
0	T	T	T	T	T	T	T
0	T	T	F	T	T	F	F
0	T	F	T	T	F	T	F
.5	T	F	F	T	F	F	F
.5	F	T	T	F	T	F	T
0	F	T	F	F	T	F	F
0	F	F	T	F	F	F	F
0	F	F	F	F	F	F	F
→				.5	.5	0	.5

3. b, c, f, and g are theorems.
   The following distribution is a counterexample to parts a, d, and e:

distribution	E	L	$E \to L$	$E \vee L$	~E	E&~L	$L \to E$	$L \leftrightarrow E$
1	T	T	T	T	F	F	T	T
0	T	F	F	T	F	T	T	F
0	F	T	T	T	T	F	F	F
0	F	F	T	F	T	F	T	T
→	1	1	1	1	0	0	1	1

4a. ϕ itself either is or isn't logically true. If it is logically true then p(ϕ) = 1. If it isn't logically true then there is a state-description in which it is false. Then the probability distribution which gives probability 1 to that state description must give probability 0 to ϕ.

⋆4b. The combinations of truth-values for ϕ and ψ are:

distrib.	SD	ϕ	ψ
a	#1	T	T
b	#2	T	F
c	#3	F	T
d	#4	F	F

If ϕ isn't logically true or logically false then one of SD #1 or SD #2 must be possible and one of SD #3 or SD #4 must be possible, and if ψ isn't logically true or logically false then one of SD #1 or SD #3 must be possible and one of SD #2 or SD #4 must be possible. Given this, either: (1) both SDs #1 and #4 are possible, or (2) both SDs #2 and #3 are possible. In case (1), if we set a = d = .5 and b = c = 0, then p(ϕ) = p(ψ) = .5, but p(ϕ&ψ) = .5, hence p(ϕ&ψ) $\neq$ p(ϕ) $\times$ p(ψ). In case (2) if we set b = c = .5 then p(ϕ) = p(ψ) = .5 but p(ϕ&ψ) = 0, and, again, p(ϕ&ψ) $\neq$ p(ϕ) $\times$ p(ψ).

Section 2.5⋆

1a. 1. Suppose that p(A&B) = p(A) × p(B).
2. A is logically equivalent to (A&B) ∨ (A&∼B). By pure logic.
3. p(A) = p((A&B) ∨ (A&∼B)). From step 2, by Theorem 4.
4. A&B and A&∼B are logically inconsistent. Pure logic.
5. p((A&B) ∨ (A&∼B)) = p(A&B) + p(A&∼B). From step 5 by axiom K4.
6. p(A) = p(A&B) + p(A&∼B). From steps 3 and 5 by algebra.
7. p(A&∼B) = p(A) − p(A&B). From step 6 by algebra.
8. p(A&∼B) = p(A) − p(A) × p(B). From steps 1 and 7 by algebra.
9. p(A&∼B) = p(A) × (1 − p(B)). From step 8 by algebra.
10. p(∼B) = 1 − p(B). By Theorem 2.
11. p(A&∼B) = p(A) × p(∼B). From steps 9 and 10 by algebra.
12. A and ∼B are probabilistically independent. From step 11. QED

⋆1c. Here is how the argument goes in the case n = 3. First, by ⋆1b, if A, B, and C are independent then so are A, B, and ∼C, and by an entirely similar argument so are A, ∼B, and C and ∼A, B, and C. Applying ⋆1b again, if ∼A, B, and C are independent, then so are ∼A, B, and ∼C, and so are ∼A, ∼B, and C. Finally, if ∼A, ∼B, and C are independent then so are ∼A, ∼B, and ∼B.
The argument for n = 4, 5, and so on is just the same, except that more steps are required to deal with each 'reversal of sign'.

2a. In the distribution below, A, B, and C are pairwise independent because p(A) = p(B) = p(C) = .1 and p(A&B) = p(A&C) = p(B&C) = .01, but p(A&B&C) = 0.

prob. dist.			truth values			formulas				
particular	variable	region	A	B	C	A	B	C	A&B	A&B&C
0	a	1	T	T	T	T	T	T	T	T
.01	b	2	T	T	F	T	T	F	T	F
.01	c	3	T	F	T	T	F	T	F	F
.08	d	4	T	F	F	T	F	F	F	F
.01	e	5	F	T	T	F	T	T	F	F
.08	f	6	F	T	F	F	T	F	F	F
.08	g	7	F	F	T	F	F	T	F	F
.73	h	8	F	F	F	F	F	F	F	F
						.1	.1	.1	.01	0

Section 2.6

1. Informal proof of theorem.
   If $\phi_1,\ldots\phi_n$ are independent in the sense of the theorem then there is an SD, α, in which all of $\phi_1,\ldots,\phi_n$ are true, and for each $i = 1,\ldots,n$ there is an SD α_i such that ϕ_j is true in α_i for all $j \neq i$, but ϕ_i is false in α_i. Now set $\mathrm{p}(\alpha_i) = \mathrm{u}_i$ for $i = 1,\ldots,n$ and set $\mathrm{p}(\alpha) = 1 - \mathrm{u}_1 - \ldots - \mathrm{u}_n$. This defines a distibution such that $\mathrm{u}(\phi_i) = \mathrm{u}_i$, for $i = 1,\ldots,n$ and $\mathrm{u}(\phi_1 \& \ldots \& \phi_n) = \mathrm{u}(\alpha) = \mathrm{u}_1 + \ldots + \mathrm{u}_n$.

Chapter 3

Section 3.2

1a. .05 1b. .05 1c. .033 1d. impossible

2a.

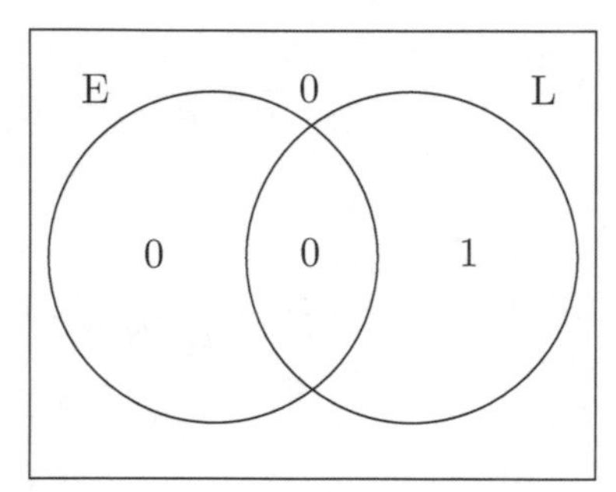

Same as:

dist.	E	L	E → L
0	T	T	T
0	T	F	F
1	F	T	T
0	F	F	T
	→0	1	1

2b. Let $\phi = \mathrm{E}$ and $\psi = {\sim}\mathrm{E}$, and take the distribution:

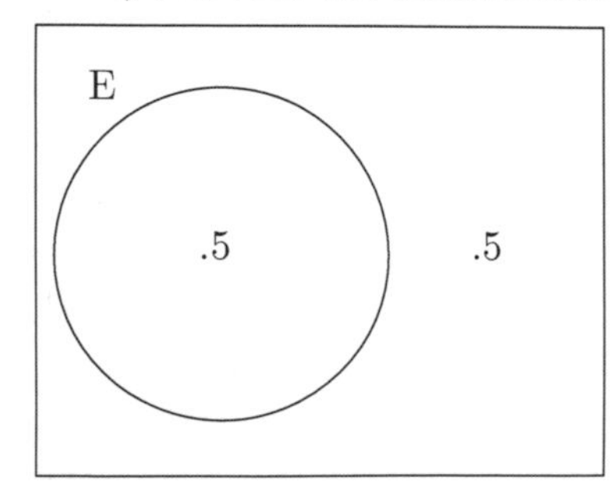

Then $u(\phi) = .5 \leq u(\psi) = .5$, but ψ is not a logical consequence of ϕ.

2c. Let ϕ, ϕ_1, and ϕ_2 all equal E&~E. Then $\phi_1 \vee \phi_2 = (E\&{\sim}E) \vee (E\&{\sim}E)$ is a logical consequence of $\phi = E\&{\sim}E$, but for any uncertainty function, u( ), $u(\phi) = u(E\&{\sim}E) = 1$, but $u(\phi_1) + u(\phi_2) = u(E\&{\sim}E) + u(E\&{\sim}E) = 1 + 1 = 2$.

2d. Let $\phi = E\&{\sim}E$, $\phi_1 = E$, and $\phi_2 = {\sim}E$, and suppose that probabilities and uncertainties are generated by the distribution in the answer to 2b. Then $\phi_1 \vee \phi_2 = E \vee {\sim}E$ is a logical consequence of $\phi = E\&{\sim}E$, but $u(\phi) = u(E\&{\sim}E) = 1$ while $u(\phi_1) = u(E) = .5$ and $u(\phi_2) = u({\sim}E) = .5$.

⋆3a. 1/7.

3b. Suppose that the distribution of probabilities is represented as follows,

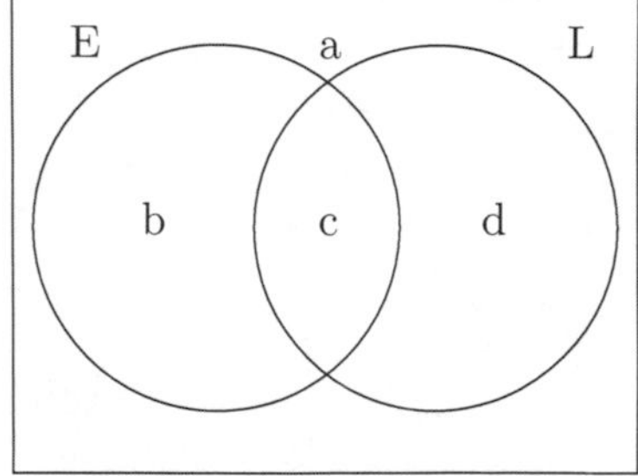

and $u(E \vee L) = a = .1$, $u({\sim}E) = b + c = .3$ and $u(L) = a + b = .4$. This implies that $a + b + c = .4$, hence $d = 1 - a - b - c = .6$, hence $a + b + d = .4 + .6 = 1$. Therefore $p(E\&L) = c = 1 - a - b - d = 0$.

3c. Special case: Suppose that the inference is $\{E, L\} \therefore E\&L$ and that $u_1 + u_2 > 1$. If probability distributions are represented as in the answer to 3b and we set $a = u_1 + u_2 - 1$, $b = 1 - u_1$, $c = 0$, and $d = 1 - u_2$, then $u(E) = u_1 = a + d$, $u(L) = u_2 = a + b$, and $u(E\&L) = a + b + d = 1 - c = 0$.

3d. Consider the inference $\{A, B\} \therefore A \leftrightarrow B$. If $u(A) = u(B) = 1$ then $p(A) = p(B) = 0$, and if this is so then $p(A \leftrightarrow B) = 1$ and $u(A \leftrightarrow B) = 0$.

Section 3.3★

1a. Minimal essential sets: {A}, {B, A → B}.
Degrees of essentialness e(A) = 1, e(B) = e(A → B) = .5.
Upper bound on uncertainty of A&B:

$$\begin{aligned} u(A\&B) &\leq e(A) \times u(A) + e(B) \times u(B) + e(A \to B) \times u(A \to B) \\ &\leq 1 \times .1 + .5 \times .2 + .5 \times .1 = .25 \end{aligned}$$

Lower bound on p(A&B): 1 − .25 = .75.

1c. Minimal essential sets: {A, B}, {A,C}
Degrees of essentialness e(A) = e(B) = e(C) = .5.
Upper bound on uncertainty of A ∨ (B&C)

$$\begin{aligned} u(A \vee (B\&C)) &\leq e(A) \times u(A) + e(B) \times u(B) + e(C) \times u(C) \\ &\leq .5 \times .1 + .5 \times .1 + .5 \times .2 = .2 \end{aligned}$$

Lower bound on p(A ∨ (B&C)): 1 − .2 = .8.

1e. Minimal essential sets: {A}, {B}, {C}
Degrees of essentialness e(A) = e(B) = e(C) = 1.
Upper bound on uncertainty of (A ∨ B ∨ C) → (A&B&C)

$$\begin{aligned} u(A\&B) &\leq e(A) \times u(A) + e(B) \times u(B) + e(C) \times u(C) \\ &\leq 1 \times .1 + 1 \times .1 + 1 \times .2 = .4 \end{aligned}$$

Lower bound on p((A ∨ B ∨ C) → (A&B&C)): 1 − .4 = .6.

2a.

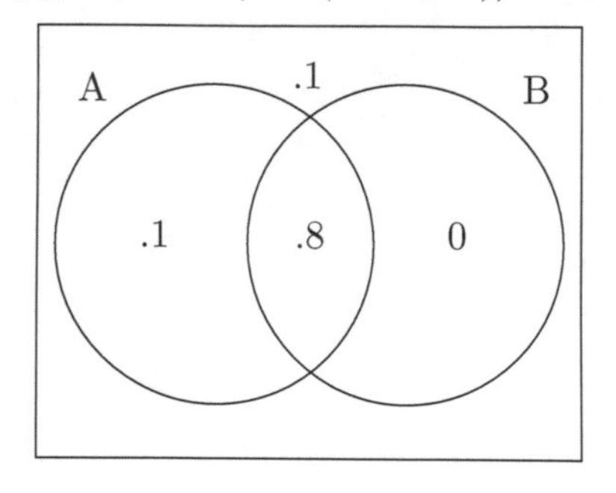

$$\begin{aligned} p(A) &= .9 \\ p(B) &= .8 \\ p(A \to B) &= .9 \\ p(A \,\&\, B) &= .8 \geq .75. \end{aligned}$$

2c.

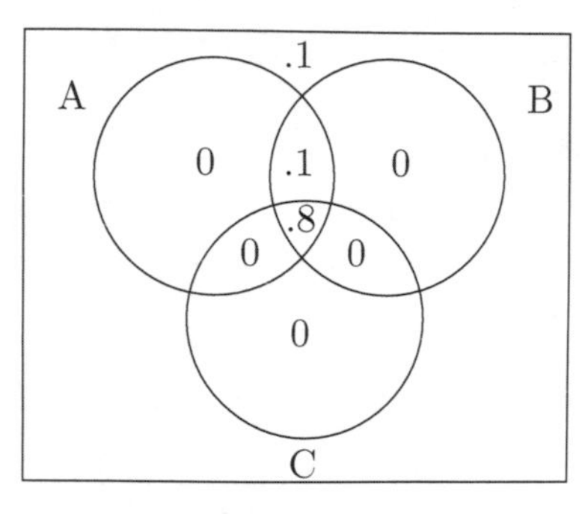

$$p(A) = .9$$
$$p(B) = .9$$
$$p(C) = .8$$
$$p(A \vee (B \,\&\, C)) = .9 \geq .8.$$

2e.

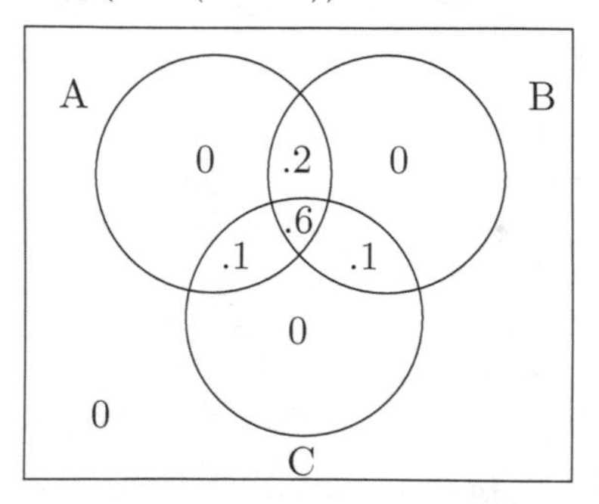

$$p(A) = .9$$
$$p(B) = .9$$
$$p(C) = .8$$
$$p((A \vee B \vee C) \rightarrow (A \,\&\, B \,\&\, C)) = .6.$$

Note that this conforms to Theorem 14, since all premises are essential, and the uncertainty of the conclusion equals the sum of the uncertainties of the premises.

3a. All of the sets $\{A_1\}, \ldots, \{A_{100}\}$ are minimal essential, and all premises have degrees of essentialness $e(A_1) = \ldots = e(A_{100}) = 1$.

3b. Since $u(A_1) = \ldots = u(A_{100}) = 1 - .99 = .01$, the maximum uncertainty of "At least 100 of $A_1, \ldots, A_{100}$ are true" is $100 \times .01 = 1$, and its minimum probability is $1 - 1 = 0$.

3c. For each A_i for $i = 1, \ldots, 100$, there is an SD ℓ_i in which A_i is false but all other A_j for $j \neq i$ are true. Assigning each ℓ_i a probability of .01 and all other SDs a probability 0 yields a probability distribution.

Each A_i is true in all ℓ_j except ℓ_i, hence the probability function generated by the foregoing distribution makes $p(A_i) = 99 \times .01 = .99$, but it gives probability 0 to "At least 100 of $A_1, \ldots, A_{100}$ are true."

$\star$4a. The minimal essential sets in this case contain 11 out of the 100 formulas $A_1, \ldots, A_{100}$; e.g., if $A_1, \ldots, A_{11}$ were omitted then "At least 90 of $A_1, \ldots, A_{100}$" wouldn't follow. All premises have degree of essentialness $e(A_i) = 1/11$. The maximum uncertainty of this conclusion is no higher than $1/11 \times .01 \times 100 = 1/11$, and its minimum probability is at least $1 - 1/11 = 10/11$.

4b. The minimal essential subsets of $A_1, \ldots, A_{100}$ that entail "At least n out of $A_1, \ldots, A_{100}$ are true" are of size $100 - n + 1$, and therefore the degree of essentialness of each premise of this inference is $e(A_i) = 1/(100 - n + 1)$. If each $u(A_i) = .01$ then the maximum uncertainty of its conclusion is $100 \times .01 \times (1/(100 - n + 1)) = 1/(100 - n + 1)$, and its minimum probability is $1 - (1/(100 - n + 1))$. If this value must be at least .75 then n cannot be greater than 97.

$\star$5a. Suppose that the premises are $\phi_1, \ldots, \phi_n$, and that ϕ is the conclusion of an inference. A subset, $\wp$, of $\phi_1, \ldots, \phi_n$ is sufficient if ϕ follows just from the premises in $\wp$, and it is essential if ϕ *doesn't* follow from the premises outside of the set. Now suppose that a set $\wp$ is sufficient and a set $\wp'$ is essential. Then ϕ doesn't follow from the premises outside of $\wp'$ but it does follow from the premises in $\wp$. Hence not all of the premises in $\wp$ can be outside of $\wp'$, i.e., the sets must intersect. QED

$\star$5b. Suppose a set of premises $\wp'$ intersected every sufficient premise set, but $\wp'$ was not essential. If $\wp'$ is not essential then ϕ must follow from the set, $\wp$, of premises outside of $\wp'$, and therefore $\wp$ must be a sufficient premise set. But since $\wp$ is the set of premises outside of $\wp'$, $\wp$ and $\wp'$ don't intersect, and therefore $\wp'$ doesn't intersect every sufficient premise set, contrary to supposition. QED

Chapter 4

Section 4.3

1. a. $p(A|B) = .25$
   c. $p(B|{\sim}A) = .3/.7 \approx .429$
   e. $p(A|A \vee {\sim}B) = .3/.7 \approx .429$
   g. $p(A|B \rightarrow A) = .3/.7 \approx .429$
   i. $p(A \vee B|B \rightarrow A) = .3/.7 \approx .429$.

2. a. $(8/52) \times (7/51) \approx .021$
   c. $4 \times (13/52) \times (12/51) \approx .072$
   e. $1 - (48/52) \times (47/51) \approx .149$.

3a.

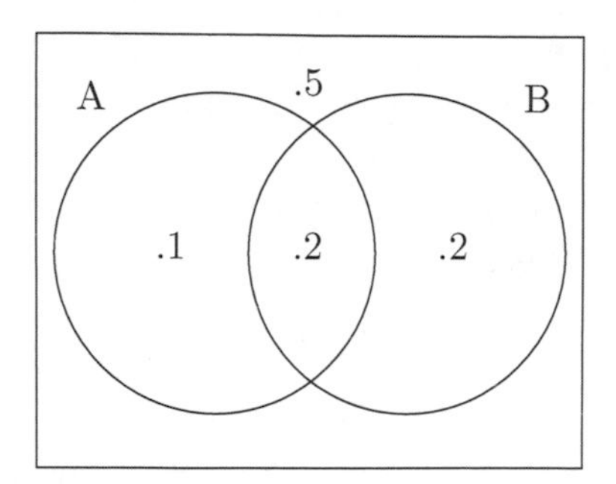

3b.

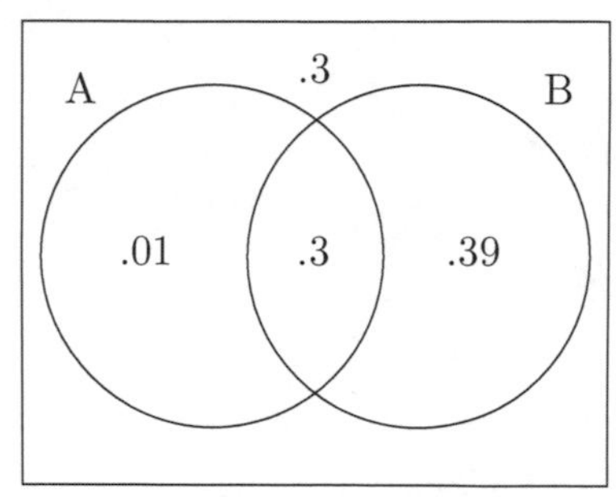

$$p(B|A) = .3/.31 \approx .968$$
$$p(B) = .69$$
$$p(B \to A|A \to B) = .6/.99 \approx .606$$
$$p(B \to A) = .61.$$

3c.

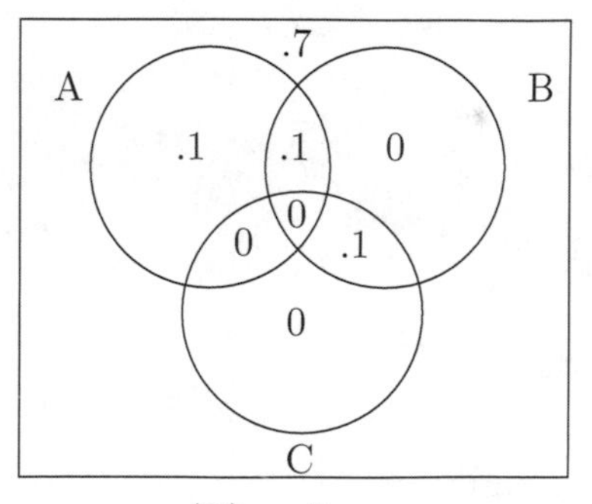

$$p(B) = .2$$
$$p(B|A) = .1/.2 = .5$$
$$p(C) = .1$$
$$p(C|B) = .1/.2 = .5$$
$$p(C|A) = 0/.2 = 0$$

3d. Let A = "There will be light rain tomorrow, and no heavy rain," let B = "Tomorrow will be cloudy," and C = "There will be heavy rain tomorrow."

★4a. If A is positively relevant to B then $p(B|A) > p(B)$, hence $p(B\&A) > p(A) \times p(B)$. Hence

$$\begin{aligned} p(A \,\&\, {\sim}B) = p(A) - p(A \,\&\, B) &< p(A) - p(A) \times p(B) \\ &< p(A)(1 - p(B)) \\ &< p(A)p({\sim}B) \end{aligned}$$

Hence A is negatively relevant to ~B. QED

4b. Suppose that A is positively relevant to B. Then, by 4a, it is negatively relevant to ~B. Hence, by the argument of 4a again, ~B is negatively relevant to A. Hence it is positively relevant to ~A. Hence ~A is positively relevant to ~B. QED

★5.

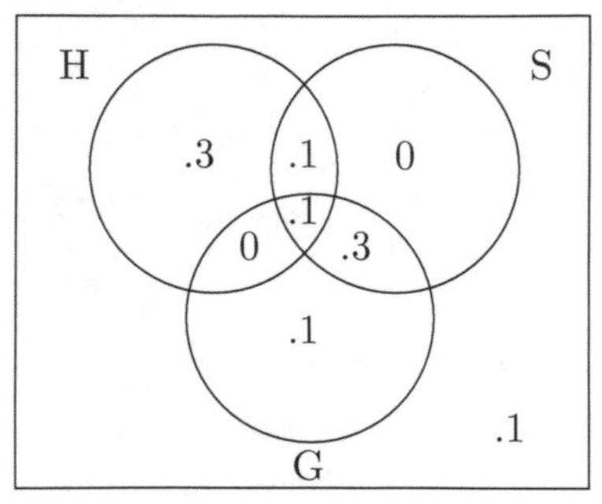

$$p(H|S \,\&\, G) = .1/.4 = .25$$
$$p(H|{\sim}S \,\&\, G) = 0/.1 = 0$$
$$p(H|S \,\&\, {\sim}G) = .1/.1 = 1$$
$$p(H|{\sim}S \,\&\, {\sim}G) = .3/.4 = .75$$
$$p(H|S) = .2/.5 = .4$$
$$p(H|{\sim}S) = .3/.5 = .6$$

★6a. $p(L) = .4/(.4 + .7) = .4/1.1 \approx .3636$
$p(R) = .7/(.4 + .7) = .7/1.1 \approx .6363$

★6b. If $p(L|L) = p(L|R)$ then $p(R|L) = 1 - p(L|L) = 1 - p(L|R)$, hence $p(R|L) + p(L|R) = 1$. Therefore $p(L) = p(L|R)/(p(L|R) + p(R|L)) = p(L|R) = p(L|L)$. QED

★6c. If $p(R|L) = p(L|R)$ then $p(L) = p(L|R)/(p(L|R)+p(L|R)) = \frac{1}{2}$, and since $p(R) = 1 - p(L)$, $p(R) = \frac{1}{2}$. QED

Section 4.5

1a. Let W = "A will win the championship." $p_0(W) = .5$, and $p_1(A) = .75$

1b. $p_0(W) = .5$, and $p_1(W) = 13/16 = .8125$ (think of it this way: team A will win the 5 game series after winning the first game if the remaining games come out either as A-A, A-B-A, A-B-B-A, B-A-A, B-A-B-A, or B-B-A-A. The chances of these are 1/4, 1/8, 1/16, 1/8, 1/16, and 1/16, respectively).

1c. $p_2(W)$ (the chance of A's winning the series after winning the first two games) is $7/8 = .875$.

2a. If $p(B|A) > p(B)$ then $p(B\&A)/p(A) > p(B)$ and $p(B\&A)/p(B) > p(A)$. Since $p(B\&A) = p(A\&B)$, $p(A\&B)/p(B) > p(A)$. And, since $p(A\&B)/p(B) = p(A|B)$, $p(A|B) > p(A)$. QED

2b. If A is a logical consequence of B then so is A&B and since B is a logical consequence of A&B, B and A&B must be logically equivalent. Therefore $p(A\&B) = p(B)$ and $p(A\&B)/p(B) = 1$. But $p(A\&B)/p(B) = p(A|B)$, hence $p(A|B) = 1$. Hence the posterior probability of A after learning B must equal 1, i.e., learning B would make A certain. Moreover, $p(A\&B) \geq p(A) \times p(B)$, hence $p(A\&B)/p(A) \geq p(B)$. Therefore $p(B|A) \geq p(B)$, so the posterior probability of B after learning A will be at least as high as the prior probability of B. QED

2c. If A and B are logically inconsistent than ~A is a logical consequence of B and ~B is a logical consequence of A. Therefore, according to 2b, learning A can't decrease the probability of ~B and learning B can't decrease the probability of ~A. But if learning A can't decrease the probability of ~B it can't increase the probability of B, and by the same argument, learning B can't increase the probability of A. QED

3. $p_1(A|C) = p_0(A \& C|B)/p_0(C|B) = p_0(A \& B \& C)/p_0(B \& C) = p_0(A|B \& C)$.

⋆4. $p_2(\phi) = p_0(\phi|B\&C)$ (where $p_2()$ is the *posterior-posterior* probability function).

⋆5. This obviously doesn't follow. Roughly, what the general statement "Each team has a 50-50 chance of winning any given game" means can be explained as follows. Suppose that the games are designated as "Game 1," "Game 2," and so on, and the teams are designated as "Team A," and "Team B." Then to say that each team has a 50-50 chance of winning any given game is to say that prior to Game 1 Teams A and B each have a 50% chance of winning, prior to Game 2 Teams A and B each have a 50% chance of winning, and so on. Supposing, however, that Game 7 is the last game of the series and Team A wins it, it doesn't follow that the team that wins the last game of the series has a 50% chance of winning it. That is, we can't substitute different names for the same thing in probability statements. Just because it *happens* that Team A is the winner of the last game and the last game is Game 7, it doesn't follow from the fact that Team A has a 50% chance of winning Game 7 that the winner of the last game has a 50% chance of winning the last game.

Section 4.7⋆⋆

1. $p(W_2|W_1) = p(W_3|W_1\&W_2) = .3$.

2a. $p(W_1) = a+b+c+d+e+f+g+h$

$$p(W_2|W_1) = \frac{a+b+c+d}{a+b+c+d+e+f+g+h}$$

$$p(W_3|W_1\&W_2) = \frac{a+b}{a+b+c+d}$$

$$p(W_4|W_1\&W_2\&W_3) = \frac{a}{a+b}$$

2b. 5 2c. 3

2d. Letting $a = p = .25$, $b = c = e = h = i = l = n = o = .04$, and $d = f = g = j = k = m = .03$ defines an order and gender-free probability distribution in which $p(W_1) = .5$, $p(W_2|W_1) = .36/.5 = .72$, $p(W_3|W_1\&W_2) = .29/.36 = .8055$, and $p(W_4|W_1\&W_2\&W_3) = .25/.29 = .8621$.

2e. In the uniform distribution in which $a = b = c = d = e = f = g = h = i = j = k = l = m = n = o = p = 1/16$, $p(W_1) = p(W_2|W_1) = p(W_3|W_1\&W_2) = p(W_4|W_1\&W_2\&W_3) = .5$.

2f. Letting $a = p = .0$, $b = c = e = h = i = l = n = o = .05$, and $d = f = g = j = k = m = .1$ defines an order and gender-free probability distribution in which $p(W_1) = .5$, $p(W_2|W_1) = .2/.5 = .4$, $p(W_3|W_1\&W_2) = .05/.2 = .4$, and $p(W_4|W_1\&W_2\&W_3) = 0/.05 = 0$.

⋆3.

$$\frac{p(D_3|W_1\&W_2\&W_3)}{p(D_2|W_1\&W_2\&W_3)} = \frac{p(D_3)}{p(D_2)} \times \frac{p(W_1\&W_2\&W_3|D_3)}{p(W_1\&W_2\&W_3|D_2)}$$

$$= \frac{p(D_3)}{p(D_2)} \times \frac{.125}{3/12} = \frac{p(D_3)}{p(D_2)} \times .5$$

★4.

independent distributions			mixtures	
D_3	D_5	D_7	$\frac{1}{2}(D_5 + D_7)$	$\frac{1}{3}(D_3 + D_5 + D_7)$y
.125	.729	.001	.385	.284
.125	.081	.009	.045	.072
.125	.081	.009	.045	.072
.125	.009	.081	.045	.072
.125	.081	.009	.045	.072
.125	.009	.081	.045	.072
.125	.009	.081	.045	.072
.125	.001	.729	.365	.284

a. $p(W_1|\frac{1}{2}(D_5 + D_7)) = .5$
$p(W_1\&W_2|\frac{1}{2}(D_5 + D_7)) = .410$
$p(W_1\&W_2\&W_3|\frac{1}{2}(D_5 + D_7)) = .365$
$p(W_2|W_1\&\frac{1}{2}(D_5 + D_7)) = .410/.5 = .820$
$p(W_3|W_1\&W_2|\frac{1}{2}(D_5 + D_7)) = .365/.410 = .890$

b. $p(W_1|\frac{1}{3}(D_3 + D_5 + D_7)) = .5$
$p(W_1\&W_2|\frac{1}{3}(D_3 + D_5 + D_7)) = .356$
$p(W_1\&W_2\&W_3|\frac{1}{3}(D_3 + D_5 + D_7)) = .284$
$p(W_2|W_1\&\frac{1}{3}(D_3 + D_5 + D_7)) = .356/.5 = .732$
$p(W_3|W_1\&W_2\&\frac{1}{3}(D_3 + D_5 + D_7)) = .284/.356 = .798$

c. Rule of succession for $\frac{1}{2}(D_5 + D_7)$:

$$p(W_{n+1}|W_1\&\ldots\&W_n) = \frac{.1^{n+1} + .9^{n+1}}{.1^n + .9^n}$$

Rule of succession for $\frac{1}{3}(D_3 + D_5 + D_7)$:

$$p(W_{n+1}|W_1\&\ldots\&W_n) = \frac{.1^{n+1} + .5^{n+1} + .9^{n+1}}{.1^n + .5^n + .9^n}$$

Chapter 5

Section 5.3

1 Consider the diagram:

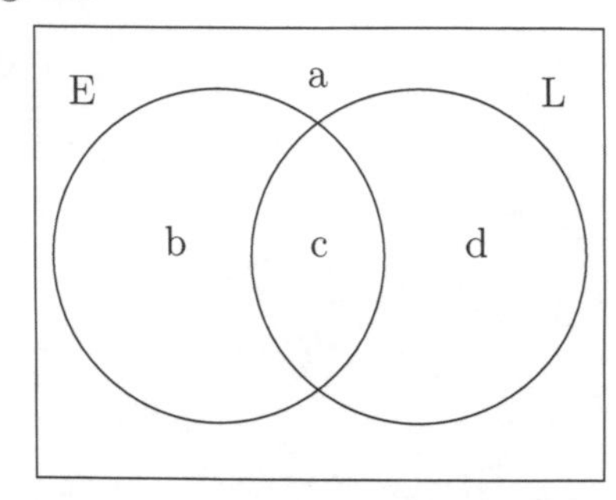

To prove that $p(\sim E) \leq u(E \vee L)/u(E \vee L|\sim E)$ note that $p(\sim E) = a+d$, $u(E \vee L) = a$, and $p((E \vee L) \& \sim E) = d$, hence $p(E \vee L|\sim E) = d/(a+d)$ and $u(E \vee L|\sim E) = a/(a+d)$. Then to prove that $p(\sim E) \leq u(E \vee L)/u(E \vee L|\sim E)$ it only has to be be sbown that $a+d \leq [a/(a/(a+d))]$. But this is trivial since $[a/(a/(a+d))] = a+d$. QED

Section 5.5

1a. Suppose that A neither entails nor is entailed by C. Then the lines #2 and #3 in the truth table below are both logical possibilities:

	dist.	A	C
#1	0	T	T
#2	.1	T	F
#3	.9	F	T
#4	0	F	F

If probabilities are distributed as shown, then $p(C) = .9$ but $p(C|A) = 0$.

1b. If A entails C then according to the Kolmogorov Axioms, $p(A) \leq p(C)$, and $p(C\&A) = p(A)$, hence $p(C|A) = p(C\&A)/p(A) = 1$.
If C entails A then $p(C\&A) = p(C)$, hence if $p(C) \geq .9$ then $p(C\&A) \geq .9$. Therefore, $p(C|A) = p(C\&A)/p(A)$ cannot equal 0.

2. Let A = Jones went to Florida,
B = he went to Hawaii,
C = he had a relaxing vacation.

Learning that Jones went either to Florida or Hawaii might not cause you to give up your belief that he had a relaxing vacation, but learning that he went to both could cause you to think otherwise.

3. Let A = The coin was tossed.
   B = It fell heads.
   C = It fell tails.

   That "The coin fell heads or it fell tails" was accepted *a priori* doesn't mean that either "It fell heads" or "It fell tails" would be accepted after it was confirmed that the coin was tossed.

4a. Let A = The first card was an ace.
   B = The second card was an ace.

   Then $A \rightarrow B$ is equivalent to "Either the first card wasn't an ace or the second card was an ace,"which would probably be probable enough to be accepted *a priori*, but would be likely to be given up if it were learned that the first card actually was an ace.[1]

4b. No.

4c.

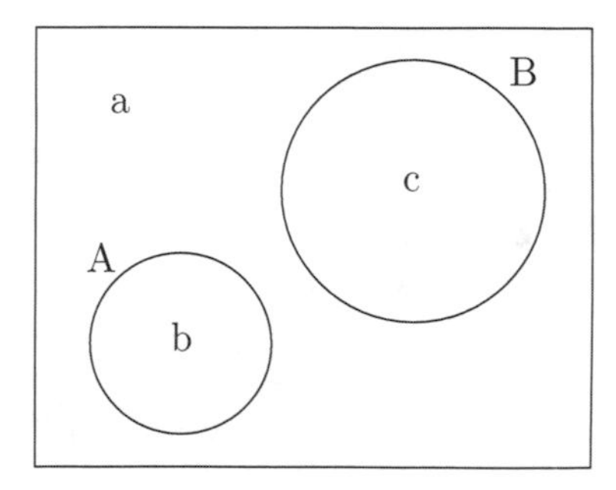

$p(A \rightarrow B) = 1 - b$
$p(A \rightarrow B|A) = p(A\&(A \rightarrow B)|A) = p(A\&B|A) = 0/b = 0.$

5. A = It was cloudy but dry yesterday.
   B = It was cloudy yesterday.
   C = It rained yesterday.

6. Let A = It rained yesterday in London.
   ~A = It didn't rain yesterday in London.
   C = The Yankees won yesterday's baseball game.

[1]But this doesn't mean that if the ordinary language conditional statement "If the first card was an ace then the second card was too" had been accepted *a priori* it would be given up if it was learned that the first card was an ace. Rather, as will be discussed in Chapters 6 and 7⋆, the material conditional $A \rightarrow B$ shouldn't be used to symbolize the ordinary language conditional.

Section 5.6⋆

1 Suppose that Jane's taking ethics is very unlikely *a priori*, and for that reason it is asserted that she won't take all three of ethics, logic, and history. On the other hand, it is conceivable that *if* she takes ethics, then she will also take both logic and history. Then the picture might look like:

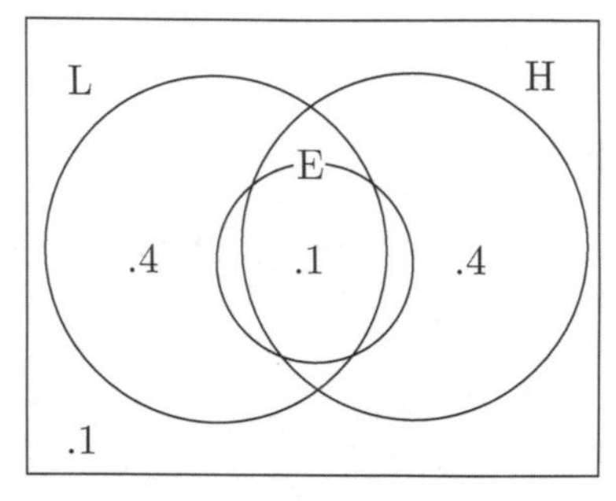

$$p(\sim(E\&L\&H)) = .9$$
$$p(\sim(E\&L\&H)|E) = 0.$$

2. Person #1 might say "Jane will take ethics or logic" on the grounds of his belief that Jane would actually take logic,[2] but hearing this, person #2 might object and say "No, Jane will take either ethics or history," on the grounds that she probably won't take logic but she could take ethics or history. From person #1's point of view, *a priori* probabilities could look like:

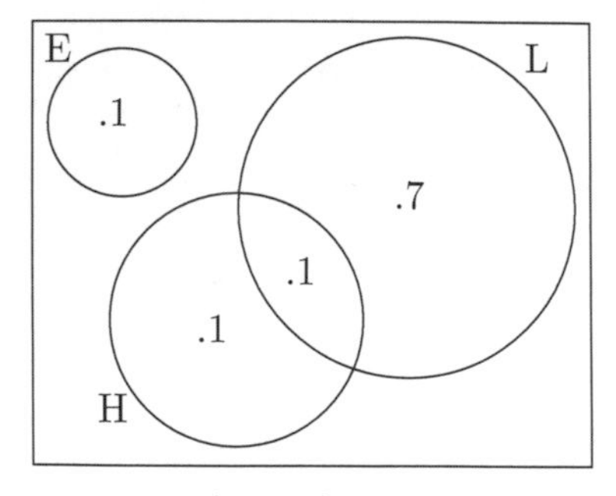

$$p(E \vee L) = .9$$
$$p(E \vee L|E \vee H) = .333$$

A more complicated analysis might put three more propositions into the picture:

[2]Though this would violate a maxim of conversational quality, not to assert a disjunction when one of its parts can be asserted.

S = person #2 asserts "No, Jane will take ethics or history."
C = Jane tosses coin C, and if it comes down heads she takes ethics and if it comes down tails she takes logic.
D = Jane tosses coin D, and if it comes down heads she takes ethics and if it comes down tails she takes history.

The 'probabilistic properties' of S, C, and D might be:

$p(E \vee L|C)$ is close to 1.
$p(E \vee H|D)$ is close to 1.
$p(C \leftrightarrow \sim D) = 1$.
$p(D|S)$ is close to 1.

A complete analysis has not been carried out, but something like it would be a contribution to probabilistic speech act theory.

Section 5.7⋆

⋆1.

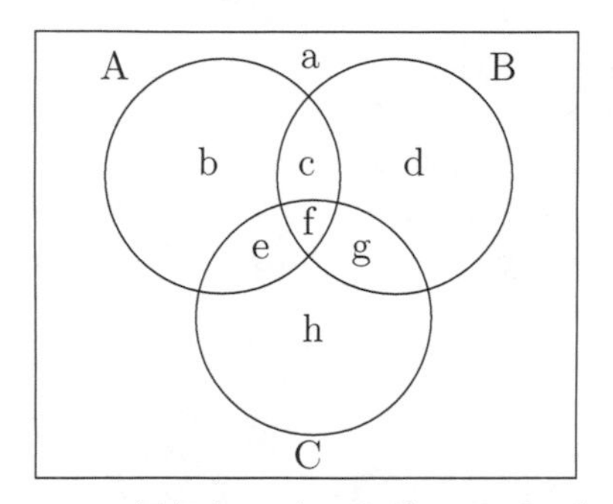

$$p(A|B) = (c+f)/(c+d+f+g)$$
$$p(A|C) = (e+f)/(e+f+g+h)$$
$$u(B|C) = (e+h)/(e+f+g+h)$$
$$u(C|B) = (c+d)/(c+d+f+g)$$

If $p(A|B) \geq p(A|C)$ then it must be shown that

$$\frac{c+f}{c+d+f+g} - \frac{e+f}{e+f+g+h} \leq \frac{e+h}{e+f+g+h} + \frac{c+d}{c+d+f+g}$$

or, equivalently,

$$\frac{f-d}{c+d+f+g} \leq \frac{2e+f+h}{e+f+g+h}$$

or, again,

$$(f-d) \times (e+f+g+h) \leq (c+d+f+g) \times (2e+f+h)$$

Multiplying out on both sides and canceling, all of the terms on the left get canceled out, hence the inequality must be true.
The same argument applies when $p(A|C) \geq p(A|B)$.

⋆2. If we suppose that the manager is perfectly truthful then if A has told him "Tell me the name someone besides myself who didn't get the job," and he has replied "B didn't get the job" it is certain that B didn't get the job; i.e., $p(\sim B|Q\&S_{\sim B}) = 1$ and $u(\sim B|Q\&S_{\sim B}) = 0$. On the other hand, if A has said "Tell me the name someone besides myself who didn't get the job," and B didn't get the job, it isn't certain that the manager will *say* "B didn't get the job." If, in fact, A got the job, then the manager could truthfully reply *either* "B didn't get the job," or "C didn't get the job," and if he chooses at random there is only 1 chance in 2 that he will say "B didn't get the job." In these circumstances, the probability that A got the job *and* that the manager doesn't reply "B didn't get the job" is $\frac{1}{3} \times \frac{1}{2} = \frac{1}{6}$. Given this, $u(S_{\sim B}|Q\&\sim B) = \frac{1}{6}$, and we can only say that

$$||p(A|Q\&S_{\sim B}) - p(A|Q\&\sim B)|| \leq 0 + \tfrac{1}{6}.$$

This is consistent with a difference of $\frac{1}{6}$ between $p(A|Q\&\sim B) = \frac{1}{2}$ and $p(A|Q\&S_{\sim B}) = \frac{1}{3}$, that is between the posterior probability of A's getting the job, given that B didn't get it, and the posterior probability of A's getting the job, given that the manager *says* that B didn't get it.

Chapter 6

Section 6.3

1a. Counterexample:

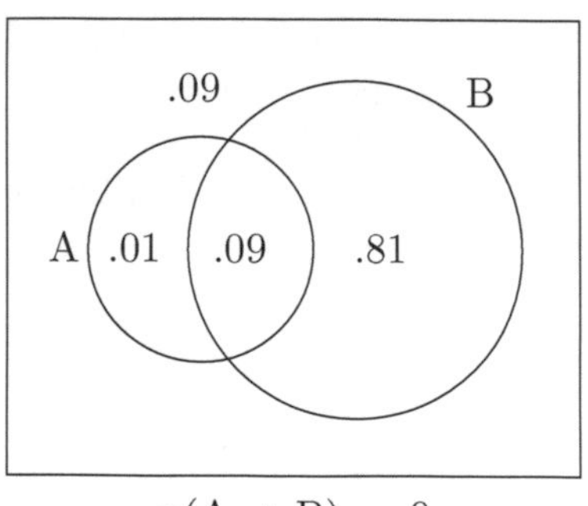

$$p(A \Rightarrow B) = .9$$
$$p(B \Rightarrow A) = .1$$

1c. No counterexample can be constructed. Argument: Suppose that

$$p(A \vee B) = b + c + d = .9$$

and

$$p(A \Rightarrow B) = c/(b + c) = .9.$$

Then $p(B) = c+d = p(A\vee B)-b$. But $b \leq 1-p(A \Rightarrow B) = 1-c/(b+c) = b/(b + c)$ because $b + c \leq 1$. Hence, since $p(B) = p(A \vee B) - b$

and $b \leq 1 - p(A \Rightarrow B)$,

$$p(B) \geq p(A \vee B) + p(A \Rightarrow B) - 1.$$

For instance, if $p(A \vee B)$ and $p(A \Rightarrow B)$ are both at least .9, then

$$p(B) \geq .9 + .9 - 1 = .8.$$

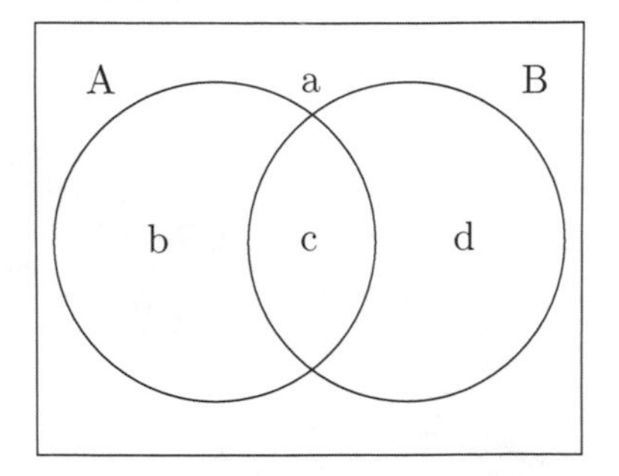

1e. Counterexample:

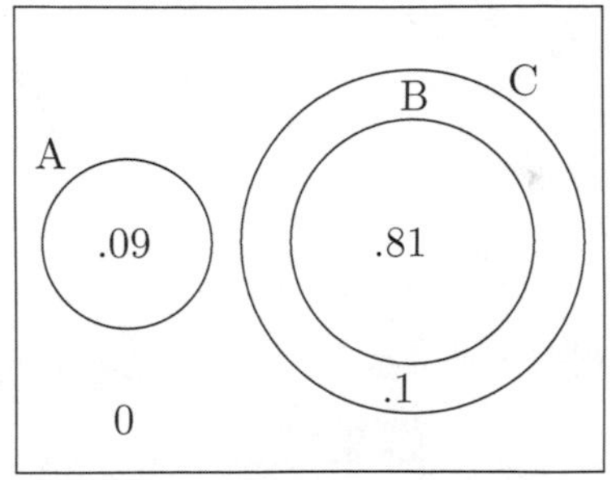

$$p((A \vee B) \Rightarrow C) = .81/.9 = .9$$

$$p(A \Rightarrow C) = 0.$$

1g. There are no counterexamples

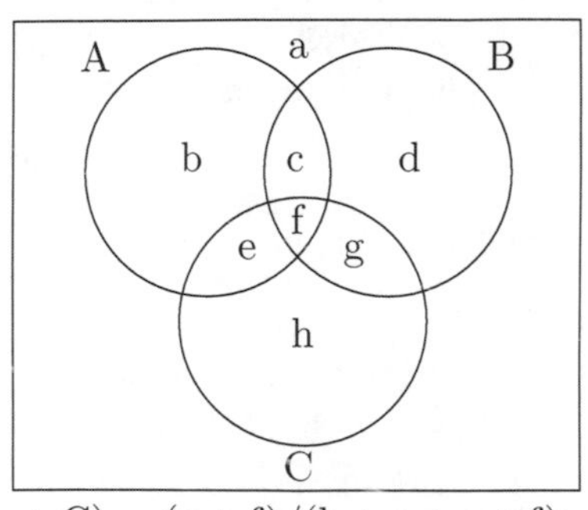

$$p(A \Rightarrow C) = (e + f)/(b + c + e + f)$$

$$p(B \Rightarrow C) = (f + g)/(c + d + f + g)$$

$$p((A \vee B) \Rightarrow C) = (e + f + g)/(b + c + d + e + f + g)$$

It is a matter of straightforward but tedious algebra to prove that

$$\frac{e+f+g}{b+c+d+e+f+g} \geq \frac{e+f}{b+c+e+f} + \frac{f+g}{c+d+f+g} - 1,$$

Hence, if p(A ⇒ C) and p(B ⇒ C) are both at least .9 then p((A ∨ B) ⇒ C) must be at least .9 + .9 − 1 = .8.

1i. Counterexample:

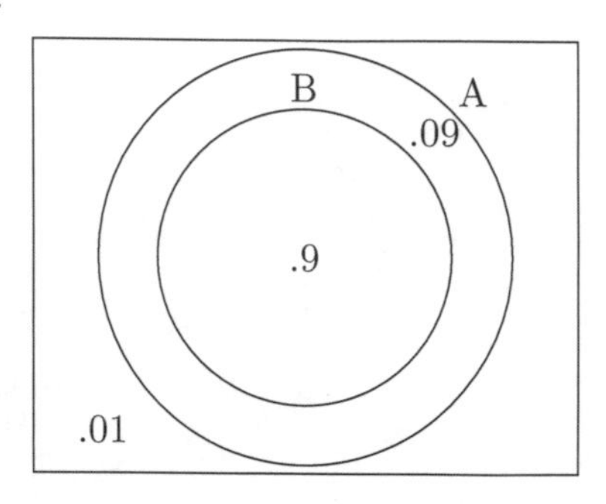

$$p(A \Rightarrow B) = .909$$
$$p(B \Rightarrow A) = 1$$
$$p(\sim A \Rightarrow \sim B) = 1$$
$$p(\sim B \Rightarrow \sim A) = .1$$

2. Parts a, e, and i. The diagrams in the answers to Exercise 1 give distributions in which the premises of the inferences have probabilities at least .9 and the conclusions have probabilities no greater than .1.

3a. If everybody goes to the party then Al will go to it. Therefore, if Al goes everybody will go.

3e. Let A = A will be the starting pitcher,
B = B will be the starting pitcher,
C = B will be the starting pitcher.[3]

Given this, (A ∨ B) ⇒ C symbolizes "If either A or B will be the starting pitcher, then B will be," from which it would be irrational to infer "If A will be the starting pitcher then B will be."

3i. Let A = The sun will rise tomorrow,
B = It will warm up during the day.

Then A ⇒ B is "If the sun rises tomorrow it will warm up during the day," which is probable, B ⇒ A is "If it warms up during the

[3]It is almost necessary to interpret B and C in the same way in order to yield an ordinary language counterexample to this inference. If they are different, as in "If either A or B is the starting pitcher the team will win," rules of conversational logic license inferring "If A is the starting pitcher the team will win."

day tomorrow then sun will rise," which is practically certain, and $\sim A \Rightarrow \sim B$ is "If the sun doesn't rise tomorrow it won't warm up during the day," which is also practically certain. But $\sim B \Rightarrow \sim A$ is "If it doesn't warm up during the day tomorrow then sun won't rise," which is very unlikely.

4. Let E = Ed wins first prize
 F = Fred wins second prize
 G = George is disappointed.

 Then the inference becomes:

$$\frac{E \Rightarrow (F \vee G), \sim F}{\sim G \Rightarrow \sim E}$$

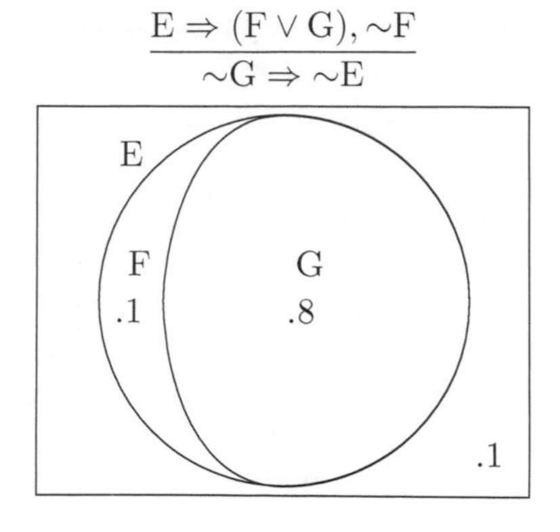

$$p(E \Rightarrow (F \vee G)) = 1$$
$$p(\sim F) = .9$$
$$p(\sim G \Rightarrow \sim E) = .5$$

 Scenario: Ed will probably win the first prize, and his father, George, hopes that his other son, Fred, will win the second prize, though that is unlikely and George will probably be disappointed. On the other hand, if he isn't disappointed it will either be because Ed didn't win the first prize after all, or he won it and Fred got the second prize.

5. Putting T's and F's in the $A \Rightarrow B$ column in the last two lines of the table and adding the probabilities of the lines that have T's in them can only total .4 or .5 or .6 or .7, none of which equals $4/7$.

6a. If an inference is classically invalid when $\Rightarrow$ is replaced by $\rightarrow$ then there is a line in the truth table for the inference in which all of its premises are true and its conclusion is false. A distribution that assigns probability 1 to that line and probability 0 to all other lines will generate probabilities equal to 1 for all of the premises and a probability of 0 for the conclusion.

6b. The foregoing implies that classical validity is a necessary condition for being 'probabilistically valid' in the sense defined, because if the inference has premises $\phi_1 \Rightarrow \psi_1, \ldots, \phi_n \Rightarrow \psi_n$, and conclusion $\phi \Rightarrow \psi$ and it is classically invalid then there is a probability function p( ) such that $p(\phi_1 \Rightarrow \psi_1) = \ldots = p(\phi_n \Rightarrow \psi_n) = 1$ but $p(\phi \Rightarrow \psi) = 0$, hence $u(\phi_1 \Rightarrow \psi_1) = \ldots = u(\phi_n \Rightarrow \psi_n) = 0$ but $u(\phi \Rightarrow \psi) = 1$, and therefore $u(\phi \Rightarrow \psi) > u(\phi_1 \Rightarrow \psi_1) + \ldots + u(\phi_n \Rightarrow \psi_n)$. Therefore the inference doesn't satisfy the Uncertainty Sum Rule.

6c. If an inference not involving probability conditionals is classically valid it satisfies the Uncertainty Sum Rule, and this implies that if its premises have probability 1 and uncertainty 0 its conclusion must have uncertainty 0 and probability 1. The fact that a probability conditional has probability 1 if and only if the corresponding material conditional has probability 1, which is easily proved, shows that this is still the case for inferences involving probability conditionals.

⋆7. The diagram below shows that Step #3 isn't 'probabilistically entailed' by Step #2. Here $p(\sim(A\&\sim B)) = .9$ but $p(A \Rightarrow B) = 0$. Note, however, that asserting ~(A&~B) in Step #2 would violate a maxim of conversational quality, which says that in circumstances in which ~A by itself is asserted, which is the case in Step #1, a more complicated and less informative proposition should not be asserted. Thus, there are two steps in the reasoning that are invalid: Step #2 is conversationally invalid, and Step #3 is probabilistically invalid. But note also that neither of these kinds of invalidity is classical, truth-conditional invalidity.

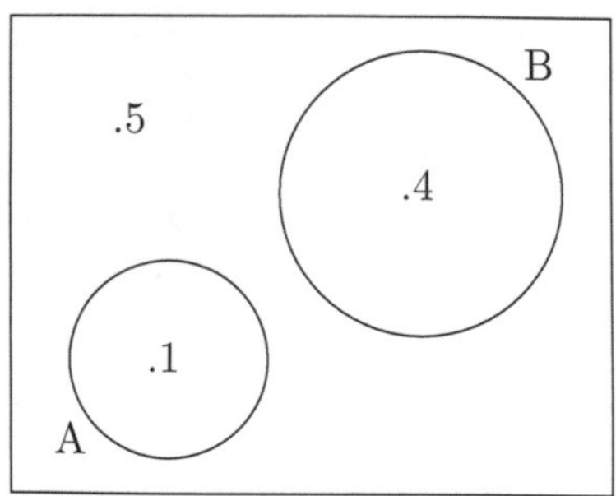

⋆8. This interesting argument has never been analyzed in detail, but all of Steps #3–#6 raise questions.

Step #3. So far, probability logic doesn't seem to apply to reasoning from assumptions or suppositions, because probability doesn't seem to apply to them. However, section 7.8⋆⋆ states a theorem that applies to the relation between an inference with premises A ∨ B ∨ C and ~B ⇒ C and another inference with premises A∨B∨C and ~B ⇒ C, *plus* ∼A, which will be returned to in commenting on Step #6.[4] Step #4 isn't 'probabilistically entailed' by Steps #1 and #3 in the sense of Exercises 5 and 6, as the diagram below shows.

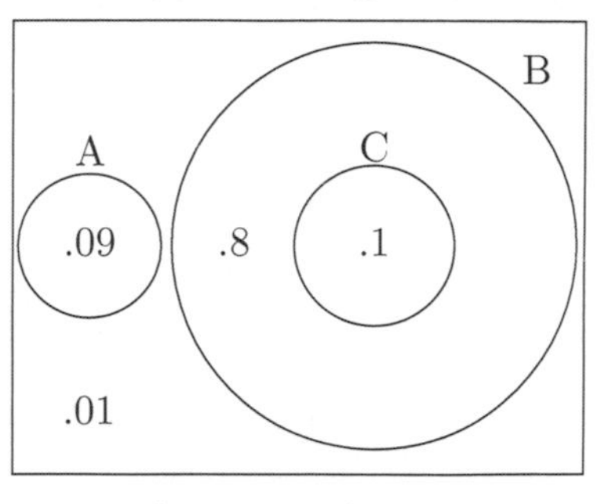

$$p(A \vee B \vee C) = .99$$
$$p(\sim A) = .91$$
$$p(\sim B \Rightarrow C) = 0$$

Step #5. In spite of the discussion in section 6.2 of the seeming inconsistency of "If the first card will be an ace then the second card will be too" and "If the first card will be an ace then the second card won't be an ace," probabilistic inconsistency won't be discussed in any detail until section 7.8⋆⋆, and we can't yet be sure of exactly how it is defined.

Step #6. Even assuming that ∼B ⇒ C and ∼B ⇒ ∼C are 'probabilistically' inconsistent, it isn't clear what follows from such a contradiction, and whether the principle of reduction to absurdity is valid. However, section 7.8⋆⋆ will state a probabilisitic version of this principle, and the Barbershop Paradox will be returned to in connection with this.

⋆9. Example: Because A follows from A together with B, it doesn't follow that you can derive B ⇒ A from A alone.

Section 6.4

⋆1. (1) {A, B} ∴ A&B is p-valid; assumed fact.

(2) {A, B, C} ∴ A is p-valid; set of premises p-entails a member.

(3) {A, B, C} ∴ B is p-valid; set of premises p-entails a member.

(4) {A, B, C} ∴ C is p-valid; set of premises p-entails a member.

[4] There is also a sense in which to formulate ∼A as a supposition can be regarded as formulating a common part of the antecedents of one or more conditionals that follow it. E.g., it is tacitly assumed that the conditional ∼B ⇒ C in Step #4 really includes ∼A in its antecendent, i.e., Step #4 really is (∼A&∼B) ⇒ C.

(5) {A, B, C} ∴ A&B is p-valid; because {A, B} ∴ A&B is p-valid and A and B are p-entailed by {A, B, C}.

(6) {A&B, C} ∴ A&B&C is p-valid; assumed fact.

(7) {A, B, C} ∴ A&B&C is p-valid; because A&B and C are p-entailed by {A, B, C}.

Section 6.5⋆

1a. OMP-ordering: $4 \succ 3 \succ 2 \approx 1 \approx \emptyset$

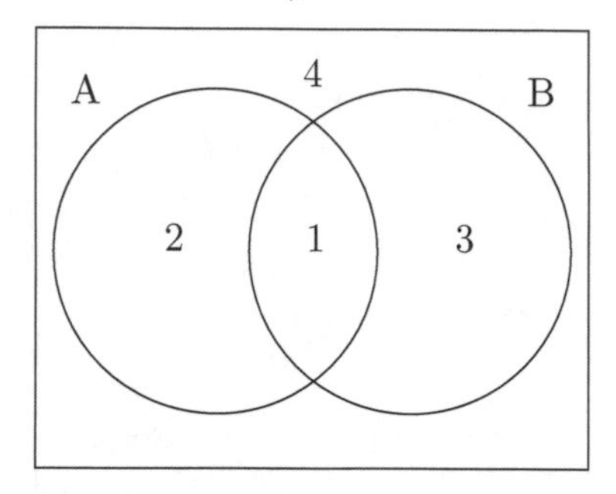

Corresponding OMP-counterexample:

	distrib.	A	B	A ⇒ B	B ⇒ A
1	0	T	T	*T*	*T*
2	0	T	F	*T*	...
3	.1	F	T	...	*F*
4	.9	F	F	...	...
		0	.1	1	0

1e. OMP-ordering counter-example: $5 \succ 2 \succ 1 \approx 3 \approx 4 \approx 6 \approx 7 \approx 8 \approx \emptyset$

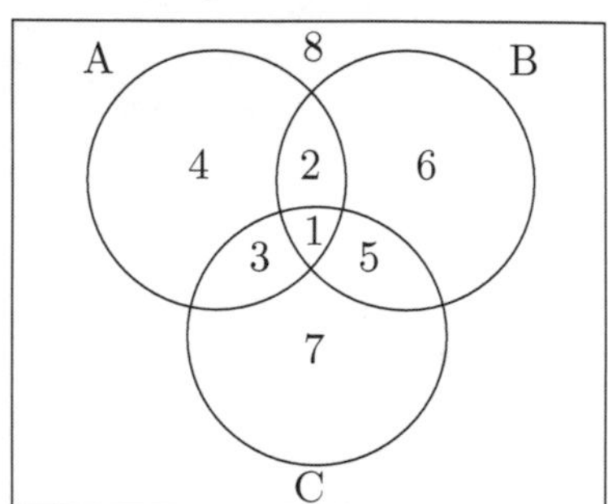

Corresponding OMP-counterexample:

regions	distribution	A	B	C	$(A \vee B) \Rightarrow C$	$A \Rightarrow C$
1	0	T	T	T	*T*	*T*
2	.10	T	T	F	*F*	*F*
3	0	T	F	T	*T*	*T*
4	0	T	F	F	*F*	*F*
5	.9	F	T	T	*T*	......
6	0	F	T	F	*F*	......
7	0	F	F	T	......	......
8	0	F	F	F	......	......
		.1	1	.9	.9	0

1i. OMP-ordering: $1 \succ 2 \succ 4 \succ 3 \approx \emptyset$

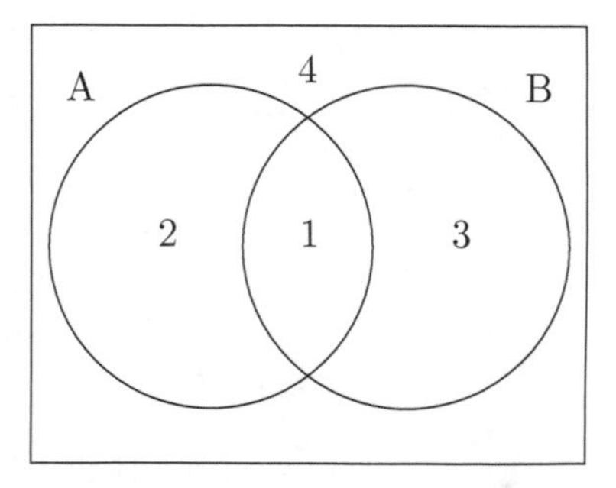

Corresponding OMP-counterexample:

	distrib.	A	B	$A \Rightarrow B$	$B \Rightarrow A$	$\sim A \Rightarrow \sim B$	$\sim B \Rightarrow \sim A$
1	.9	T	T	*T*	*T*	...	...
2	.09	T	F	*F*	...	...	*F*
3	0	F	T	...	*F*	*F*	...
4	.01	F	F	...	...	*T*	*T*
		.99	.9	.909	1	1	.1

2a.

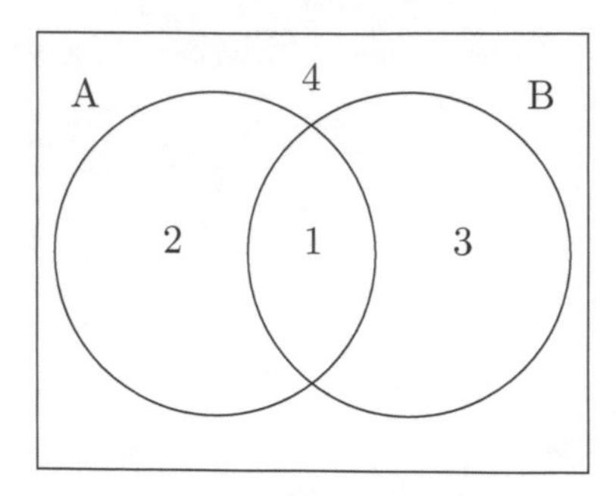

If A holds in an OMP-ordering then $\{1,2\} \succ \{3,4\}$, and if A $\Rightarrow$ B holds then $\{1\} \succ \{2\}$. If $\{1,2\} \succ \{3,4\}$ and $\{1\} \succ \{2\}$ then clearly $\{1\} \succ \{3,4\}$, and since $\{1\} \succ \{2\}$ and $\{2\} \succ \{3,4\}$, it follows that $\{1\} \succ \{1,3,4\}$. Hence $\{1\} \succ \{2,4\}$ and therefore $\{1,3\} \succ \{2,4\}$, which is what is required for B to hold in the ordering.

2c. If A $\Rightarrow$ B holds in the ordering then $\{1\} \succ \{2\}$, hence $\{1,3,4\} \succ \{2\}$. But that is what is required for A $\rightarrow$ B to hold in the ordering.

2e.

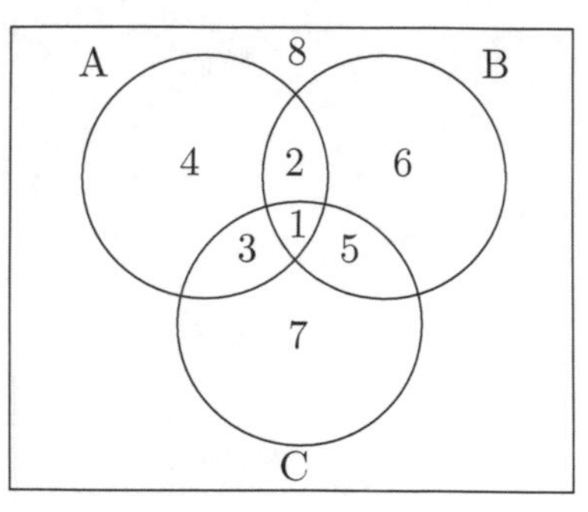

1. For A $\Rightarrow$ B to hold $\{1,2\} \succ \{3,4\}$
2. For B $\Rightarrow$ A to hold $\{1,2\} \succ \{5,6\}$
3. For A $\Rightarrow$ C to hold $\{1,3\} \succ \{2,4\}$
4. Either $\{1\} \succ \{2,4\}$ or $\{3\} \succ \{2,4\}$
5. If $\{1\} \succ \{2,4\}$ then $\{1\} \succ \{2\}$, hence, by (1), $\{1\} \succ \{3,4\}$, so $\{1\} \succ \{3\}$.
6. If $\{1\} \succ \{2\}$ and $\{1\} \succ \{3\}$ then $\{1\} \succ \{2,3\}$, hence $\{1,5\} \succ \{2,3\}$.
7. That $\{1,5\} \succ \{2,3\}$ means that B $\Rightarrow$ C holds in the ordering.
8. If $\{3\} \succ \{2,4\}$ then it is not the case that $\{2\} \succ \{3,4\}$, hence, by (1), $\{1\} \succ \{3,4\}$.
9. If $\{1\} \succ \{3,4\}$ but not $\{2\} \succ \{3,4\}$ then $\{1\} \succ \{2\}$.

10. If $\{1\} \succ \{2\}$ and $\{1\} \succ \{3,4\}$ then $\{1\} \succ \{2,3\}$ and $\{1,5\} \succ \{2,3\}$.
11. If $\{1,5\} \succ \{2,3\}$ then $\mathrm{B} \Rightarrow \mathrm{C}$ holds in the ordering.

⋆3. If an inference is classically invalid then, replacing all probability conditionals by material conditionals, there is a line in the truth table for the inference in which all of its premises are true but its conclusion is false. If the SD corresponding to that line is assigned probability 1 and all other SDs are assigned probability 0, this distribution generates a probability function in which all of the premises of the original inference have probability 1 and the conclusion has probability 0.

⋆4. Suppose that $p((A \vee B) \Rightarrow C)$, $p(A \Rightarrow C)$, and $p((A \vee B) \Rightarrow B)$ are generated by the distribution in the diagram below:

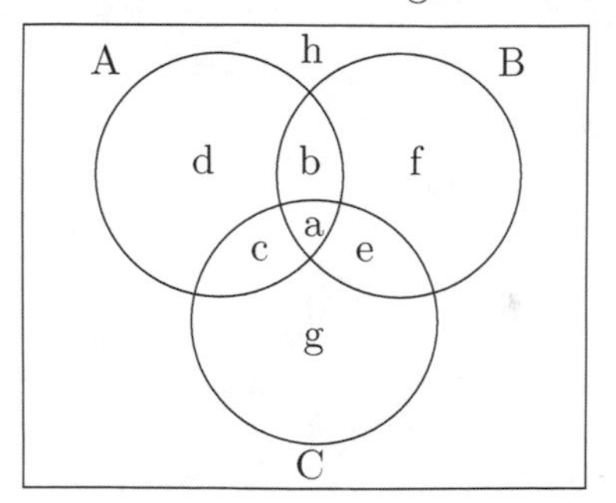

$$u((A \vee B) \Rightarrow C) = (b+d+f)/(a+b+c+d+e+f)$$
$$u(A \Rightarrow C) = (b+d)/(a+b+c+d)$$
$$u((A \vee B) \Rightarrow B) = (c+d)/(a+b+c+d+e+f),$$

If $u(A \Rightarrow C) \times u((A \vee B) \Rightarrow B) > u((A \vee B) \Rightarrow C)$ then

$$\frac{b+d}{a+b+c+d} \times \frac{c+d}{a+b+c+d+e+f} > \frac{b+d+f}{a+b+c+d+e+f},$$

hence

$$\frac{b+d}{a+b+c+d} \times (c+d) > b+d+f,$$

hence

$$(b+d) \times (c+d) > (b+d+f) \times (a+b+c+d),$$

hence

$$bc+bd+cd+d^2 > ab+b^2+bc+bd+ad+bd+cd+d^2+af+bf+cf+df.$$

But canceling equal terms on the left and right side above yields:

$$0 > ab+b^2+ad+d^2+af+bf+cf+df,$$

which is impossible.

Section 6.6⋆

2a. Given the distribution

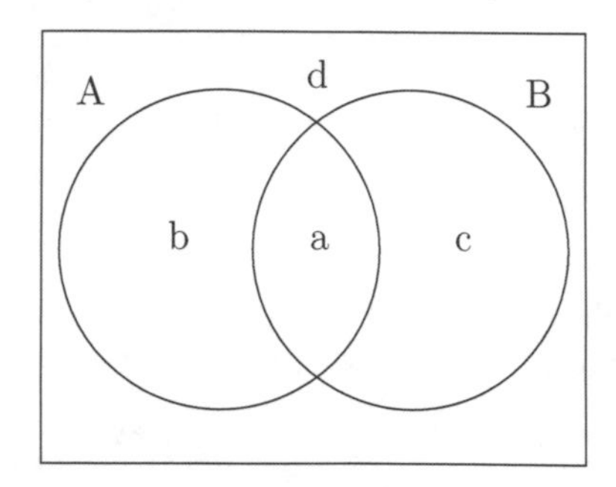

$$p(A) = a + b$$
$$\text{and}$$
$$p(B \Rightarrow A) = a/(a + c).$$

It follows that $u(B \Rightarrow A) = c/(a+c) \leq (c+d)/(a+c) = u(A)/p(B)$. Hence if $p(A) \geq .99$ and $u(A) \leq .01$, then $u(B \Rightarrow A) \leq .01/p(B)$. Hence, if $p(B) \geq .1$ then $u(B \Rightarrow A) \leq .01/.1 = .1$ and $p(B \Rightarrow A) \geq .9$. In other words, if $p(A) \geq .99$ and we suppose that, statistically, there are 9 chances in 10 that B has a probability of at least .1, then there are 9 chances in 10 that $p(B \Rightarrow A) \geq .9$. Roughly, most of the time when A is probable, $B \Rightarrow A$ will also be probable.[5]

2b. If B is a 'random proposition' with a high probability then ~B is a random proposition with a low one, and there will only be a random chance that most of the region corresponding to it will lie inside the region corresponding to another proposition, A; i.e., given that B is highly probable, there is only a random chance that $\sim B \Rightarrow A$ will be highly probable. Therefore, this 'inference' doesn't have a 'statistical justification' like that of the other fallacy of material implication, to infer $B \Rightarrow A$ from A.

---

[5]This statistical argument is closely related to D. Bamber's geometrical argument outlined in Appendix 5 for the 'statistical reasonableness' of inferences of the form $A \vee B \mathrel{.\cdot.} \sim A \Rightarrow B$.

An interesting side issue concerns a connection between the p-invalid inference $A \mathrel{.\cdot.} B \Rightarrow A$ and the rather strange inference $\{A, B\} \mathrel{.\cdot.} B \Rightarrow A$, which is nevertheless easily shown to be p-valid (the reader will have no difficulty showing that $A \Rightarrow B$ holds in all OMP-orderings in which both A and B hold). But the second premise, B, is *marginally essential* in the sense that A and B are not both required to be probable to guarantee that $A \Rightarrow B$ is probable; all that is required is that A should be probable while B is not too improbable. Thus, we saw above that if $p(A) \geq .99$ and $p(B) \geq .1$, it follows that $p(A \Rightarrow B) \geq .1$. Adams (1981) gives an extended discussion of the phenomenon of marginal essentialness, which only arises in inferences that involve probability conditionals.

Section 6.7⋆

1 The object of working on these essay type questions is to lead the student to think about foundational matters that don't have cut and dried 'solutions'. Chapter 8⋆ discusses issues related to part b.

Chapter 7⋆

Section 7.2

1a.
1. A&B G.
2. T $\Rightarrow$ (A&B) From 1, by CF.
3. (T&(A&B)) $\Rightarrow$ (B&A) By LC.
4. T $\Rightarrow$ (B&A) From 2 and 3 by RT.
5. B&A From 4, by CF. QED

1c.
1. A $\Rightarrow$ ∼A G.
2. (A&∼A) $\Rightarrow$ B By LC.
3. A $\Rightarrow$ B From 1 and 2 by RT. QED

1e.
1. A $\Rightarrow$ B G.
2. (A&B) $\Rightarrow$ (C $\rightarrow$ B) By LC.
3. A $\Rightarrow$ (C $\rightarrow$ B) From 1 and 2 by LC. QED

1g.
1. A $\Rightarrow$ B G.
2. A $\Rightarrow$ C G.
3. (A&B) $\Rightarrow$ C From 1 and 2 by AR.
4. ((A&B)&C) $\Rightarrow$ (B&C) By LC.
5. (A&B) $\Rightarrow$ (B&C) From 3 and 4 by RT.
6. A $\Rightarrow$ (B&C) From 1 and 5 by RT. QED

1i.
1. A $\Rightarrow$ B G.
2. B $\Rightarrow$ A G.
3. (A&B) $\Rightarrow$ (A&B) By LC.
4. A $\Rightarrow$ (A&B) From 1 and 3 by RT.
5. (B&A) $\Rightarrow$ (A&B) By LC.
6. B $\Rightarrow$ (A&B) From 2 and 5 by RT.
7. (A $\vee$ B) $\Rightarrow$ (A&B) From 4 and 6 by DA. QED

⋆2a
1. A G.
2. B G.
3. T $\Rightarrow$ A From 1 by CF.
4. T $\Rightarrow$ B From 2 by CF.
5. (T&A) $\Rightarrow$ B From 3 and 4 by AR.
6. ((T&A)&B) $\Rightarrow$ (A&B) By LC.

7. (T&A) $\Rightarrow$ (A&B) From 5 and 6 by RT.
8. T $\Rightarrow$ (A&B) From 3 and 7 by RT.
9. A&B From 8 by CF. QED

$\star$2c.
1. (A $\vee$ B) $\Rightarrow$ ~A G.
2. ((A $\vee$ B)&~A) $\Rightarrow$ B By LC.
3. (A $\vee$ B) $\Rightarrow$ B From 1 and 2 by RT.
4. ((A $\vee$ B)&B) $\Rightarrow$ ~A From 1 and 3 by AR.
5. B $\Rightarrow$ ~A From 4 by EA, because (A $\vee$ B)&B and B are logically equivalent. QED

$\star$2e.
1. A $\Rightarrow$ B G.
2. B $\Rightarrow$ ~A G.
3. (A&B) $\Rightarrow$ (A $\rightarrow$ B) LC.
4. A $\Rightarrow$ (A $\rightarrow$ B) From 1 and 3 by RT.
5. ~A $\Rightarrow$ (A $\rightarrow$ B) By LC.
6. (A $\vee$ ~A) $\Rightarrow$ (A $\rightarrow$ B) from 4 and 5 by DA.
7. T $\Rightarrow$ (A $\rightarrow$ B) From 6 by EA.
8. (B&~A) $\Rightarrow$ (B $\rightarrow$ ~A) By LC.
9. B $\Rightarrow$ (B $\rightarrow$ ~A) From 2 and 8 by RT.
10. ~B $\Rightarrow$ (B $\rightarrow$ ~A) By LC.
11. (B $\vee$ ~B) $\Rightarrow$ (B $\rightarrow$ ~A) From 9 and 10 by DA.
12. T $\Rightarrow$ (B $\rightarrow$ ~A) From 11 by EA.
13. (T&(A $\rightarrow$ B)) $\Rightarrow$ (B $\rightarrow$ ~A) From 7 and 12 by AR.
14. [(T&(A $\rightarrow$ B))&(B $\rightarrow$ ~A)] $\Rightarrow$ ~A By LC.
15. (T&(A $\rightarrow$ B)) $\Rightarrow$ ~A From 13 and 14 by RT.
16. T $\Rightarrow$ ~A From 7 and 15 by RT.
17. ~A From 16 by CF. QED

Section 7.3

1a.
1. A
2. B $\Rightarrow$ B LC.
3. B $\rightarrow$ B From 2, by $\Re$12. QED

1c.
1. A G.
2. A $\Rightarrow$ B G.
3. B $\Rightarrow$ C G.
4. B From 1 and 3 by MP (Rule $\Re$8). QED

2. Proof of $\Re$12.
   1. $\phi \Rightarrow \psi$
   2. $(\phi \& \psi) \Rightarrow (\phi \rightarrow \psi)$ LC.

3. $\phi \Rightarrow (\phi \rightarrow \psi)$ From 1 and 2 by RT.
4. $\sim\phi \Rightarrow (\phi \rightarrow \psi)$ LC.
5. $(\phi \vee \sim\phi) \Rightarrow (\phi \rightarrow \psi)$ From 3 and 4 by DA.
6. $\mathrm{T} \Rightarrow (\phi \rightarrow \psi)$ From 5 by EA.
7. $\phi \rightarrow \psi$ From 6 by CF. QED

Proof of $\Re 13$.

1. $\phi \Rightarrow \psi$
2. $\phi \Rightarrow \eta$
3. $(\phi \,\&\, \eta) \Rightarrow \psi$ From 1 and 2 by AR.
4. $((\phi \,\&\, \eta) \,\&\, \psi) \Rightarrow (\psi \,\&\, \eta)$ LC.
5. $(\phi \,\&\, \eta) \Rightarrow (\psi \,\&\, \eta)$ From 3 and 4 by RT.
6. $\phi \Rightarrow (\psi \,\&\, \eta)$ From 2 and 5 by RT. QED

Section 7.4

1. Proof of $\Re 17$.

Suppose that $\wp_1$ and $\wp_2$ are the factual formulas ϕ and ψ, and these are replaced by their trivial equivalents $\mathrm{T} \Rightarrow \phi$ and $\mathrm{T} \Rightarrow \psi$.

1. $(\mathrm{T} \Rightarrow \phi) \bigwedge (\mathrm{T} \Rightarrow \psi)$
2. $(\mathrm{T} \vee \mathrm{T}) \Rightarrow ((\mathrm{T} \rightarrow \phi)\&(\mathrm{T} \rightarrow \psi))$ From 1, by the definition of $\bigwedge$.
3. $\mathrm{T} \Rightarrow ((\mathrm{T} \rightarrow \phi)\&(\mathrm{T} \rightarrow \psi))$ From 2 by EA.
4. $[\mathrm{T}\&((\mathrm{T} \rightarrow \phi)\&(\mathrm{T} \rightarrow \psi))] \Rightarrow (\mathrm{T} \rightarrow \phi)$ LC.
5. $\mathrm{T} \Rightarrow (\mathrm{T} \rightarrow \phi)$ From 3 and 4 by RT.
6. $(\mathrm{T}\&(\mathrm{T} \rightarrow \phi)) \Rightarrow \phi$ LC.
7. $\mathrm{T} \Rightarrow \phi$ From 5 and 6 by RT.
8. ϕ From 7 by CF.
9. $[\mathrm{T}\&((\mathrm{T} \rightarrow \phi)\&(\mathrm{T} \rightarrow \psi))] \Rightarrow (\mathrm{T} \rightarrow \psi)$ LC.
10. $\mathrm{T} \Rightarrow (\mathrm{T} \rightarrow \psi)$ From 3 and 9 by RT.
11. $(\mathrm{T}\&(\mathrm{T} \rightarrow \psi)) \Rightarrow \psi$ LC.
12. $\mathrm{T} \Rightarrow \psi$ From 10 and 11 by RT.
13. ψ From 12 by CF. QED

Proof of $\Re 19$

Suppose that $\wp_1$ and $\wp_2$ are the formulas $\phi_1 \Rightarrow \psi_1$ and $\phi_2 \Rightarrow \psi_2$, respectively.

1. $(\phi_1 \Rightarrow \psi_1) \bigwedge (\phi_2 \Rightarrow \psi_2)$
2. $(\phi_1 \vee \phi_2) \Rightarrow [(\phi_1 \rightarrow \psi_1)\&(\phi_2 \rightarrow \psi_2)]$ From 1 by the definition of $\bigwedge$.
3. $[(\phi_1 \vee \phi_2)\&[(\phi_1 \rightarrow \psi_1)\&(\phi_2 \rightarrow \psi_2)]] \Rightarrow [(\phi_2 \rightarrow \psi_2)\&(\phi_1 \rightarrow \psi_1)]$ LC.
4. $(\phi_1 \vee \phi_2) \Rightarrow [(\phi_2 \rightarrow \psi_2)\&(\phi_1 \rightarrow \psi_1)]$ From 2 and 3 by RT.
5. $(\phi_2 \vee \phi_1) \Rightarrow [(\phi_2 \rightarrow \psi_2)\&(\phi_1 \rightarrow \psi_1)]$ From 4 by EA.
6. $(\phi_2 \Rightarrow \psi_2) \bigwedge (\phi_1 \Rightarrow \psi_1)$ From 5 by definition of $\bigwedge$. QED

2.

	A	B	C	D	$A \Rightarrow B$	$C \Rightarrow D$	$(A \vee C) \Rightarrow [(A \rightarrow B)\&(C \rightarrow D)]$
1	T	T	T	T	*T*	*T*	*T*
2	T	T	T	F	*T*	*F*	*F*
3	T	T	F	T	*T*	...	*T*
4	T	T	F	F	*T*	...	*T*
5	T	F	T	T	*F*	*T*	*F*
6	T	F	T	F	*F*	*F*	*F*
7	T	F	F	T	*F*	...	*F*
8	T	F	F	F	*F*	...	*F*
9	F	T	T	T	...	*T*	*T*
10	F	T	T	F	...	*F*	*F*
11	F	T	F	T	...	...	....
12	F	T	F	F	...	...	....
13	F	F	T	T	...	*T*	*T*
14	F	F	T	F	...	*F*	*F*
15	F	F	F	T	...	...	....
16	F	F	F	F	...	...	....

Section 7.5

1.

	A	B	$A \rightarrow B$	$A \Rightarrow B$	$A \Rightarrow {\sim}B$	$B \Rightarrow {\sim}A$	$(A \vee B) \Rightarrow {\sim}A$	$(B \vee {\sim}A) \Rightarrow A$
1	T	T	T	*T*	*F*	*F*	*F*	*T*
2	T	F	F	*F*	*T*	...	*F*	......
3	F	T	T	...	...	*T*	*T*	*T*
4	F	F	T	...	...	...	......	*F*

All of these are classically valid, but a, b, c, and f are p-invalid. d and e are p-valid, and in each case the set of all of the inference's premises yields the conclusion.[6]

[6]Curiously, in both inferences d and e, it is not possible to verify either premise without falsifying the other, and given this, it follows that the inference is p-valid if and only if it is classically valid.

2.

	A	B	C	$A \Rightarrow C$	$B \Rightarrow \sim C$	$A \Rightarrow \sim B$	$A \Rightarrow B$	$A \rightarrow \sim B$	$\sim A$	$(A \vee B) \Rightarrow \sim A$	$(A \vee B) \Rightarrow B$
1	T	T	T	*T*	*F*	*F*	*F*	F	F	*F*	*T*
2	T	T	F	*F*	*T*	*F*	*T*	F	F	*F*	*T*
3	T	F	T	*T*	...	*T*	*F*	T	F	*F*	*F*
4	T	F	F	*F*	...	*T*	*F*	T	F	*F*	*F*
5	F	T	T	...	*F*	...	...	T	T	*T*	*T*
6	F	T	F	...	*T*	...	...	T	T	*T*	*T*
7	F	F	T	...	...	...	...	T	T	......	...
8	F	F	F	...	...	...	...	T	T	......	...

	A	B	C	$A \Rightarrow \sim B$	$B \Rightarrow \sim C$	$C \Rightarrow \sim A$	$\sim A \Rightarrow \sim A$	$(A \vee B \vee C) \Rightarrow (A \leftrightarrow (B \leftrightarrow C))$
1	T	T	T	*F*	*F*	*F*	...	*T*
2	T	T	F	*F*	*T*	...	...	*F*
3	T	F	T	*T*	...	*F*	...	*F*
4	T	F	F	*T*	...	...	...	*T*
5	F	T	T	...	*F*	*T*	*T*	*F*
6	F	T	F	...	*T*	...	*T*	*T*
7	F	F	T	...	...	*T*	*T*	*T*
8	F	F	F	...	...	...	*T*	....

No subsets of the premises of inferences a, b, and c yield these inferences' conclusions. The set of the one premise of inference d yields its conclusion, and the first three premises of inference e yields that inference's conclusion, but the whole set of its premises does not.

3. Diagrammatic counterexample to inference a of exercise 2:

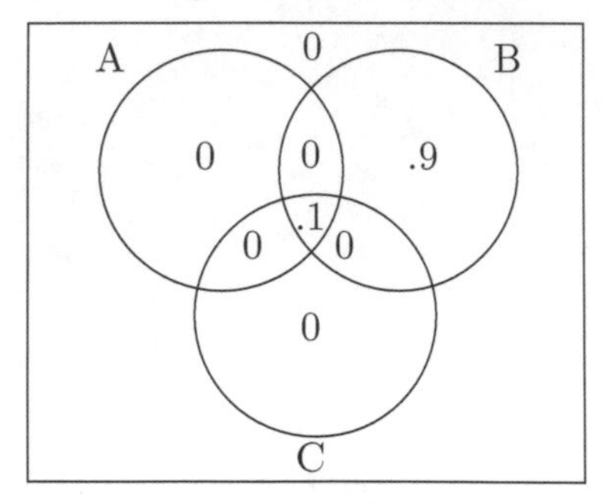

$$p(A \Rightarrow C) = 1$$
$$p(B \Rightarrow \sim C) = .9$$
$$p(A \Rightarrow \sim B) = 0$$

Diagrammatic counterexample to inference c of exercise 2:

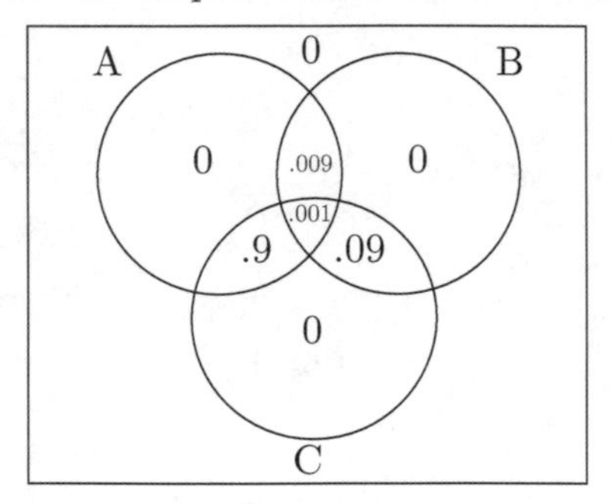

$$p(A) = .91$$
$$p(B \Rightarrow C) = .91$$
$$p((A\&B) \Rightarrow C) = .1$$

Derivation of the conclusion of inference e of exercise 2 from its premises (using derived rules of inference):

1. $A \Rightarrow \sim B$ G.
2. $B \Rightarrow \sim C$ G.
3. $C \Rightarrow \sim A$ G.
4. $(A \vee B) \Rightarrow [(A \to \sim B)\&(B \to \sim C)]$ From 1 and 2 by $\Re$15.
5. $(A \vee B \vee C) \Rightarrow [(A \to \sim B)\&(B \to \sim C)\&(C \to \sim A)]$
   From 3 and 4 by $\Re$15.
6. $(A \vee B \vee C) \Rightarrow [A \leftrightarrow (B \leftrightarrow C)]$ From 5 by $\Re$14. QED
   Note in particular that $A \leftrightarrow (B \leftrightarrow C)$ is a logical consequence $(A \to \sim B)\&(B \to \sim C)\&(C \to \sim A)$.

Section 7.7

1a. 1. $A \Rightarrow B$ G.
   2. $A \Rightarrow \sim(A \rightarrow B)$ Assumption.
   3. $A \Rightarrow \sim B$ From 2 by $\Re 9$, since $\sim B$ is a logical consequence of $\sim(A \rightarrow B)$.
   4. $A \Rightarrow (B\&\sim B)$ From 1 and 3, by $\Re 13$.
   5. $A \Rightarrow F$ From 4, by $\Re 9$, since $B\&\sim B$ entails F.
   6. $A \Rightarrow (A \rightarrow B)$ By RAA, since $A \Rightarrow F$ follows from assumption $A \Rightarrow \sim(A \rightarrow B)$ plus premise. QED

1c. 1. $(A \vee B) \Rightarrow \sim A$ G.
   2. $(A \vee B) \Rightarrow \sim B$ Assumption.
   3. $(A \vee B) \Rightarrow (\sim A\&\sim B)$ From 1 and 2 by $\Re 13$.
   4. $[(A \vee B)\&(\sim A\&\sim B)] \Rightarrow F$ LC.
   5. $(A \vee B) \Rightarrow F$ From 3 and 4 by RT.
   6. $(A \vee B) \Rightarrow B$ By RAA since $(A \vee B) \Rightarrow F$ follows from assumption $(A \vee B) \Rightarrow \sim B$, plus the premise of the inference. QED

⋆2. If $\{\wp_1, \ldots, \wp_n\} \therefore \phi \Rightarrow \psi$ is valid then some subset of $\{\wp_1, \ldots, \wp_n\}$ must yield $\phi \Rightarrow \psi$; suppose that is $\{\wp_1, \ldots, \wp_m\}$ for some $m \leq n$. Then by the part of the argument already given for the conditional RAA rule, $\{\wp_1, \ldots, \wp_m, \phi \Rightarrow \sim\psi\}$ must yield $\phi \Rightarrow F$ and therefore $\{\wp_1, \ldots, \wp_m, \phi \Rightarrow \sim\psi\}$ must p-entail $\phi \Rightarrow F$. Therefore $\{\wp_1, \ldots, \wp_n, \phi \Rightarrow \sim\psi\}$ must p-entail $\phi \Rightarrow F$.
Conversely, if $\{\wp_1, \ldots, \wp_n, \phi \Rightarrow \sim\psi\}$ p-entails $\phi \Rightarrow F$ then there is a subset of $\{\wp_1, \ldots, \wp_n, \phi \Rightarrow \sim\psi\}$ that yields $\phi \Rightarrow F$. If the subset that yields $\phi \Rightarrow F$ is $\{\wp_1, \ldots, \wp_m, \phi \Rightarrow \sim\psi\}$, which includes $\phi \Rightarrow \sim\psi$, then by the part of the argument already given $\{\wp_1, \ldots, \wp_m\}$ yields and therefore p-entails $\phi \Rightarrow \psi$. If the subset that yields $\phi \Rightarrow F$ is $\{\wp_1, \ldots, \wp_m\}$, which doesn't include $\phi \Rightarrow \sim\psi$, it nevertheless p-entails $\phi \Rightarrow \psi$ because $\phi \Rightarrow F$ p-entails $\phi \Rightarrow \psi$ and $\{\wp_1, \ldots, \wp_m\}$ p-entails $\phi \Rightarrow F$.

⋆3. If both $A \Rightarrow B$ and $A\&\sim B$ p-entail $\phi \Rightarrow \psi$ but $\phi \Rightarrow \psi$ weren't a conditional tautology then it is easy to see that $A \Rightarrow B$ and $A\&\sim B$ would both have to yield $\phi \Rightarrow \psi$. Now, the possible combinations of *ersatz* truth values of $A \Rightarrow B$, $A\&\sim B$, and $\phi \Rightarrow \psi$ can be represented in a nine-line truth table below:

	$A \Rightarrow B$	$\phi \Rightarrow \psi$	$A \& {\sim}B$
1	T	T	F
2	~~T~~	~~...~~	F
3	~~T~~	~~F~~	F
4	...	T	T
5	...	~~...~~	~~T~~
6	...	~~F~~	~~T~~
7	F	T	T
8	F	~~...~~	~~T~~
9	F	~~F~~	~~T~~

All the lines in the table in which $\phi \Rightarrow \psi$ has either an *F*or no *ersatz* truth value have 'strikouts' through them, because $\phi \Rightarrow \psi$ has a 'lower' value in these lines than values of either A ⇒ B or A&~B. For instance, line 6 has a strikout because ~A&B is true in that line while $\phi \Rightarrow \psi$ is *F*, which is inconsistent it being yielded by A&~B, and line 2 has a strikeout because A ⇒ B is *T*but $\phi \Rightarrow \psi$ is truth-valueless, which is inconsistent with it being yielded by A ⇒ B. Therefore, assuming that $\phi \Rightarrow \psi$ is yielded by both A ⇒ B and A&~B, it must always have the value *T*, and therefore it must be a conditional tautology.[7]

Section 7.8⋆⋆

⋆4. The argument might be formalized as follows:

1. (A ∨ B ∨ C) (i.e., one of A, B, or C must be in the shop; rule 1)
2. ~B ⇒ ~C (If B goes out C must go with him; rule 2)
3. ~A Assumption
4. ~B ⇒ C From 1 and 3.
5. Step 4 contradicts step 2.
6. Since assumption ~A leads to contradiction, A must be true.

In effect, this argument assumes that if the premises {A∨B∨C, ~B ⇒ ~C} plus the assumption ~A entail ~B ⇒ C, and that contradicts the premise ~B ⇒ ~C, then the {A ∨ B ∨ C, ~B ⇒ ~C} must entail A. But that doesn't follow by conditional RAA.

[7]Together with the fact that {A ⇒ B,A&~B} p-entails F, that only tautologies can be p-entailed by both A ⇒ B and A&~B shows that A&~B satisfies essential requirements for being the logical negation of A ⇒ B. However, that A&~B isn't a 'true negation' in all respects is shown by the fact that it doesn't satisfy the Reduction to Absurdity principle. Thus, we can't prove that A ⇒ B follows from A → B by showing that A → B together with A&~B p-entails a contradiction. In fact, we are close to showing that just as our language doesn't contain 'true conjunctions' of conditional statements, it doesn't contain 'true negations' of them either.

Chapter 8⋆

Section 8.3

1a.

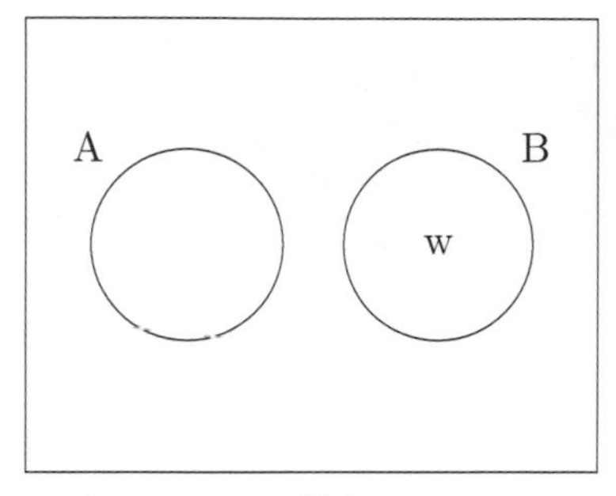

B true

$A \Rightarrow B$ false

1c.

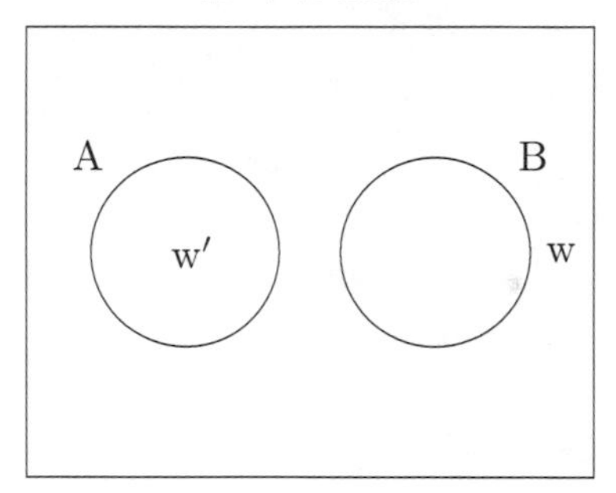

$A \leftrightarrow B$ true

$A \Rightarrow B$ false

1e.

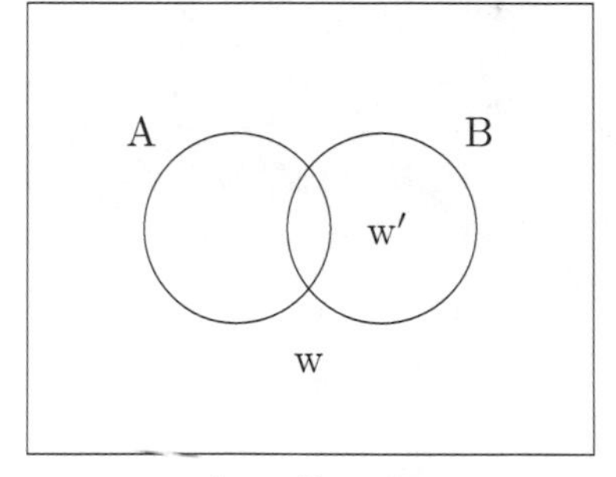

$A \Rightarrow B$ true

$\sim A \Rightarrow \sim B$ true

$B \Rightarrow A$ false

2a. Should be valid. If the nearest A world isn't a B world, i.e., if $A \Rightarrow \sim B$ is true, it should not be the case that the nearest A world *is* a B world, i.e., $A \Rightarrow B$ is true.

⋆2b. Should be valid.

2c. Should be invalid. The nearest A ⇒ B world to w is w (because A and B are both true in w), and w is also a B world. Hence, since both A ⇒ B and B are true in w, so is (A ⇒ B) ⇒ B. But ∼A ⇒ B is false in w because ∼A is false in w but B is false in nearest world to w in which ∼A is true.

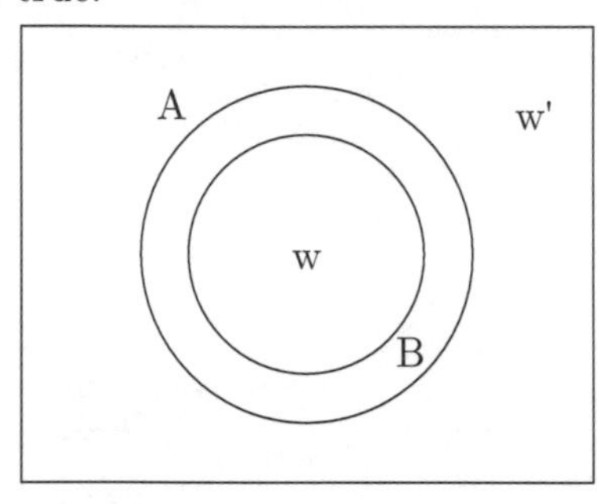

2e. Invalid.

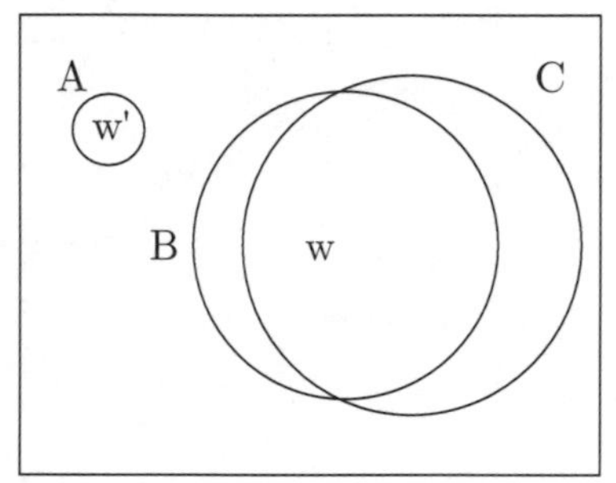

2g. Invalid.

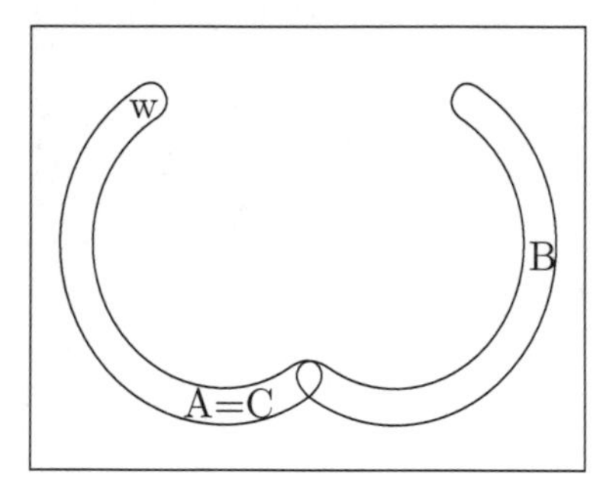

Appendix 2

1a. $p(\Pi_2 \& {\sim}\Pi_1) = \iota + 0 = \iota$.

1c. $p(P \Rightarrow \Pi_2) = \dfrac{p(P\&\Pi_2)}{p(\Pi_2)} = \dfrac{0+0}{0+0+0+\iota} = 0$.

1e. $p((P \leftrightarrow \Pi_1) \Rightarrow P) = \dfrac{p((P\leftrightarrow\Pi_1)\&P)}{p(P\leftrightarrow\Pi_1)} = \dfrac{p(P\&\Pi_1)}{p(P\&\Pi_1)+p(\sim P\&\sim\Pi_1)}$
$= \dfrac{p(P\&\Pi_1)}{p(P\&\Pi_1)+p(\sim P\&\sim\Pi_1)} = \dfrac{\iota+0}{(\iota+0)+(\iota+.6+\iota)} = \dfrac{\iota}{.6+3\iota} < \dfrac{\iota}{.6} < 2\iota.$

2a. $p(P \vee \Pi_2 \vee \Pi_3) = b + \iota$

2c. $p(\Pi_1|\Pi_1 \vee \Pi_2) = \dfrac{\iota}{\iota+\iota} = .5$

2e. $p(B \vee \Pi_1|C \vee \Pi_2) = \dfrac{\iota}{c+\iota} = \dfrac{\iota}{c} - \dfrac{\iota^2}{c^2} + \dfrac{\iota^3}{c^3} - \ldots\ldots$

3a.

	A	B	$A \Rightarrow B$
0	T	T	*T*
ι	T	F	*F*
$1-\iota$	F	T	*N*
0	F	F	*N*
	ι	$1-\iota$	0

3c.

	A	B	$\sim(A\&B)$	$A \Rightarrow \sim B$
ι	T	T	F	*F*
0	T	F	T	*T*
$1-\iota$	F	T	T	*N*
0	F	F	T	...
	ι	1	$1-\iota$	0

3e.

	A	B	C	$A \Rightarrow B$	$B \Rightarrow C$	$A \Rightarrow C$
0	T	T	T	*T*	*T*	*T*
ι	T	T	F	*T*	*F*	*F*
0	T	F	T	*F*	...	*T*
0	T	F	F	*F*	...	*F*
$1-2\iota$	F	T	T	...	*T*	...
ι	F	T	F	...	*F*	...
0	F	F	T	...	...	...
0	F	F	F	...	...	...
	ι	1	$1-\iota$	1	$1-2\iota$	0

Appendix 4

⋆1. Assuming that the new information is ~B and $p_1(\phi) = p_0(\phi|B)$, we can write:

$$P_1(A\Rightarrow B) = \frac{p_1(A\&B)}{p_1(A)} = \frac{p_0(A\&B|\sim B)}{p_0(A|\sim B)} = \frac{p_0(A\&B\&\sim B)/p_0(\sim B)}{p_0(A\&\sim B)/p_0(\sim B)}$$

If $p_0(A\&\sim B) > 0$, hence $p_0(\sim B) > 0$, this implies:

$$p_1(A\Rightarrow B) = \frac{p_0(A\&B\&\sim B)}{p_0(\sim B)} = 0.$$

But $p_1(A\Rightarrow B) = p_0(A\Rightarrow B|\sim B) = \dfrac{p_0(\sim B\&(A\Rightarrow B))}{p_0(\sim B)}$

If $p_1(A \Rightarrow B) = 0$ but $p_0(\sim B) > 0$, this implies that $p_0(\sim B\&(A \Rightarrow B)) = 0$.

⋆2.

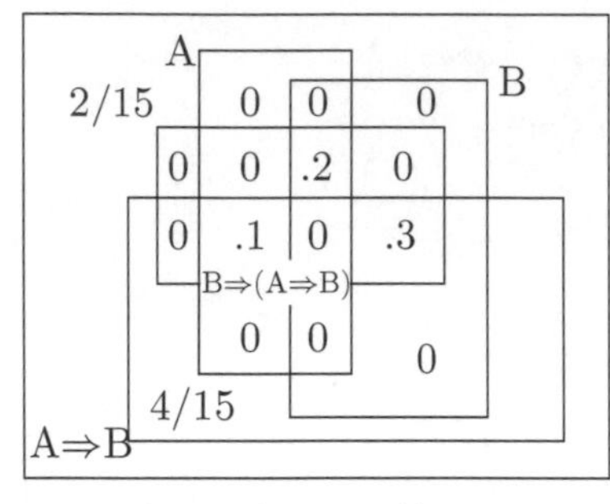

$$p(B \Rightarrow (A \Rightarrow B)) = .6$$

Appendix 7

1a. $(\exists x)Px \Rightarrow (\forall x)(Px \to Sx)$ ∴ $(\forall x)(Px \to Sx)$ is valid. Trivially, verifying $(\exists x)Px \Rightarrow (\forall x)(Px \to Sx)$ entails verifying $(\forall x)(Px \to Sx)$. That falsifying $(\forall x)(Px \to Sx)$ entails falsifying $(\exists x)Px \Rightarrow (\forall x)(Px \to Sx)$ follows from the fact that the former is a classical consequence of the latter.

1b. That Pc ∴ $(\forall x)Px \Rightarrow (\exists x)Px$ is valid follows from the fact that its conclusion is logically true, hence $(\forall x)Px \Rightarrow (\exists x)Px$ cannot be falsified and therefore it is yielded by the empty subset of the premise set {Pc}.

1c. $(\exists x)(Fx \vee Sx)$ ∴ $\sim(\exists x)Fx \Rightarrow (\forall x)Sx$ is invalid because verifying its premise does not entail verifying its conclusion, and furthermore its conclusion can be falsified. Diagrammatic counterexample:

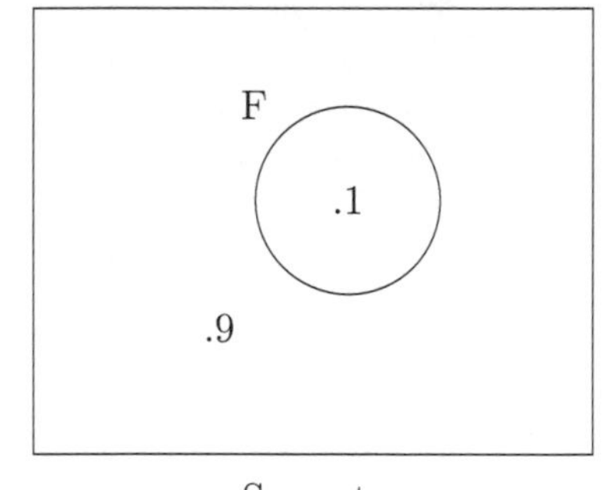

S empty
$p(\Gamma_1) = .9$

1

F and S empty
$p(\Gamma_2) = 1$

$$p((\exists x)(Fx \vee Sx)) = .9$$
$$p(\sim(\exists x)Fx \Rightarrow (\forall x)Sx) = 0.$$

Real-life counterexample: F = freshmen, S = sophomores, domain = persons at a party. It is 90% probable that 10% of the persons at the party were freshmen and no one at it was a sophomore, and that is the case in world Γ_1. But there is a 10% probability that there were no freshmen or sophomores at the party, which is the case in world Γ_2. In that case there is a 90% probability that *someone* at the party was either a freshman or a sophomore, but a 0% probability that if *no one* at the party was a freshman then someone at it was a sophomore.

1c. $\sim$(Fc&Sc) $\therefore$ Fc $\Rightarrow$ $\sim$Sc is invalid: verifying $\sim$(Fc&Sc) doesn't entail verifying Fc $\Rightarrow$ $\sim$Sc and the latter isn't yielded by the empty set of premises. Diagrammatic counterexample:

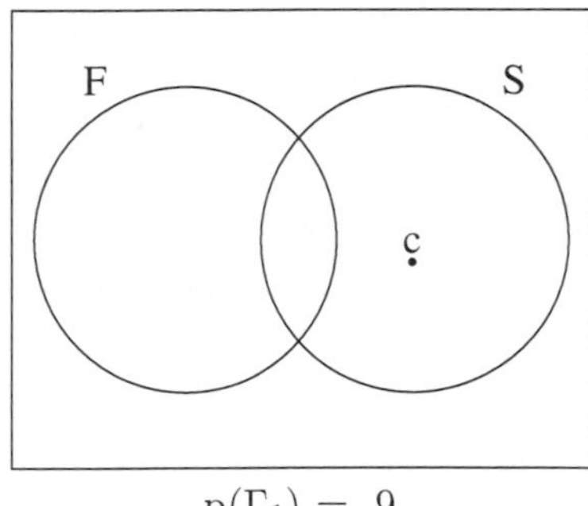

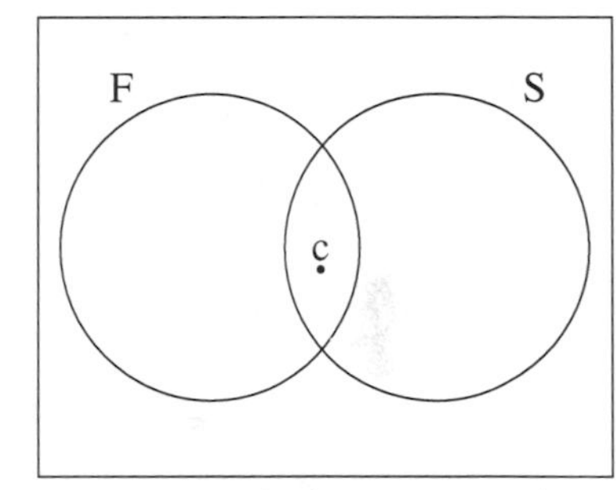

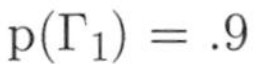
$p(\Gamma_1) = .9$ $\qquad$ $p(\Gamma_2) = .1$

$$p(\sim(\text{Fc\&Sc})) = .9$$
$$p(\text{Fc} \Rightarrow \sim\text{Sc}) = 0$$

Real-life counterexample: Domain = persons at a university, F = freshmen at the university, S = students at the university, c = Charles, who is a student but probably not a freshman. Then $\sim$(Fc&Sc) symbolizes "It is not the case that Charles is a freshman and a student," which is probable since Charles is probably not a freshman (the 'F' region should have been drawn as a subset of the 'S' region, since presumably freshman are a subclass of the class of students), but Fc $\Rightarrow$ $\sim$Sc symbolizes "If Charles is a freshman then he is not a student," which would be absurd.

⋆2a. Assuming that $p(B_1) = .99$, $p(m_{i+1}Hm_i) = 1$ for $i = 1, \ldots, n-1$, and $p(Bm_{i-1}\&m_iHm_{i-1}) \rightarrow Bm_i) = .9999$ for $i = 2, \ldots, n$, n must be at least 1,000 for it to be possible that $p(Bm_n)$ should equal 0.

⋆2b. Replacing '$\rightarrow$' by '$\Rightarrow$' and assuming that $p(Bm_{i-1}\ \&\ m_iHm_{i-1}) \Rightarrow Bm_i) = .9999$, while it is still the case that $p(B_1) = .99$, $p(m_{i+1}Hm_i) = 1$ for $i = 1, \ldots, n-1$, $p(Bm_n)$ must be greater than 0 no matter how large n is.

⋆3. a. $a < b$, $b < c$, and $c < d$ have essentialness 1 and $a < c$, $a < d$, and $b < d$ have essentialness 0.

b. If all premises have probability at least .9, hence uncertainty no more than .1, the maximum uncertainty of the conclusion is $1 \times .1 + 1 \times .1 + 1 \times .1 + 0 \times .1 + 0 \times .1 + 0 \times .1 = .3$, hence the minimum probability of $a < b < c < d$ is .7.

c. A possible probabilistic state of affairs in which all premises have probability at least .9 while $a < b < c < d$ has probability .7 involves four possible orderings of a, b, c, and d:

$$\Gamma_1 = a < b < c < d$$
$$\Gamma_2 = b < a < c < d$$
$$\Gamma_3 = a < c < b < d$$
and $$\Gamma_4 = a < b < d < c.$$

The reader can easily calculate that if $p(\Gamma_1) = .7$ and $p(\Gamma_2) = p(\Gamma_3) = p(\Gamma_4) = .1$, then $p(a < b) = p(b < c) = p(c < d) = .9$, $p(a < c) = p(a < d) = p(b < d) = 1$, and $p(a < b < c < d) = .7$.

⋆4. We can diagram the formulas involved as follows:

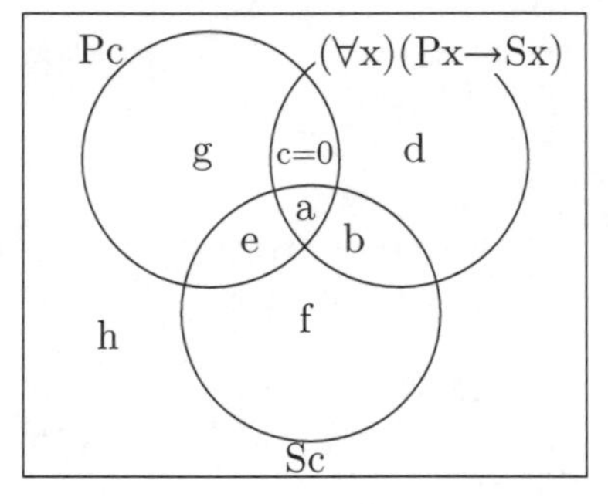

Given that $(\forall x)(Px \rightarrow Sx)$ and Pc logically entail Sc, region c should be empty, i.e., $c = 0$. Then $p((\forall x)(Px \rightarrow Sx)) = a + b + d$, $p((\forall x)(Px \rightarrow Sx)|Pc) = a/(a + g + e)$, and $p((\forall x)(Px \rightarrow Sx)|Pc\&Sc) = a/(a + e)$. Since, trivially, $a/(a + e) \geq a/(a + g + e)$, it follows that $p((\forall x)(Px \rightarrow Sx)|Pc\&Sc) \geq p((\forall x)(Px \rightarrow Sx)|Pc)$.

Appendix 8

1a. $(b = c) \Rightarrow Ab \therefore (c = b) \Rightarrow Ac$ is valid. Verifying $(b = c) \Rightarrow Ab$ entails verifying Ab and $b = c$, and since these logically entail $c = b$ and Ac they entail verifying $(c = b) \Rightarrow Ab$. Conversely, falsifying $(c = b) \Rightarrow Ab$ entails verifying $c = b$ and falsifying Ac. But if $c = b$ is true but Ac is false then $b = c$ is true and Ab is false, hence $(b = c) \Rightarrow Ab$ is falsified. Hence $(b = c) \Rightarrow Ab$ yields $(c = b) \Rightarrow Ac$, and therefore the inference is valid.

1c. $(\forall x)(Ax \rightarrow (x = f)) \therefore Af \Rightarrow As$ is invalid, since verifying $(\forall x)(Ax \rightarrow (x = f))$ doesn't entail verifying $Af \Rightarrow As$. Here is a probabilistic counterexample:

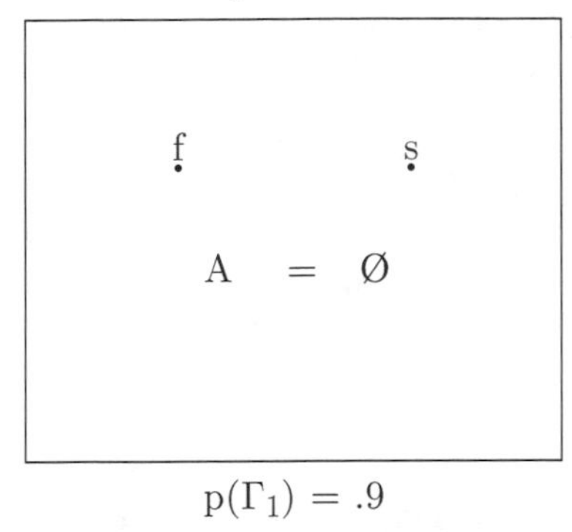

$p(\Gamma_1) = .9$

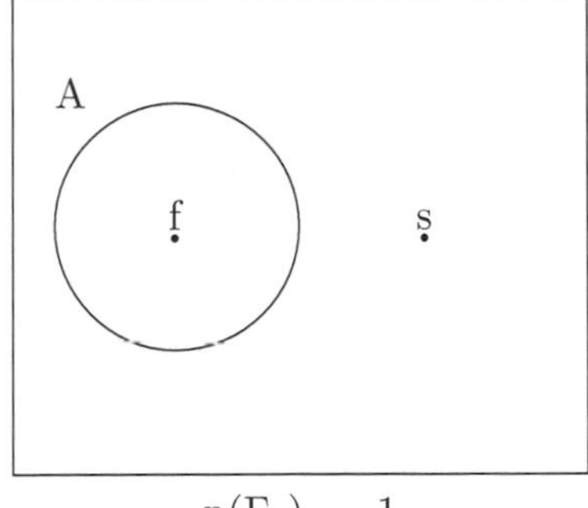

$p(\Gamma_2) = .1$

$p((\forall x)(Ax \rightarrow (x = f))) = .9$ (because Ax is false and $Ax \rightarrow (x = f)$ is true for all x in world Γ_1, hence $(\forall x)(Ax \rightarrow (x = f))$ is true in it.)
$p(Af \Rightarrow As) = p(Af\&As)/p(As) = 0/.1 = 0$.
Real-life counterexample: Letting A symbolize 'is a first-arrival', f stand for the first person to arrive at the party, and s stand for the second person to arrive, $(\forall x)(Ax \rightarrow (x{=}f))$ symbolizes "Anyone who is a first-arrival is the first person to arrive, and $Af \Rightarrow As$ symbolizes "If the first person to arrive is a first-arrival then the second person to is a first-arrival," which is absurd.

2a. The minimal essential sets in this inference are $\{a{=}b, b{=}c\}$, $\{a{=}b, a = c\}$, and $\{b = c, a = c\}$, therefore each premise has degree of essentialness $\frac{1}{2}$ because this is the reciprocal of the size of the smallest essential set to which it belongs.

2b. If $p(a{=}b) = p(a{=}c) = p(b{=}c) = .9$ then $u(a{=}b) = u(a{=}c) = u(b{=}c) = .1$, and the maximum uncertainty of $(a = b)\&(b = c)$ is $\frac{1}{2}(.1{+}.1{+}.1) = .15$, hence the minimum probability of $(a{=}b)\&(b{=}c)$ is .85.

2c. Suppose that there are four worlds, Γ_1–Γ_4, as follows:

In Γ_1 a, b, and c are all equal;
In Γ_2 a and b are equal, but they are different from c;
In Γ_3 a and c are equal, but they are different from b;
In Γ_4 b and c are equal, but they are different from a.

Then if $p(\Gamma_1) = .85$ and $p(\Gamma_2) = p(\Gamma_3) = p(\Gamma_4) = .05$, it follows that $p(a{=}b) = p(a{=}c) = p(b{=}c) = .9$, while $p((a{=}b)\&(b{=}c)) = .85$.

2d. A real-life possibility might be one in which a = Albert, b = the banker, and c = the chairman (of a committee), and we know that at least two of Albert, the banker, and the chairman are one and the same, and very probably they are *all* the same—Albert happens to be both the banker and the chairman of the committee. If the latter is 85% probable and each of the other possibilities is 5% probable then p(a = b), p(a = c), and p(b = c) will all equal .9, while p((a = b)&(b=c)) will equal .85.

⋆3. The four-member set {a=b, a=c, a=d, a=e} is essential, and there is no smaller essential set that contains a = b. Therefore the degree of essentialness of a=b is $\frac{1}{4}$. Assuming that each of the 10 premises has the same degree of essentialness and its uncertainty is .1, the maximum uncertainty of the conclusion must be $\frac{1}{4} \times 10 \times .1 = .25$. Therefore the minimum probability of the conclusion must be .75.
A probabilistic model in which each premise has probability .9 and the conclusion has probability .75 involves a world, Γ, and five other worlds $\Gamma_a, \ldots, \Gamma_e$. Γ has probability .75, and all five of a, b, c, d, and e are equal in it, while each of $\Gamma_a, \ldots, \Gamma_e$ has probability .05 and all but one of a, b, c, d, and e are equal in them, b, c, d, and e being equal and a being different from the rest in Γ_a, b being different from the rest in Γ_b, and so on. Then the probability of all of a, b, c, d, and e being equal is the same as the probability of Γ, which is .75, and the probability that a equals b is the sum of the probabilities of worlds Γ, Γ_c, Γ_d, and Γ_e, which is $.75 + .05 + .05 + .05 = .9$.

Appendix 9

1a. (Px ∨ Qx) ⇒ Rx ∴ Px ⇒ Rx is m-invalid.

Truth-table:	Px	Qx	Rx	(Px ∨ Qx) ⇒ Rx	Px ⇒ Rx
1	T	T	T	*T*	*T*
2	T	T	F	*F*	*F*
3	T	F	T	*T*	*T*
4	T	F	F	*F*	*F*
5	F	T	T	*T*	...
6	F	T	F	*T*	...
7	F	F	T	......	...
8	F	F	F	......	...

(Px ∨ Qx) ⇒ Rx doesn't yield Px ⇒ Rx because it is verified in line 5 while Px ⇒ Rx is neither verified nor falsified in that line. Also, Px ⇒ Rx isn't yielded by the empty subset of {(Px ∨ Qx) ⇒ Rx}, since it is falsified in line 4 but nothing in the empty subset is falsified in that line.

Proportionality diagram:

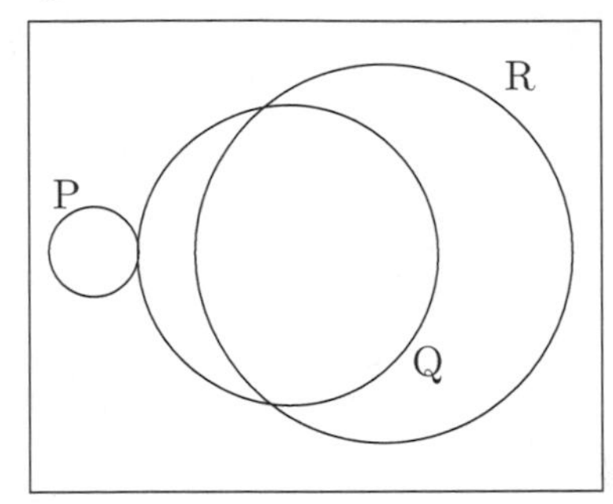

A high proportion of the union of P and Q is in R, but none of P is in R.

1c. Px $\leftrightarrow$ Qx $\therefore$ Px $\Rightarrow$ Qx is m-invalid.

Truth-table:	Px	Qx	Px $\leftrightarrow$ Qx	Px $\Rightarrow$ Qx
1	T	T	T	*T*
2	T	F	F	*F*
3	F	T	F	...
4	F	F	T	...

Px $\Rightarrow$ Qx isn't yielded by Px $\leftrightarrow$ Qx because it isn't verified in line 4, where Px $\leftrightarrow$ Qx is verified, and it isn't yielded by the empty set since it is falsified in line 2 but nothing in the empty set is falsified in that line.

Proportionality diagram:

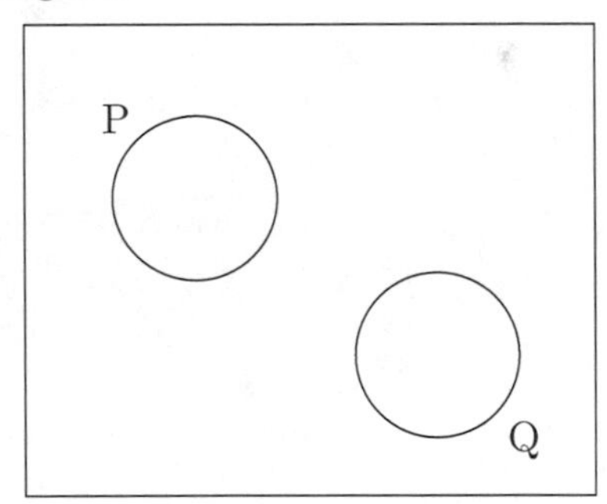

Almost every point in the large rectangle is in the region corresponding to Px $\leftrightarrow$ Qx (which is everything outside the circles P and Q), but none of P is inside Q.

1e. {Px, Px → Qx} ∴ Px ⇒ Qx is m-valid.

Truth-table:	Px	Qx	Px → Qx	Px ⇒ Qx
1	T	T	T	*T*
2	T	F	F	*F*
3	F	T	T	...
4	F	F	T	...

{Px, Px → Qx} yields Px ⇒ Qx. Line 1 is the only one in which no premise is false and at least one is true, but the conclusion is true in that line, and line 2 is the only line in which the conclusion is false, but Px → Qx is falsified in that line.

1g. Px ⇒ (Qx&Rx) ∴ (Px&Qx) ⇒ Rx is m-valid.

Truth-table	Px	Qx	Rx	Px ⇒ (Qx&Rx)	(Px&Qx) ⇒ Rx
1	T	T	T	*T*	*T*
2	T	T	F	*F*	*F*
3	T	F	T	*F*	......
4	T	F	F	*F*	......
5	F	T	T	......	......
6	F	T	F	......	......
7	F	F	T	......	......
8	F	F	F	......	......

Px ⇒ (Qx&Rx) yields (Px&Qx) ⇒ Rx because line 1 is the only line in which the premise is verified, and the conclusion is verified in that line, and line 2 is the only line in which the conclusion is falsified, and the premise is falsified in that line.

2a. xHy ∴ yHx is m-valid. τ(xHy) is the proportion of pairs ⟨x, y⟩ that stand in the relation H, but that is the same as τ(yHx). Hence, τ(yHx) must be high if τ(xHy) is high.

2c. Px∨Qx ∴ Px∨Qy isn't m-valid because τ(Px∨Qx) can equal 1 while τ(Px ∨ Qy) equals $\frac{3}{4}$. To see this, suppose that Px = "x is an even number" and Qx = "x is an odd number" in the domain of numbers from 1 to 100. Then τ(Px ∨ Qx) = 1 because all numbers from 1 to 100 are either even or odd, but τ(Px∨Qy) is false in the 50 instances in which x is an odd number and the 50 instances in which y is an even number. Therefore there are 50×50 = 2,500 out of 10,000 pairs of numbers from 1 to 100 in which Px∨Qy is false, hence τ(Px∨Qy) = $\frac{3}{4}$.

3a. xHy ∴ (∀x)xHy is m-invalid. To see this, suppose that xHy is the relation x≠y among numbers from 1 to 100. Then xHy is true for all pairs x and y except the 100 pairs in which x=y. Since there are

$100 \times 100 = 10{,}000$ pairs in all, and xHy is true in all but 100 of them, $\tau(\text{xHy}) = (10,000\text{-}100)/10,000 = .99$. On the other hand, $(\forall x)$xHy is true for any given value of y if and only if x is different from it for *all* values of x. But that is never the case, since y is never different from itself. Therefore, $\tau((\forall x)\text{xHy}) = 0$.

3b. xHy $\therefore$ $(\exists x)$xHy is m-valid. You can see this intuitively if you picture the relation H as in proportionality diagram 2:

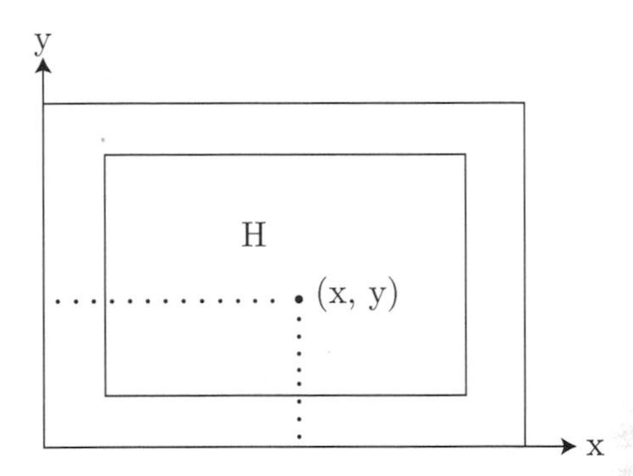

Here the 'universe' of pairs $\langle x, y\rangle$ lie inside the larger rectangle and the the pairs that stand in the relation H lie inside the smaller one. Then $\tau(\text{xHy})$ is the proportion of the larger rectangle that the smaller one occupies, which is pictured as being large. But for every pair $\langle x, y\rangle$ inside the rectangle H, *some* object x must stand in relation xHy to y, and since most pairs $\langle x, y\rangle$ are inside H most objects y must have some x related to them. I.e., for most y it must be the case that $(\exists x)$xHy is true, hence $\tau((\exists x)\text{xHy})$ must be high. [Note: In fact, it is not hard to show that $\tau((\exists x)\text{xHy}) \geq \tau(\text{xHy})$.]

4. $\tau((\text{xHy\&yHz}) \Rightarrow \text{xHz})$ could be high and $\tau((\text{xHy\&yHz\&}zHw) \Rightarrow \text{xHw})$ low if almost all sequences of 'relatives' came in chains of three, in which an x is related to a y and y is related to a z, to which x is also related, but in the rare chains of four, in which x is related to a y, y is related to a z, and z is related to a w, x *isn't related to w.*

5. This would have to be both a very polygamous and a very gender-unbalanced society. Suppose, for instance, that there were one wife who was shorter than all of her 10,000 husbands, but there were 100 wives who were taller than their one and only husbands. Then 100 out of 101 wives would be taller than their husbands, and 10,000 out of 10,100 husbands would be taller than their wives.

References

Ackermann, Wilhelm, 1954, *Solvable Cases of the Decision Problem*, Amsterdam: North-Holland Publishing Company.

Adams, Ernest W., 1962, "On rational betting systems", *Archiv für Mathematische Logik und Grundlagensforschung*, 6(1–2): 7–29, and 6/34, pp. 112–128.

———, 1964, "On the reasonableness of inferences involving conditionals", *Memorias des XIII Congresso International de Filosofia, Communicaciones Libres*, 5: 1–10.

———, 1965, "The logic of conditionals", *Inquiry*, 8: 166–197.

———, 1966, "Probability and the logic of conditionals", in Hintikka and Suppes (1966), pp. 265–316.

———, 1970, "Subjunctive and indicative conditionals", *Foundations of Language*, 6: 89–94.

———, 1974, "The logic of 'Almost All'", *Journal of Philosophical Logic*, 3: 3–17.

———, 1975, *The Logic of Conditionals*, Dordrecht: D. Reidel Publishing Company.

———, 1977, "A note on comparing probabilistic and modal semantics of conditionals", *Theoria*, XLIII(3): 186–194.

———, 1981, "Transmissible improbabilities and marginal essentialness of premises in inferences involving indicative conditionals", *Journal of Philosophical Logic*, 10(2): 149–177.

———, 1983, "Probabilistic enthymemes", *Journal of Pragmatics*, 7: 283–295.

———, 1986, "On the logic of high probability", *Journal of Philosophical Logic*, 15: 255–279.

———, 1987, "On the meaning of the conditional", *Philosophical Topics*, XV(1): 5–22.

———, 1988a, "Consistency and decision: Variations on Ramseyan themes", in Harper and Skyrms (1988), pp. 49–69.

———, 1988b, "Confirming approximate generalizations", in Fine, Arthur and Jarrett Leplin (eds.), *PSA 1988*, volume I, pp. 10–16, Anne Arbor: Edwards Brothers.

———, 1992, "On the empirical status of measurement axioms: The case of subjective probability", in Savage, C. Wade and Philip Ehrlich (eds.), *Philosophical and Foundational Issues in Measurement Theory*, p. 53–73, Hillsdale, NJ: Lawrence Erlbaum Associates, Inc.

———, 1994, "Updating on conditional information", *Special Issue of IEEE Transactions on Systems, Man and Cybernetics*, 24(12): 1708–1713.

———, 1995, "Remarks on a theorem of mcgee", *Journal of Philosophical Logic*, 24(4): 343–348.

———, 1996, "Four probability–preserving properties of inferences", *Journal of Philosophical Logic*, 25: 1–24.

———, 1997, "Practical possibilities", *Pacific Philosophical Quarterly*, 78: 113–127.

Adams, Ernest W. and Ian F. Carlstrom, 1979, "Representing approximate ordering and equivalence relations", *Journal of Mathematical Psychology*, 19(2): 182–207.

Adams, Ernest W. and Howard P. Levine, 1975, "On the uncertainties transmitted from premises to conclusions in deductive inferences", *Synthese*, 30(3): 429–460.

Adams, Ernest W. and Roger D. Rosenkrantz, 1980, "Applying the Jeffrey decision model to rational betting and information acquisition", *Theory and Decision*, 12: 1–20.

Appiah, Anthony, 1985, *Assertion and Conditionals*, Cambridge: Cambridge University Press.

Ayer, A. J., 1959, *Logical Positivism*, Glencoe, Illinois: Free Press.

———, 1968, *The Origins of Pragmatism*, San Francisco: Freeman, Cooper and Company.

Bacon, Francis, 1612 [1965], "Essays", in Warhaft, Sidney (ed.), *Francis Bacon, a Selection of his Works*, Toronto: Macmillan Publishing.

Baird, John C. and Elliot Noma, 1978, *Fundamentals of Scaling and Psychophysics*, New York: John Wiley and Sons.

Bamber, Donald, 1997, "Entailment with near surety of generalizations having nonuniform exception rate thresholds", SPAWAR Systems Center, San Diego, California.

Bayes, Thomas, 1940, "An essay towards solving a problem in the doctrine of chances", in Deming, William Edwards (ed.), *Facsimiles of Two Papers by Bayes*, Washington, D.C.: U. S. Department of Agriculture, originally published in *Transactions of the Royal Society*, London, 1763, 53: 370–418.

Bennett, Jonathan, 1995, "Farewell to the phlogiston theory of conditionals", *Mind*, 97(388): 509–527.

Buchler, Justus (ed.), 1955, *Philosophical Writings of Peirce*, New York: Dover Publications Inc.

Calabrese, Philip G., 1994, "A theory of conditional information with applications", *Special Issue of IEEE Transactions on Systems, Man and Cybernetics*, 24(12): 1676–1684.

Carlstrom, Ian F., 1975, "Truth and entailment for a vague quantifier", *Synthese*, 30: 461–495.

———, 1990, "A truth-functional logic for near-universal generalizations", *Journal of Philosophical Logic*, 19(4): 379–405.

Carnap, Rudolf, 1945, "Two concepts of probability", *Philosophy and Phenomenological Research*, 5: 513–32.

———, 1947, *Meaning and Necessity*, Chicago: University of Chicago Press.

———, 1950, *Logical Foundations of Probability*, Chicago: University of Chicago Press.

Chihara, Charles S., 1987, "Some problems for Bayesian confirmation theory", *British Journal for the Philosophy of Science*, 38: 551–560.

Church, Alonzo, 1936a, "A note on the *Entscheidungsproblem*", *Journal of Symbolic Logic*, 1(1): 40–41.

———, 1936b, "Correction to *A note on the Entscheidungsproblem*", *Journal of Symbolic Logic*, 1(3): 101–102.

———, 1956, *Introduction to Mathematical Logic*, volume I, Princeton, New Jersey: Princeton University Press.

Churchland, Paul M., 1981, "Eliminative materialism and propositional attitudes", *Journal of Philosophy*, 78: 67–90.

Cooper, William S., 1978, *Foundations of Logico-Linguistics*, Holland: D. Reidel Publishing Company.

Copi, Irving M., 1965, *Symbolic Logic*, New York: The Macmillan Company, 2nd edition.

Davidson, Donald, 1980, *Essays on Actions and Events*, The Clarendon Press: Oxford.

Davidson, Donald and Patrick Suppes, 1956, "A finitistic axiomatization of subjective probability and utility", *Econometrica*, 24: 264–275.

Diaconis, Persi and David Friedman, 1980, "de Finetti's generalizations of exchangeability", in Jeffrey, Richard C. (ed.), *Studies in Inductive Logic and Probability*, pp. 233–250, Berkeley: University of California Press.

Dodgson, C. (Lewis Carroll), 1894, "A logical paradox", *Mind*, 3: 436–438, N.S.

———, 1895, "What the tortoise said to Achilles", *Mind*, 4: 278–280, N.S.

Doob, Joseph L., 1953, *Stochastic Processes*, New York: John Wiley & Sons.

Dubois, Didier and Henri Prade, 1994, "Conditional objects as nonmonotonic consequence relationships", *IEEE Transactions on Systems, Man, and Cybernetics*, 24(12): 1724–1740.

Dudman, V. H., 1994, "On conditionals", *Journal of Philosophy*, 91(3): 113–128.

Dummett, Michael, 1977, *Elements of Intuitionism*, Oxford: Oxford University Press.

Edgington, Dorothy, 1991, "On the mystery of the missing matter of fact", *Aristotelian Society Supplementary*, 65: 1850–209.

———, 1995, "Conditionals", *Mind*, 104: 235–329.

———, 1997, "Vagueness by degrees", in O'Keefe, Rosanna and Peter Smith (eds.), *Vagueness: A Reader*, Cambridge, Massachusetts: MIT Press.

Eells, Ellery, Brian Skyrms, and Ernest W. Adams (eds.), 1994, *Probability and Conditionals: Belief Revision and Rational Decision*, Cambridge: Cambridge University Press.

Ellis, Brian, 1973, "On the logic of subjective probability", *Philosophy of Science*, 24: 125–152.

Feller, William, 1957, *Introduction to Probability Theory and its Applications*, volume I, New York: John Wiley and Sons Inc., second Edition.

Fine, Terrence L., 1973, *Theories of Probability*, New York: Academic Press.

de Finetti, Bruno, 1937, "Foresight, its logical laws, its subjective sources", in Kyburg, Henry E. and Howard E. Smokler (eds.), *Studies in Subjective Probability*, New York: John Wiley and Sons.

———, 1975, *Theory of Probability, an Introductory Treatment*, New York: Wiley and Sons, translated from the Italian by A. Machi and A. Smith.

Fisher, Ronald A., 1959, *Smoking: the Cancer Controversy*, Edinburgh and London: Oliver and Boyd.

Fodor, Jerry, 1983, *The Modularity of Mind*, Cambridge, Massachusetts: MIT Press.

van Fraassen, Bas, 1976, "Probabilities of conditionals", in Harper, W.L. and C. Hooker (eds.), *Foundations of probability theory, statistical inference, and statistical theories of science*, pp. 261–308, Dordrecht-Holland: D. Reidel.

Frege, Gottlob, 1892, "Sense and reference", in Feigl, Herbert and Wilfrid Sellars (eds.), *Readings in Philosophical Analysis*, New York: Appleton–Century Crofts.

Gärdenfors, Peter, 1988, "Causation and the dynamics of belief", in Harper and Skyrms (1988), pp. 85–104.

Gardner, Martin, 1961, *Mathematical Puzzles and Diversions, a New Selection*, New York: Simon and Schuster.

Gibbard, Allan, 1981, "Two recent theories of conditionals", in Harper et al. (1981), pp. 211–247.

Goguen, Joseph A., 1968–9, "The logic of inexact concepts", *Synthese*, 19: 325–373.

Goldman, Alvin I., 1970, *A Theory of Human Action*, Englewood Cliffs, New Jersey: Prentice–Hall.

Goodman, Irwin R., Hung T. Nguyen, and Elbert A. Walker, 1991, *Conditional Inference and Logic for Intelligent Systems: a Theory of Measure–free Conditioning*, Amsterdam: North Holland Publishing Company.

Goodman, Nelson, 1947, "The problem of counterfactual conditionals", *Journal of Philosophy*, 44: 113–128.

———, 1955, *Fact, Fiction, and Forecast*, Cambridge, Massachusetts: Harvard University Press.

Grice, H. Paul, 1975, "Logic and conversation", in Davidson, Donald and G. Harman (eds.), *The Logic of Grammar*, Encino, California: Dickinson Publishing Company.

———, 1989, *Studies in the Way of Words*, Cambridge, Massachusetts: Harvard University Press.

Hacking, Ian, 1967, "Slightly more realistic personal probability", *Philosophy of Science*, 34(4): 311–325.

Hailperin, Theodore, 1965, "Best possible inequalities for the probability of a logical function of events", *American Mathematical Monthly*, 72: 343–359.

———, 1997, *Sentential Probability Logic*, Bethlehem, PA: Lehigh University Press.

Hájek, Alan, 1994, "Triviality on the cheap?", in Eells et al. (1994), pp. 113–140.

Halmos, Paul R., 1950, *Measure Theory*, New York: Van Nostrand.

Harper, William L. and Brian Skyrms (eds.), 1988, *Causation in Decision, Belief Change, and Statistics: Proceedings of the Irvine Conference on Probability and Causation*, Dordrecht: Kluwer Academic Publishers.

Harper, William L., Robert Stalnaker, and Glenn Pearce (eds.), 1981, *Ifs: Conditionals, Belief, Decision, Chance, and Time*, Dordrecht: D. Reidel

Hawthorne, James, 1996, "On the logic of nonmonotonic conditionals and conditional probabilities", *Journal of Philosophical Logic*, 25: 185–218.

Hempel, Carl G., 1965, *Aspects of Scientific Explanation*, New York: The Free Press.

Hintikka, Jaakko, 1970, "Surface information and depth information", in Hintikka, Jaakko and Patrick Suppes (eds.), *Information and Inference*, pp. 263–297, Dordrecht: D. Reidel Publishing.

Hintikka, Jaakko and Patrick Suppes (eds.), 1966, *Aspects of Inductive Logic*, Dordrecht: North-Holland Publishing Company.

Hoover, Douglas N., 1982, "A normal form theorem for $\mathcal{L}_{\omega_1 p}$, with applications", *Journal of Symbolic Logic*, 47(3): 605–624.

Horwich, Paul, 1990, *Truth*, Oxford: Basic Blackwell.

Hume, David, 1739, *A Treatise of Human Nature*, Oxford: Clarendon Press, edition of 1888, edited by L. A. Selby-Bigge.

Jackson, Frank, 1987, *Conditionals*, Oxford: Basil Blackwell.

Jeffrey, Richard C., 1964, "If", *Journal of Philosophy*, 65: 702–3.

———, 1983, *The Logic of Decision*, Chicago and London: University of Chicago Press, second edition.

———, 1992, *Probability and the Art of Judgment*, Cambridge: Cambridge University Press.

Jennings, R. E., 1994, *The Genealogy of Disjunction*, Oxford: Oxford University Press.

Kant, Immanuel, 1781, *Kritik der Reinen Vernunft*, Macmillan & Co., Ltd, english translation by N. K. Smith as *Immanuel Kant's Critique of Pure Reason*, 1929.

Kemeny, John G., 1955, "Fair bets and inductive probabilities", *Journal of Symbolic Logic*, 20(3).

Kendall, Maurice G., 1948, *Advanced Theory of Statistics*, volume II, London: Charles Griffin and Co.

Kneale, William and Martha Kneale, 1962, *The Development of Logic*, The Clarendon Press: Oxford.

Kolmogorov, A. N., 1950, *The Foundations of Probability*, New York: Chelsea Publishing Company.

Kraft, Charles H., John W. Pratt, and A. Seidenberg, 1959, "Intuitive probability on finite sets", *Annals of Mathematical Statistics*, 30: 408–419.

Krantz, David H., R. Duncan Luce, Patrick Suppes, and Amos Tversky, 1971, *Foundations of Measurement*, volume I, New York: Academic Press.

Krauss, Sarit, Daniel Lehmann, and Menachem Magidor, 1990, "Nonmonotonic reasoning, preferential models, and cumulative logics", *Artificial Intelligence*, 44: 167–207.

Kripke, Saul, 1963, "Semantical analysis of modal logic, i", *Zeitschrift fur Mathematische Logik und Grundlagen der Mathematik*, 9: 67–96.

———, 1965, "Semantical analysis of modal logic, ii", in Addison, John West, Leon Henkin, and Alfred Tarski (eds.), *Theory of Models, Proceedings of the 1963 International Symposium at Berkeley*, pp. 206–220, Amsterdam: North-Holland Publishing Company.

———, 1980, *Naming and Necessity*, Cambridge, Massachusetts: Harvard University Press.

Kyburg, Henry E., 1965, "Probability, rationality, and the rule of detachment", in Bar-Hillel, Yehoshua (ed.), *Proceedings of the 1964 International Congress for Logic, Methodology, and Philosophy of Science*, pp. 301–310, Amsterdam: North-Holland Publishing Co.

Laplace, Pierre-Simon, 1812, *Traité Analytique des Probabilités*, Paris.

Lehman, Daniel and Menachem Magidor, 1988, "Rational logics and their models; a study in cumulative logics", *Technical Report #TR-88-16.*

Lehman, R. Sherman, 1955, "On confirmation and rational betting", *Journal of Symbolic Logic*, 20(3).

Lewis, C. I. and Cooper H. Langford, 1932, *Symbolic Logic*, New York: Dover Publications.

Lewis, David, 1973, *Counterfactuals*, Oxford: Basil Blackwell.

———, 1976, "Probabilities of conditionals and conditional probabilities", *Philosophical Review*, 85: 297–315.

———, 1981, "Counterfactuals and comparative possibility", in Harper et al. (1981), pp. 57–86.

Locke, John, 1689, *An Essay Concerning Human Understanding*, New York: Dover Publications, with biographical, critical, and historical prolegomena by Alexander Campbell Fraser.

Luce, R. Duncan and Howard Raiffa, 1957, *Games and Decisions*, New York: Wiley and Company.

Luce, R. Duncan, David H. Krantz, Patrick Suppes, and Amos Tversky, 1989, *Foundations of Measurement*, volume II, New York: Academic Press.

Makinson, David and Peter Gärdenfors, 1991, "Relationships between the logic of theory change and nonmonotonic logic", in Fuhrmann, André and Michael Morreau (eds.), *The Logic of Theory Change: Workshop, Konstanz, FRG, October 13–15, 1989 Proceedings*, volume 465, pp. 183–205, Springer Verlag.

Manders, Kenneth L., 1979, "Theory of all substructures of a structure", *Journal of Symbolic Logic*, 44: 583–598.

Mates, Benson, 1965, *Elementary Logic*, New York: Oxford University Press.

McGee, Vann, 1981, "Finite matrices and the logic of conditionals", *Journal of Philosophical Logic*, 10: 349–351.

———, 1985, "A counterexample to *Modus Ponens*", *Journal of Philosophy*, 82: 462–471.

———, 1989, "Probabilities of conditionals", *Philosophical Review*, 98: 485–542.

———, 1991, *Truth, Vagueness, and Paradox*, Indianapolis/Cambridge: Hackett Publishing Company.

———, 1994, "Learning the impossible", in Eells et al. (1994), pp. 179–200.

McKeon, Robert (ed.), 1941, *The Basic Works of Aristotle*, New York: Random House.

Mill, John Stuart, 1895 [1843], *A System of Logic*, New York: Harper and Brothers, eighth edition.

von Mises, Richard, 1957, *Probability, Statistics, and Truth*, London: George Allen and Unwin Ltd., second revised English edition.

Montague, Richard and Leon Henkin, 1956, "On the definition of 'Formal Deduction'", *Journal of Symbolic Logic*, 21(2): 129–136.

Nilsson, Nils J., 1986, "Probabilistic logic", *Artificial Intelligence*, 28(1): 71–87.

Northrop, Eugene P., 1944, *Riddles in Mathematics*, London: Penguin Books.

Pearl, Judea, 1988, *Probabilistic Reasoning in Intelligent Systems*, San Mateo, California: Morgan Kaufmann Publishers.

Pearl, Judea and M. Goldzmidt, 1991, "System-z+: a formalism for reasoning with variable strength defaults", *Proceedings of the Ninth International Conference on Artificial Intelligence*, 1: 399–404.

Peirce, Charles S., 1878, "How to make our ideas clear", *Popular Science Monthly*, reprinted as Chapter 3, J. Buchler, *Philosophical Writings of Peirce*, Dover Publications, New York, 1955.

———, 1903, "Abduction and induction", Reprinted as Chapter 11 in J. Buchler (ed) *Philosophical Writings of Peirce*, Dover Publications, New York, 1955.

Popper, Karl R., 1959, *The Logic of Scientific Discovery*, Hutchinson of London.

Quine, W. V. 0., 1950, *Methods of Logic*, New York: Holt, Rinehart, and Winston, first edition.

———, 1953, *From a Logical Point of View*, Cambridge, Massachusetts: Harvard University Press.

———, 1958, *Mathematical Logic*, Cambridge, Massachusetts: Harvard University Press, revised edition.

———, 1959, *Methods of Logic*, New York: Henry Holt and Company, revised edition.

Ramsey, Frank Plumpton, 1926, "Truth and probability", Printed in Ramsey (1931), 156–198.

———, 1927, "Facts and propositions", *Proceedings of the Aristotelian Society (Supplementary)*, 7(1): 153–170, reprinted in Ramsey (1931), 138–155.

———, 1931, *The Foundations of Mathematics and other Essays*, New York: The Humanities Press, r. B. Braithwaite (editor), reprinted in 1950.

Rand McNally, 1986, *Rand McNally Road Atlas MCMLXIV*, Rand McNally & Co.

Reichenbach, Hans, 1949, *The Theory of Probability*, Berkeley and Los Angeles: University of California Press.

Robinson, Abraham, 1996, *Non-standard Analysis*, Princeton: Princeton University Press, revised edition.

Rosser, J. B. and A. R. Turquette, 1952, *Many-valued Logics*, Amsterdam: North-Holland Publishing Company.

Russell, Bertrand, 1912, *The Problems of Philosophy*, London: Oxford University Press.

———, 1914, "The relation of sense-data to physics", *Scientia*, 16(4): 1–27, reprinted in Russell (1917), pages 145–179.

———, 1917, *Mysticism and Logic and Other Essays*, London: George Allen & Unwin.

———, 1945, *A History of Western Philosophy*, New York: Simon and Schuster.

Salmon, Nathan, 1986, *Frege's Puzzle*, Massachusetts: MIT Press.

Savage, Leonard J., 1954, *The Foundations of Statistics*, New York: John Wiley and Sons. Inc.

Schmitt, Frederick F., 1995, *Truth, a Primer*, Boulder, San Francisco, Oxford: Westview Press.

Schurz, Gerhard, 1994, "Probabilistic justification for default reasoning", in Nebel, Bernhard and Leonie Dreschler-Fisher (eds.), *KI-94: Advances in Artificial Intelligence*, pp. 248–259, Berlin: Springer.

Scott, Dana, 1964, "Measurement structures and linear inequalities", *Journal of Mathematical Psychology*, 1(2): 233–247.

Scott, Dana and Peter Krauss, 1966, "Assigning probabilities to logical formulas", *Studies in Logic and the Foundations of Mathematics*, 43: 219—264.

Scott, Dana and Patrick Suppes, 1958, "Foundational aspects of theories of measurement", *Journal of Symbolic Logic*, 23: 113–128.

Searle, John, 1969, *Speech Acts*, Cambridge: Cambridge University Press.

Shannon, Claude E. and Warren Weaver, 1949, *The Mathematical Theory of Communication*, Urbana, Ill: University of Illinois Press.

Shimony, Abner, 1955, "Coherence and the axioms of confirmation", *Journal of Symbolic Logic*, 20(1): 1–28.

———, 1970, "Scientific inference", in Colodny, Robert (ed.), *The Nature and Function of Scientific Theories: Essays in Contemporary Science and Philosophy*, volume 4 of *University of Pittsburgh series in the philosophy of science*, pp. 79–172, Pittsburgh, PA: University of Pittsburgh Press.

Simpson, E. H., 1951, "The interpretation of interaction in contingency tables", *Journal of the Royal Statistical Society, Ser. B*, 13: 238–241.

Skyrms, Brian, 1981, "The prior propensity account of subjunctive conditionals", in Harper et al. (1981), p. 259–265.

———, 1986, *Choice and Chance, an Introduction to Inductive Logic*, Belmont, California: Wadsworth Publishing Company, 3rd ed.

———, 1990, *The Dynamics of Rational Deliberation*, Cambridge, Massachusetts: Harvard University Press.

Stalnaker, Robert C., 1968, "A theory of conditionals", in Rescher, Nicholas (ed.), *Studies in Logical Theory*, pp. 98–112, Blackwell, reprinted in Harper et al. 1981: 41–56.

———, 1981, "A defense of conditional excluded middle", in Harper et al. (1981), pp. 87–104.

Stich, Stephen, 1983, *From Folk Psychology to Cognitive Science, The Case Against Belief*, Massachusetts: MIT Press.

Suppes, Patrick, 1966, "Probabilistic inference and the concept of total evidence", in Hintikka and Suppes (1966), p. 49–65.

———, 1970, *A Probabilistic Theory of Causality*, Amsterdam: North-Holland Publishing Company.

Suppes, Patrick, David H. Krantz, R. Duncan Luce, and Amos Tversky, 1990, *Foundations of Measurement*, volume III, New York: Academic Press.

Tarski, Alfred, 1930-1, "O pojęciu prawdy w odniesieniu do sformalizowanych nauk dedukcyjnych", *Ruch Filozoficzny*, xii: 210–211, though date is 1930-1, actually published in 1933, title in English is "On the notion of truth in reference to formalized deductive sciences".

———, 1936, "Der Wahrheitsbegriff in den formalisierten Sprachen", *Studia Philosophica*, 1: 261–405, reprint dated 1935.

———, 1944, "The semantic conception of truth and the foundations of semantics", *Philosophy and Phenomenological Research*, 4(3): 341—-376.

———, 1956, "The concept of truth in formalized languages", in *Logic, Semantics, Metamathematics, Papers from 1923 to 1938*, pp. 152–278, Oxford: Oxford University Press, edited by J. H. Woodger, first English translation.

Vickers, John M., 1976, *Belief and Probability*, Dordrecht, Holland: D. Reidel Publishing Company.

———, 1988, *Chance and Structure, an Essay on the Logical Foundations of Probability*, The Clarendon Press: Oxford.

Vineberg, Susan, 1996, "Eliminative induction and Bayesian confirmation", *Canadian Journal of Philosophy*, 26(2): 257–266.

Weingartner, Paul and Gerhard Schurz, 1986, "Paradoxes solved by simple relevance criteria", *Logique et Analyse*, 29: 3–40.

Wittgenstein, Ludwig, 1922, *Tractatus Logic-Philosophicus*, London: Routledge and Kegan Paul.

Zadeh, Lotfi A., 1965, "Fuzzy sets", *Information and Control*, 8: 338–353.

Name Index

Subject Index